Environmental Hydraulics

Environmental Hydraulics

Contributors

Mansoor Zoveidavianpoor, Ariffin Samsuri et al.

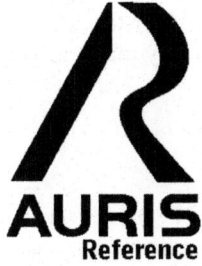

AURIS
Reference

www.aurisreference.com

Environmental Hydraulics

Contributors: Mansoor Zoveidavianpoor, Ariffin Samsuri et al.

Published by Auris Reference Limited

www.aurisreference.com

United Kingdom

Environmental Hydraulics

ISBN: 978-1-78154-910-0

British Library Cataloguing in Publication Data
A CIP record for this book is available from the British Library

Printed in the United Kingdom

Exclusively distributed by CBS Publishers & Distributors Pvt. Ltd.

Sales & Distribution Rights only for India, Pakistan, Bangladesh, Sri Lanka, Nepal and Bhutan.This book is not to be sold outside these territories.

Contents

List of Abbreviations

CAM	Crassulacean Acid Metabolism
CCA	Chromated Copper Arsenate
DBC	Dissolved Black Carbon
DBH	Diameter At Breast Height
DCRI	Drill Cutting Re-Injection
DOC	Dissolved Organic Carbon
FC	Field Capacity
HF	hydraulic Fracturing
HRU	Hydrologic Response Units
HSSC	High Sediment Concentrations
KPK	Khyber Pakhtunkhwa
LCC	Leaf Carbon Concentration
LDMC	Leaf Dry Matter Content
LES	Leaf Economics Spectrum
LNC	Leaf Nitrogen Concentration
MAGICC	Model for Assessment of Greenhouse gas Induced Climate Change
MCL	Maximum Contaminant Level
OBMs	Oil Based Muds
PCA	Principal Components Analysis
PKN	Perkins-Kern-Nordgren
PTFs	Pedotransfer Functions
PWP	Permanent Wilting Point
REV	Representative Elementary Volume
RMSE	Root Mean Squared Errors
SLA	Specific Leaf Area
TDN	Total Dissolved Nitrogen
TDP	Thermal Dissipation Probes
USA	United States
USDA	United States Department of Agriculture
UTM	Universal Transverse Mercator
VPD	Vapor Pressure Deficit
WBMs	Water Based Muds
WHO	World Health Organization

List of Contributors

Mansoor Zoveidavianpoor
Universiti Teknologi, Faculty of Petroleum & Renewable Energy Engineering, Malaysia

Ariffin Samsuri
Universiti Teknologi, Faculty of Petroleum & Renewable Energy Engineering, Malaysia

Seyed Reza Shadizadeh
Petroleum University of Technology, Abadan Faculty of Petroleum Engineering, Iran

Lixin Chen
Key Laboratory Soil and Water Conservation and Desertification Combating, Ministry of Education, College of Soil and Water Conservation, Beijing Forestry University, Beijing, People's Republic of China

Zhiqiang Zhang
Key Laboratory Soil and Water Conservation and Desertification Combating, Ministry of Education, College of Soil and Water Conservation, Beijing Forestry University, Beijing, People's Republic of China

Brent E. Ewers
Program in Ecology, Department of Botany, University of Wyoming, Laramie, Wyoming, United States of America

Stefan Wirtz
Department of Physical Geography, Trier University, Trier, Germany

Manuel Seeger
Department of Physical Geography, Trier University, Trier, Germany
Department of Land Degradation and Development, Wageningen University, Wageningen, The Netherlands

Andreas Zell
Department 7.3- Technical Physics, Saarland University, Saarbru¨ cken, Germany

Christian Wagner
Department 7.3- Technical Physics, Saarland University, Saarbru¨ cken, Germany

Jean-Frank Wagner
Department of Geology, Trier University, Trier, Germany

Johannes B. Ries
Department of Physical Geography, Trier University, Trier, Germany

Stanislaus J. Schymanski
Department of Environmental Systems Sciences, ETH Zurich, Zurich, Switzerland

Dani Or
Department of Environmental Systems Sciences, ETH Zurich, Zurich, Switzerland

Maciej Zwieniecki
Department of Plant Sciences, University of California Davis, Davis, California, United States of America

Rebecca T. Barnesa
Department of Earth Science, Rice University, Houston, Texas, United States of America

Morgan E. Gallagher
Department of Earth Science, Rice University, Houston, Texas, United States of America

Caroline A. Masiello
Department of Earth Science, Rice University, Houston, Texas, United States of America

Zuolin Liu
Department of Earth Science, Rice University, Houston, Texas, United States of America

Brandon Dugan
Department of Earth Science, Rice University, Houston, Texas, United States of America

Yifeng Chen
State Key Laboratory of Water Resources and Hydropower Engineering Science, Key Laboratory of Rock Mechanics in Hydraulic Structural Engineering, Wuhan University, P. R. China

Chuangbing Zhou
State Key Laboratory of Water Resources and Hydropower Engineering Science, Key Laboratory of Rock Mechanics in Hydraulic Structural Engineering, Wuhan University, P. R. China

Wenzel Kröber
Institute of Biology, Geobotany and Botanical Garden, Martin-Luther-University Halle-Wittenberg, Halle (Saale), Germany

Shouren Zhang
State Key Laboratory of Vegetation and Environmental Change, Institute of Botany, the Chinese Academy of Sciences, Beijing, China

Merten Ehmig
Institute of Biology, Geobotany and Botanical Garden, Martin-Luther-University Halle-Wittenberg, Halle (Saale), Germany

Helge Bruelheide
Institute of Biology, Geobotany and Botanical Garden, Martin-Luther-University Halle-Wittenberg, Halle (Saale), Germany
German Centre for Integrative Biodiversity Research (iDiv) Halle-Jena-Leipzig, Leipzig, Germany

Julia L. Barringer
U.S. Geological Survey, USA

Pamela A. Reilly
U.S. Geological Survey, USA

Preface

Engineering soundness and economic feasibility are no longer sufficient criteria for construction of hydraulic works. As a result, environmental considerations have become very much a part of hydraulic analyses. In response to growing environmental concerns, the field of hydraulics has expanded and a new branch, called Environmental Hydraulics, has emerged. The focus of this book, Environmental Hydraulics, is on hydraulic analyses of those environmental issues that are important for protection, restoration, and management of environmental quality. The motivation for this book grew out of the desire to provide a hydraulic discussion of some of the key environmental issues. The focus of first chapter is injection of wastes related to the drilling process, which involve processing cuttings into small particles, mixing them with water and other additives to make slurry, and injecting it into a subsurface geological formation at pressure high enough to fracture the rock. Second chapter highlights on urban tree species show the same hydraulic response to vapor pressure deficit across varying tree size and environmental conditions. Third chapter presents the comparison of experimental results with assumptions used in numerical models. In fourth chapter, the effect of rapid environmental fluctuations (e.g. irradiance due to moving sunflecks) on the heat and mass exchange of the leaf and resulting changes in leaf temperature and hydration status were simulated. Fifth chapter presents biochar-induced changes in soil hydraulic conductivity and dissolved nutrient fluxes constrained by laboratory experiments and sixth chapter presents the stress/strain-dependent hydraulic properties of fractured rock masses under mechanical loading or engineering disturbance achieved. Seventh chapter highlights the potential use of well-known leaf traits from the leaf economics spectrum to predict plant species' drought resistance. Last chapter presents a brief overview of the history of groundwater As contamination and summarizes information about the sources, occurrence and mobility of As in groundwater

Chapter 1

OVERVIEW OF ENVIRONMENTAL MANAGEMENT BY DRILL CUTTING RE-INJECTION THROUGH HYDRAULIC FRACTURING IN UPSTREAM OIL AND GAS INDUSTRY

Mansoor Zoveidavianpoor[1], Ariffin Samsuri[1] and
Seyed Reza Shadizadeh[2]

[1]Universiti Teknologi, Faculty of Petroleum & Renewable Energy Engineering,
Malaysia

[2] Petroleum University of Technology, Abadan Faculty of Petroleum Engineering,
Iran

INTRODUCTION

For the reason of worldwide increased activities of upstream oil and gas industry for future energy demands which will be associated with more waste generation, zero discharge is considered an environmentally friendly approach of complying with environmental legislations. Drilling is one of the major operations in upstream oil and gas industry that can potentially impact the environment through generation of different types of wastes. The drilling process generates millions of barrels of drilling waste each year; primarily used drilling fluids and drill cuttings especially oil-contaminated drill cuttings. In the early years of the oil industry, little attention was given to environmental management of drilling wastes. The rapid development of drilling operation in order to fulfill the global energy demands and so the drilling environmental regulatory requirements have become stricter, drilling and mud system technologies have advanced, and many companies have voluntarily adopted waste management options with more benign environmental impacts that those used in the past. Moreover, it is crucial to find out why drilling wastes are important nowadays, how they generated and by which means those waste could be disposed off with higher efficiency and acceptable HSE and economically concerns. Drill Cutting Re-Injection (DCRI) is one of the processes that developed as an environmentally friendly and zero discharge technology in upstream oil and gas industry.

A variety of oil field wastes are disposed of through injection, such as produced water that re-injected through tens of thousands of wells for enhanced recovery or disposal. Other oil field wastes that are injected at some sites include work over and completion fluids, sludge, sand, scale, contaminated soils, and storm water, among others. The focus of this chapter is injection of wastes related to the drilling process, which involve processing cuttings into small particles, mixing them with water and other additives to make slurry, and injecting it into a subsurface geological formation at pressure high enough to fracture the rock. DCRI has been given other terms by different authors such as fracture slurry injection, grind and inject, and drill cuttings injection.

The most critical aspect in waste injection through hydraulic fracturing (HF) in upstream oil and gas industry, which is DCRI, will be reviewed in this chapter. The subject of this chapter, DCRI, is a specialized area in upstream petroleum industry; even though many brilliant papers presented on various environmental areas, overview papers that present a context for those more specific studies are needed. This chapter will presents in an effort to review the environmental management of DCRI in upstream petroleum industry. The aims are firstly, to review the drilling process and different types of drilling fluid. Afterwards, because it's considered as a key in identifying containment formations to prevent waste migration to water resources and environment in DCRI operations, HF technology will be introduced in the second part of this chapter. Finally, after reviewing the essential parts of DCRI, drilling wastes and HF, the nature of DCRI and its role in environmental management will be presented in details.

OVERVIEW OF DRILLING OPERATION

Oil and gas wells are drilled to depths of several hundred to more than 5,000 meters. Figure 1 shows a schematic of typical drilling rig, which uses a rotating drill bit attached to the end of a drill pipe. Drilling fluids (muds) are pumped down through the hollow drill pipe, through the drill bit nozzles and up the annular space between the drill pipe and the hole. Drilling mud mixture is particularly related to site and hole condition; it used to lubricate and cool the drill bit, maintains pressure control of the well as it is being drilled, and helps to removes the cuttings from the hole to the surface, among other functions. In fact, the technology of mud mixing and treatment has been recognized as a source of pollutants.

Mud and drill cuttings are separated by circulating the mixture over vibrating screens called shale shakers. As the bit turns, it generates fragments of rock (cuttings), which will be separated from the mud by shale shakers that will moves the accumulated cuttings over the screen to a point for further treatment or management. Consequently, additional lengths of pipe are added to the drill string as necessary. As a common practice in drilling of oil and gas wells, when a target depth has been reached according to the drilling plan, the drill string is removed and the exposed section of the borehole is permanently stabilized and lined with casing that is slightly smaller than the diameter of the hole. The main function is to maintain well-bore stability and pressure integrity. (Three sizes of casing depicted inFigure 1). Cement is then is pumped into the space between the wall of the drilled hole and the outside of the casing to secure the casing and seal off the upper part of the borehole. Each new portion of casing is smaller in diameter than the previous portion through which it is installed. The final number of casing strings depends on the total depth of the well and the sensitivity of the formations through which the well passes. The process of drilling and adding sections of casing continues until final well depth is reached.

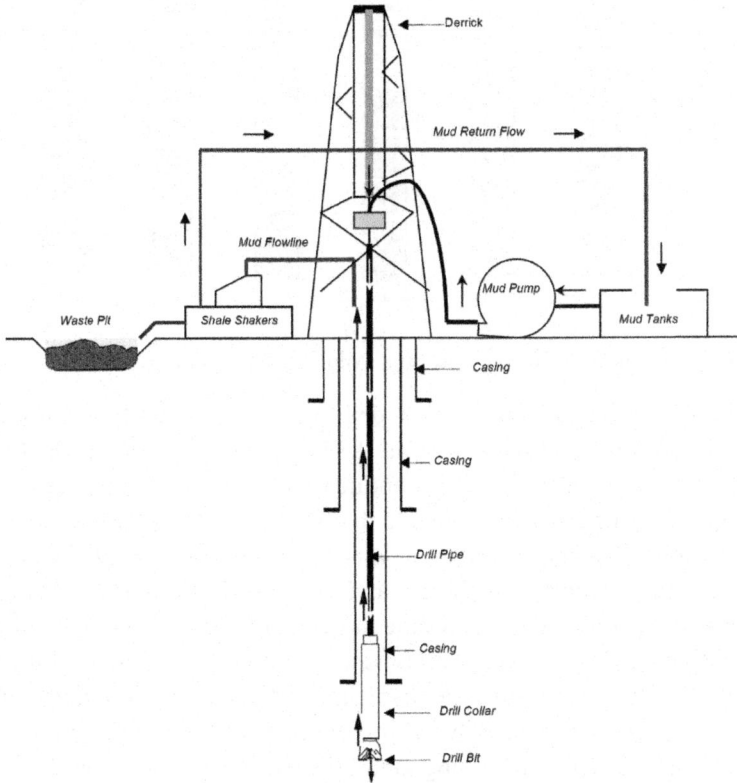

Figure 1. A schematic of a drilling rig (not to scale).

Two primary types of wastes are generated in drilling of oil and gas wells; drill cuttings and drilling fluids. Most drilling fluids contain bentonite clay, water, barite, specialized additives, and some types of muds also contain hydrocarbons. Large volumes of drilling muds are stored in aboveground tanks or pits. The liquid muds pass through the screen and are recycled into the mud system, which is continuously treated to maintain the desired properties for a successful drilling operation. Depending on the depth and diameter of the well bore, the volume of drilling wastes generated from each well varies; typically, several thousand barrels of drilling waste are generated per well. Figure 2 is a demonstration of the generated drilling waste from a 2400 meters well depth that comprises of four different borehole sizes.

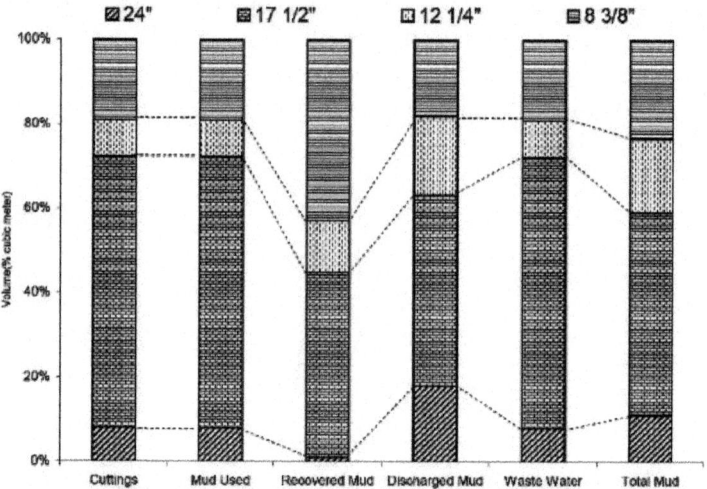

Figure 2. A typical drill cutting and mud volumes for a 2400 meters well depth.

The generation of wastes from drilling fluid and drill cuttings could be recognized at different stages of the drilling operation. When drilling at the first few hundred meters to run conductor casing or surface casing, higher quantities of cuttings are produced; that's because borehole diameter is the largest during this stage. Substantial waste fluid must be handled when drilling deep wells that encountered shale's and/or unstable formations. So, oil based muds (OBMs) is utilized to overcome those problems which will be mixed with other drilling fluids in waste pit and disposed to the environment. Furthermore, higher volume of wastes must be displaced in the completion phase of drilling operation which is replaced by completion fluids and equipment. Physical condition of a waste pit during and after drilling operation is illustrated in Figures 3 and 4, respectively. More details could be found by Shadizadeh and Zoveidavianpoor, (2008).

Figure 3. Mud pit condition during drilling operation.

Figure 4. Mud pit condition after drilling operation.

ENVIRONMENTAL IMPACTS OF DRILLING MUDS

In upstream petroleum industry, drilling is the major operation that can potentially impact the environment. Drilling operation generates a significant volume of wastes. The composition of drilling fluid constituents is depicted in Table 1. Environmentally responsible actions require an understanding of the characteristics of these wastes and how they are generated in order to minimize their environmental impacts by known environmental protection methods. In this section, environmental impacts of a drilling mud will be presented along with a case study on mud pit samples for heavy metals (Cd, Cr, Ni, and Al) concentrations during and after the drilling operation. For more details please consult Shadizadeh and Zoveidavianpoor, 2008 and 2010.

Table 1. Elemental composition of drilling fluid constituents (ppm) (Bleier et al., 1993).

Elements	Water	Cuttings	Barite	Clay	Chrome-lig-nosulfonate	Lignite	Caustic
Aluminum	0.3	40,400	40,400	88,600	6,700	6,700	0.013
Arsenic	0.0005	3.9	34	3.9	10.1	10.1	0.039
Barium	0.01	158	590,000	640	230	230	0.26
Cadmium	0.0001	0.08	6	0.5	0.2	0.2	0.0013
Chromium	0.001	183	183	8.02	40,030	65.3	0.00066
Cobalt	0.001	183	183	8.02	40,030	65.3	0.00066
Copper	0.0002	2.9	3.8	2.9	5	5	0.00053
Iron	0.003	22	49	8.18	22.9	22.9	0.039
Lead	0.5	21,900	12,950	37,500	7,220	7,220	0.04
Magnesium	0.003	37	685	27.1	5.4	5.4	0.004
Mercury	4	23,300	3,900	69,800	5,040	5,040	17,800
Nickel	0.0001	0.12	4.1	0.12	0.2	0.2	5
Potassium	0.0005	15	3	15	11.6	11.6	0.09
Silicon	2.2	13,500	660	2,400	3,000	460	51,400
Sodium	7	206,000	70,200	271,000	2,390	2,390	339
Strontium	6	3,040	3,040	11,000	71,000	2,400	500,000
Cobalt	0.07	312	540	60.5	1030	1030	105

A potential source of heavy metals in drilling fluid is from crude itself. Crude oil naturally contains widely varying concentrations of various heavy metals. In the selected well a combination of water based muds (WBMs) and OBMs had used. As shown in Table 2, the major components of WBMs in the investigated site were barite, salt, starch, bentonite, and lime. The metals of greatest concern, because of their potential toxicity and/or abundance in drilling fluids, include chromium, cadmium, and nickel (Neff, 2002). Some of these metals are added intentionally to drilling muds as metal salts or organometallic compounds. Others are present as trace impurities in major mud ingredients, particularly barite and bentonite. One of the major drilling mud additives used in both WBMs and OBMs in the investigated well is barite. The amount of barite used in the investigated well as shown in Table 2 is 702 tonnes. Barite contains variable amounts of heavy metals and it is the main source of heavy metals in the investigated site. Metals concentrations in mud pit of selected well during and after drilling operation are presented in Figure 5. Chromium concentration was detected in the samples at 0–0.08 ppm. Other heavy metals were also at high levels and showed significantly higher values specially by using OBMs: cadmium 0–0.006 ppm, nickel 0–0.024 ppm, and aluminum 0–341 ppm. However, these heavy metal levels are generally above toxic levels. As shown in Figures 5, the concentrations of cadmium, chromium, and nickel increased progressively in the fourth sampling periods because of the contamination of the mud pit with OBMs that was initiated in the fourth sampling period. Concentration of aluminum increased from the first to the third sampling periods, whereas in the fourth period it shows decreased values from 0.05 ppm to 0.006 ppm. Aluminum was not observed in the fifth and sixth sampling periods but maintained an increased value from the seventh to the end of the sampling periods. In the entire study area, chromium levels ranged from 0 to 0.08 ppm but no concentration was observed after the seventh period of the sampling. This can be explained by the storm runoff water at the investigated well site that washes away all these wastes, especially in the mud pits to other locations or seepage from the discharge pits into the surrounding soils. The statistics of the investigated heavy metals are shown in Table 3.

Table 2. Drilling fluid used in the selected well (Shadizadeh and Zoveidaviampoor, 2010).

Properties		24" hole @ 60 m	17½" hole @ 1510 m	12¼" hole @ 2158 m	8½" hole @ 2330 m
Mud Properties	Mud system	WBM	WBM	WBM	OBM
	pH	10-10.5	10.5-9.8	8-10	9-9.5
	Average salt concentration (mg/l)	2000	185600	297600	380100
	Average calcium concentration (mg/l)	464	2404	3320	231
	YP	11	4-7	6-78	19-27
	PV	35	5-10	8-58	8-12
	Initial Gel	22	3-6	1-13	2
	10 Min. Gel	30	4-8	2-6	3
	Mud lost @ unit (bbl)	0	2588	1252	802
	Density (pcf)	70-62	68-79	79-146	69.5
	Barite (t)	0	27	674.4	0

Mud Material	2	166	168	15
Salt (t)				
Starch (sx)	0	30	727	0
Bentonite (t)	160	750	0	0
Lime (sx)	123	69	222	130
CMS H.V (sx)	0	0	0	17
IRSATROL(sx)	0	0	0	140
Diesel (bbl)	0	0	0	615

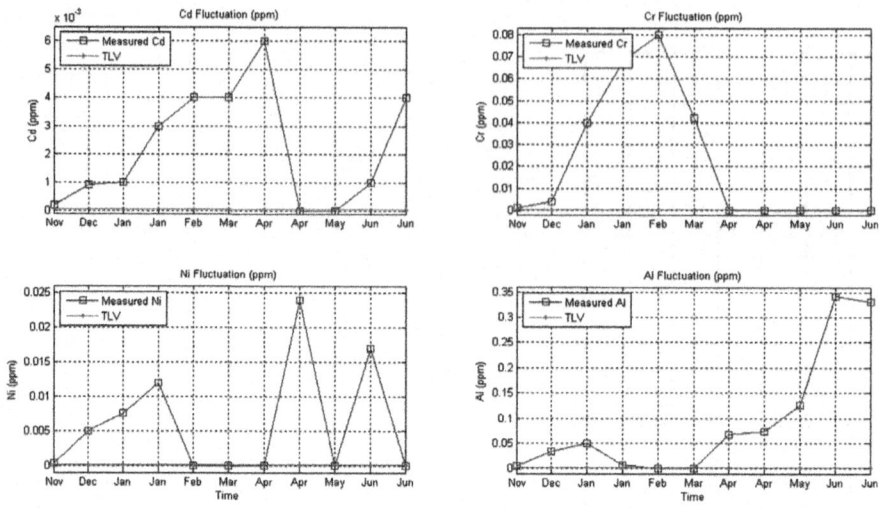

Figure 5. Heavy metals fluctuation during and after drilling operation.

Table 3. Heavy metals statistics in the case study.

Statistics	Heavy Metals (ppm)			
	Cd	Cr	Ni	Al
Max	0.0060	0.0800	0.024	0.341
Mean	0.0022	0.0214	0.005991	0.09396
Median	1.0000e-003	1.0000e-003	0.0003	0.05
Mode	0.0040	0	0	0
Std	0.0021	0.0306	0.008349	0.1255
Range	0.0060	0.0800	0.024	0.341

POTENTIAL EFFECTS ON NATURAL RESOURCES, AND MINIMIZATION STRATEGIES

Drilling wastes can harm ecosystems, plants, and animals and cause health problems in humans. Many materials that are released into reserve mud pits also release drilling wastes into the environment, which calls for public awareness as well. When released heavy metals are discharged into unlined pits the toxic substances in the pits can leach directly into the soil and may contaminate groundwater. Additionally, there is no evidence of zero discharge in lined pits. In contrast to most organic pollutants, trace metals are not usually eliminated from aquatic ecosystems by natural processes due to their non-biodegradability. Both toxic and nontoxic heavy metals tend to accumulate in bottom sediments, from which they may be released by various processes of remobilization. Frequently, these metals can move up the biological chain, eventually reaching humans, where they can cause chronic and acute ailments (Ankley et al., 1993). As presented in the previous sections, routine drilling wastes such as drilling muds and cuttings contain a variety of toxic chemicals; they are known to be hazardous to wildlife, livestock, and human health. If pollutants from oil well drilling build up in the food chain, people who consume those natural resources from the contaminated drilled well area could be at risk of health problems such as genetic defects and cancer. For environmental protection, different strategies are considered; (1) restoring the well site to its natural state after drilling, (2) let the liquid to be evaporated, (3) Bioremediation, (4) multi-pit system, and (5) DCRI, which is the focus of this chapter. Because DCRI deal with the initiation and propagation of a fracture in a rock matrix by means of hydraulic pressure, HF will briefly be discussed in the next section.

HYDRAULIC FRACTURING

Initially, fracturing was a low technology operation consisting of the injection, at low temperature, of a few thousand gallons of napalm into low-pressure reservoirs. Substantially, HF has evolved into a highly engineering and complex procedure. As a technology has improved, so has the number of wells, formations, and fields that can be successfully fractured, increased. The development of high pressure pump units, high strength proppant, and sophisticated fracturing fluids, has meant that deep, low permeability, high temperature, reservoirs can now be fractured (Veatch et al., 1989). This technology is a well-known process, which was originally applied to overcome near wellbore skin damage (Smith, 2006). Since then, it has been expanded to such applications as (1) reservoir stimulation for increase hydrocarbon deliverability, (2) increase drainage area, and decrease pressure drop around the well to minimize problems with asphaltene and/or paraffin deposition, (3) geothermal reservoir recovery, (4) waste disposal, (5) control of sand production, (6) to measure the in-situ stress field and (7) heat extraction (geothermal energy) from deep formations. Obviously, there could be other uses of HF, but the majority of the treatments are performed for the mentioned reasons. HF has made significant contributions to the petroleum industry since its inception (Veatch et al., 1989). By 2009 HF activity has increased 5-fold compared to the investment of a decade earlier and has become the second largest outlay of petroleum companies after drilling (Economides, 2010).

HF is the pumping of fluids at high rates and pressures in order to break the rock. A typical chart of fracturing which shows the common treatment stages is shown in Figure 6. The operation begins with injection of a mixed acid and water named Pre-pad. A mixture of water and a polymer, named Pad, will follows. The fracture will initiated in this stage but contains no proppant. To make the fracture open for fluid flow, a mixture of proppant and the fracturing fluid, which called Slurry will have injected. For more details please consult Daneshy, 2010.

As it clear from section 2, the need has been arises to treat/manage the drill cuttings toward zero discharge by utilization of HF.

Figure 6. A typical fracturing chart illustrates the steps to HF a well (Daneshy, 2010).

There are both similarities and distinct differences between HF and DCRI which shown in Table 4. More details could be found from Arthur (2010).

Table 4. Comparison between DCRI and HF.

Issue	Drill Cutting Re-injection	Hydraulic Fracturing
Target interval	Non Reservoir	Reservoir
Pumping period	Long-term	Short-term
Pumping pressure	Fracture	Fracture
Slurry mixture	Cuttings and fracturing fluids	Proppant and fracturing fluids
Fracture containment study	Essential	Essential

WASTE MANAGEMENT BY DCRI

An Overview

Even though the generation of drill cuttings is a certain result of drilling, those wastes can be treated and/or managed in a number of ways. A summary chart on different drilling wastes management options are presented in Figure 7. As mentioned earlier, the focus of this chapter will be on DCRI.

Valuable literature available regarding the disposal options including: lessons learned concerning biotreating exploration and production wastes (McMillen et al. 2004), successful cases of fixation (Zimmerman and Robert, 1991), converting cuttings into a valuable sources by using vermicomposting (Paulse, 2004), and thermal treatment (Bansal and Sugiarto, 1999).

As summarized in Figure 7, environmental management of drilling wastes may be categorized in three options; waste minimization, recycle/reuse, and disposal. The first and second options are not addressed here. Table 5 shows a comparison among disposal methods which may classified into fixation, thermal treatment, DCRI, and bioremediation/composting. Among the four methods for disposal option that may be considered when deciding on waste management options, the focus of this chapter is on DCRI.

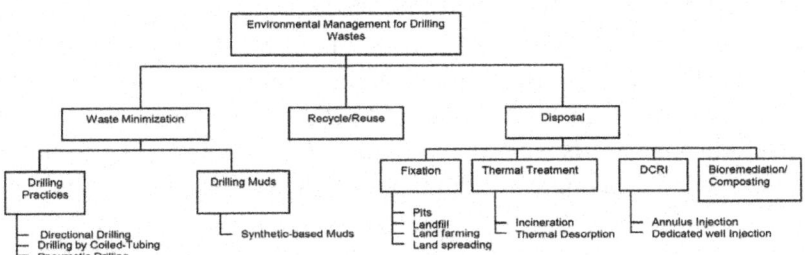

Figure 7. Different approaches in environmental management for drilling wastes.

Comparative assessments on alternative disposal options are outlined in Table 5. As clearly shown, environmental impacts and safety risks, which are the most important factors among others, have low level degree and therefore its vulnerability as the best option increases to be adopted as the environmentally friendly drilling waste disposal process. in addition to zero discharge, other advantages of DCRI include; no transportation concerns, no future cleanup responsibilities by the operator, full control over the waste management process, world wide applicability, and its favorable economics. According to Reddoch, (2008): "DCRI is simply the lowest cost, easiest course of action for most drilling operations."

Table 5. Qualitative and quantitative comparison in disposal approaches.

Comparison Factors	Fixation	Thermal Treatment	DCRI	Bioremediation/ Composting
Environmental Impact	Low	High	Low	Medium
Cost	$9-10/bbl a	$90/metric ton a	$5/bb'b	$500/cubic meter c
Safety Risks	High	High	Low	Medium
Technical	Low	Medium	High	Medium

The question is raised that what is the relationship between environmental management and DCRI? It's clear that DCRI process will maintain waste containment in a target interval with zero discharge and consequently low HSE risks. Other goals such as cost management and asset management are not covered in this chapter. For more details please consult Bruno et al. (2000).

We can visualize DCRI to loss of circulation of drilling fluids in conventional under balanced drilling operation. Also, it's quite similar to HF operation, because we need to propagate the fractures in the selected horizon and this goal will be achieved by utilization of fracture propagation models which conventionally employed in HF treatment.

Cuttings may be re-injected into the annulus of a well being drilled or into a dedicated well. In annulus injection, cutting would be stored until the desired formation is reached. Whereas in dedicated disposal well, one or more dedicated disposal wells would be drilled and drill waste systems put in place in those wells. A schematic of both types of DCRI is shown in Figure 8.

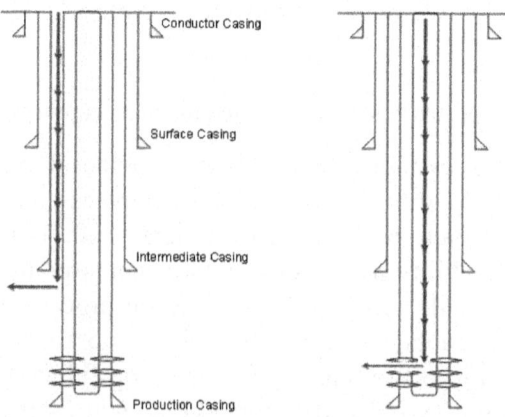

Figure 8. Two major types of DCRI; annulus injection (left) and dedicated well (right).

Drill cuttings may be injected into subsurface geological formations at the drilling site, offshore or onshore and would provide a complete disposal solution. Its worth to note that onshore operations have a wider range of options than offshore operations.

Readers may be asks why this process is called drill cutting re-injection? That's because drill cuttings will be returned back to their origin, deep beneath the Earth's surface.

A sketch of basic setup and flow of DCRI process is shown in Figure 9. Drill cuttings and other oilfield wastes are slurried by being milled and sheared in the presence of water. The resulting slurry is then disposed of by pumping it into a dedicated disposal well, or through the open annulus of a previous well into a fracture created at the casing shoe set in a suitable formation.

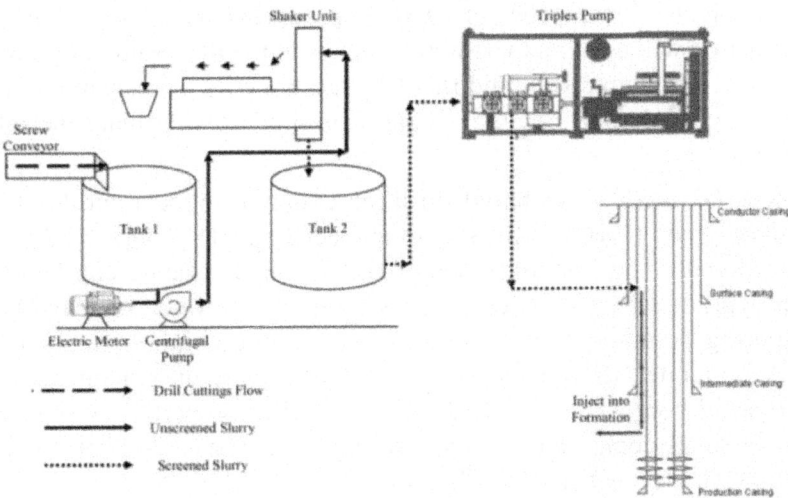

Figure 9. A sketch of basic setup and flow of DCRI.

A Case Study

In addition to the drill cuttings and drilling fluids, various waste streams need to be handled and disposed of properly include: produced water, contaminated rainwater, scales, and produced sand. DCRI provides a secure operation by injecting cuttings and associated fluids up to several thousand meters below the surface into hydraulically created fractures. In order to guarantee containment within the selected underground formation and perform sufficient design of surface facilities, simulations are performed for the anticipated downhole waste domain.

In this regard, a feasibility study was performed to show the possibility of DCRI in Ahwaz oilfield located in southern Iranian oilfields. The possibility of annular injection and dedicated injection wells was investigated in this study. The objectives were to (1) estimate the volume of drilling waste produced from drilling of each wellbore of the field, (2) select the most appropriate disposal formation in the field, and (3) determine whether the drill wastes can be safely injected into a dedicated well or annular space. Numerous scenarios were considered in the feasibility studies to ensure safe containment of any injected drilling waste. More details could be found by Shadizadeh and Zoveidavianpoor, (2011).

The volumes of drill cuttings and muds, type of utilized mud, and geological information are shown inTable 6. The required data to conduct this study is depicted in Table 7.

In particular, the expense of DCRI requires that the operator knows how the formation will respond to treatment, and whether the treatment design such as selection of pump rates, fluid rheology, accurate rock mechanic properties, pumping schedule and fracture propagation model, will create the intended fracture.

Most 2D models are based on three common models entitles Perkins-Kern-Nordgren (PKN), Khristianovic-Geertsma de Klerk (KGD), and Radial models. The first and second models which assume constant height, are appropriate when the stress contrasts are high between the pay layer and neighboring formations and these contrasts follow lithologic boundaries. For Radial model, its better works in a setting where the fracture grows in a formation of homogeneous stress and mechanical properties so that fracture height is small compared to formation layer thickness. A brief comparison among 2D models is listed in Table 8.

Table 6. Generalized geologic data along with drill cuttings and mud volumes.

Depth (m)	Formation Name	Column	Setting depth (inch m)	Lithology	Hole Size (inch)	Cutting volume (bbl)	Mud volume (bbl)	Mud type
1550	Aghajari			Marl with Sandston bonds	26	132	4400	WBM
1660	Mishan			Marl with Limestone basement	17 1/2	2040	2800	WBM
2332	Gachsaran			Marl, Salt, Anhydrate.	9 5/8	219	3500	WBM
3590	Asmari			Limestone with Sandstone	8 1/2	73	800+400	WBM+OBM

Table 7. Explanation of required data for DCRI simulation.

Required data	Description
Injection batch volumes and injection rates	Injection of the slurry is often conducted intermittently in batches into the selected disposal formation, followed by a period of shut-in. depending upon the batch volume and the injection rate, each batch injection may last from less than an hour to several days or even longer.
Minimum in situ stress	Most important in fracture simulation that controls fracture-height growth, fracture azimuth and vertical and horizontal orientation, fracture width, treatment pressures, fracture conductivity, and wastes containment in disposal horizon.
Pore pressure	Very critical parameter to planning and carrying out successful DCRI, because the stress state of the poro-elastic medium is directly influenced by pore pressure or reservoir pressure.
Young's modulus	Is the ration of longitudinal stress to longitudinal strain, which has significant effect on fracture geometry, especially on fracture width
Poisson's ratio	Is a measure of the compressibility of material perpendicular to applied stress that has significant effect on fracture geometry
Casing setting depths and injection point	The target which the slurry has to be injected via annulus or dedicated well.
Fluid leak-off data	Means the leaking of fluid from the surface of a fracture into the surrounding rock formation. It's an important parameter controlling the size and geometry of the hydraulically induced fracture.
Slurry rheology	The study of the deformation and flow of matter, that crucial for maintaining zonal isolation.
Fracture toughness	Is an important parameter in fracture modeling and is a measure of a material's resistance to fracture propagation

The main advantage of a more advanced method such as pseudo 3D (P3D) over 2D models is that it does not require estimating fracture height, but it does require input of the magnitude of minimum horizontal stress in the zone to be fractured and in the zones immediately above and below.

Table 8. Comparison of 2D fracture models.

Model Name	Plan View	Cross Section View	Pressure-Time Trend	Description
PKN				Cross section= Elliptical Width ∝ height Width < KGD Length > KGD Suitable when: length>height
KGD				Cross section: Rectangular Width ∝ height Suitable when: length<height
Radial				Cross section= Elliptical Suitable when: length=height

Simulation Study

Based on the petrophysical logs, from lithological point of view, the relevant formations are fairly marl, sandstone and limestone with an average rock density 2.33gr/cm3. The vertical stress was calculated by integrating the available bulk density with respect to depth. Vertical stress gradient is calculated as Eq. (1):

$$\sigma_v = 0.433\rho_{OB} = 0.433 \times 2.33 = 1psi \: / \: ft \tag{1}$$

The values of minimum horizontal stress of Aghajari, Mishan, and Gachsaran formations were 1693, 3847, and 4489, respectively which calculated from Eq. (2) is:

$$\sigma = \frac{\upsilon(\sigma_v D - 2p) + p}{1 - F\upsilon} \tag{2}$$

Elasticity of the formations is determined with the sonic log. Table 9 lists the values of the static elastic Young's modulus, Poisson's ration, leak-off coefficient for the different formation zones shown inTable 5. These values are based on the dynamic elastic Young's module obtained from sonic and density logs. Static elastic Young's module values are often two times smaller than dynamic values derived from sonic logs. The elastic Young's module values that are listed in Table 9 are arbitrarily one-half of their dynamic equivalents. The larger than usual values were used in the analysis for these shallower formations.

Table 9. Formation properties used in fracture simulations.

Zone Name	Zone Height (ft)	Poisson's Ratio*	Pore Pressure* (psi)	Fracture Gradient* (psi/ft)	In-situ Stress (psi)	Young's Modulus (MM psi)	Leak-off Coefficient (ft.min$^{-0.5}$)	Toughness (psi.min$^{-0.5}$)
Aghajari	5250	0.29	1050	0.650	1693	2	0.00081	1000
Mishan	330	0.31	2567	0.714	3847	2	0.00087	1000
Gachsaran	330	0.36	2878	0.780	4489	2	0.00089	1000

Table 10. Physical properties of injected cuttings slurry.

Density	1.26 SG
Particle Loading	80/100 mesh proppant at a consternation of 2 PPG
Apparent Viscosity	161 cp @ 170 1/S
Non-Newtonian power law indices	N=0.26; k=0.15

Slurry rheology design did not performed in this paper and is beyond the scope of this article; however by considering the cuttings brought out of the wellbore and the drilling muds used in Ahwaz oil field, a reasonable result was earned of rheology characteristics of the injection slurry. It was assumed that the cuttings slurry with final rheological condition would behave in a manner similar to the drilling muds used in Ahwaz oil field. Slurry and solid properties are selected from past DCRI operation in literature (Abou-Sayed et al., 2002), which is also near the nature of selected drilling fluids and cuttings lithology of the Ahwaz oilfield and are presented in Table 10.

For the scenario of casing injection into a dedicated injection well, the intermediate casing can be set on top of Gachsaran formation. The casing is assumed to perforate at a depth about 50 m under the Aghajari formation and the center of the Mishan formation. The initial fracture is assumed to be at the center of the perforated interval.

Simulation Results

After determining all required data, a fracture geometry model was selected for use in the simulation. As described previously, the dedicated injection mechanism is more suitable for the Mishan formation because it is deep enough and consists of limestone lithology in a base that is appropriate for reinjection. In each case, the geometry reported indicates the maximum fracture achieved when slurry is pumped continuously. The simulation study is represented for both dedicated wells that consist of two cases and annulus injection well mechanisms.

Dedicated Well Injection Mechanism

Two cases will be presented in this section, which differs in the magnitude of two parameters; Young's modulus and leak-off coefficient.

Case 1: For a case like Ahwaz oilfield in which the vertical distribution of the minimum in situ stress is uniform, a circular fracture is expected. The formations had Young's modulus and leak-off coefficients as shown in Table 9. For this simulation, the fracturing would initiate from the Mishan formation and broke through the Aghajari formation but was still 4,700 ft. below the surface when 50,000 bbl of slurry had been injected continuously. Table 11 summarizes the results of this simulation. Figures 10 and 11show predicted the fracture shape plot after injection of 50,000 bbl continuously at 5 bbl/min.

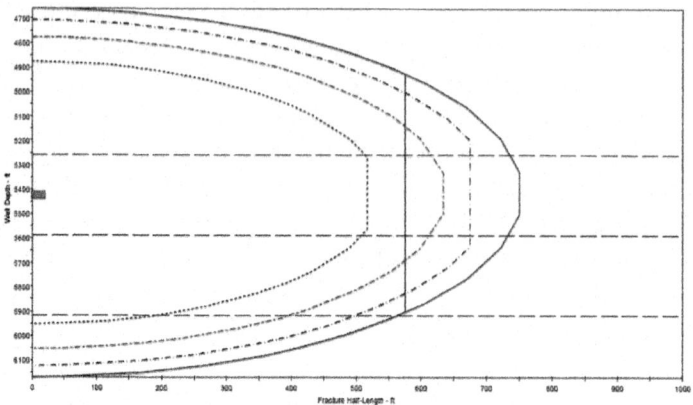

Figure 10. Fracture geometry history- Radial model (case 1).

Case 2: The formations were assumed to have Young's modulus that was twice those listed in Table 9. Also, the leak-off coefficient for formations used was specified as one half of the value listed in Table 9. This extremely large modulus and small leak-off resulted in a much larger fracture. Consequently, this is a very conservative analysis. Even for this very conservative case, the fracture that broke through the Aghajari formation was still 4,550 ft. below the surface when almost 50,000 bbl of slurry had been injected continuously at 5 bbl/min. Table 11 summarizes the results of the fractures created.

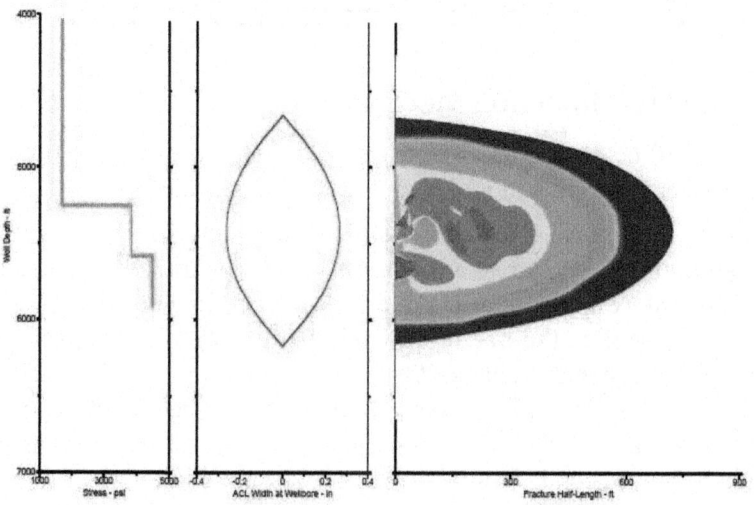

Figure 11. Fracture profile and cuttings concentration- Radial model (Case 2).

Table 11. Simulation's results of dedicated well injection.

Parameters	Case 1	Case 2
Slurry volume (bbls)	50000	50000
Fracture half-length (ft)	576	795
Fracture width at well (in)	0.276	0.237
Net pressure (psi)	71	89
Max surface pressure (psi)	1755	1807
Shut-in time (hrs)	13	26

Annulus Well Injection Mechanism

Annulus injection is only possible if the annulus of an intermediate casing string in an existing well is open to a suitable subsurface formation and this well satisfies a range of screening criteria. The allowable injection pressures for annulus injectors are often lower than the allowable pressures for dedicated wells because of casing burst and collapse limitations for annulus injectors. By considering the lithology and casing design of Ahwaz oilfield, it is concluded that the planned slurry injection would occur in an 18 5/8-in./13 3/8-in. annulus. Other annuli are not possible for injection because they are open to unsuitable subsurface formations. To prevent the upward migration of injected wastes to the surface, the 18 5/8-in. casing string should set at about 1,000 ft. and cement back to the surface, and the 13 3/8-in. string should cement back to 1,500 ft. below the previous casing shoe. This provides a window across the Upper Miocene marl and sandstone of Aghajari formation. For this simulation, the fracturing initiated from the Aghajari formation and grew toward the surface but was still 500 ft. below the surface when 15,000 bbl of slurry had been injected continuously. Table 12 presents the different parameters of the fracture created. Figure 12 shows the predicted fracture shape plot after injection of 15,000 bbl continuously at 5 bbl/min.

Table 12. Simulation results of annular well injection.

Parameters	Radial Model
Slurry volume (bbls)	15000
Fracture half-length (ft)	230
Fracture width at well (in)	643
Net pressure (psi)	2.39
Max surface pressure (psi)	968

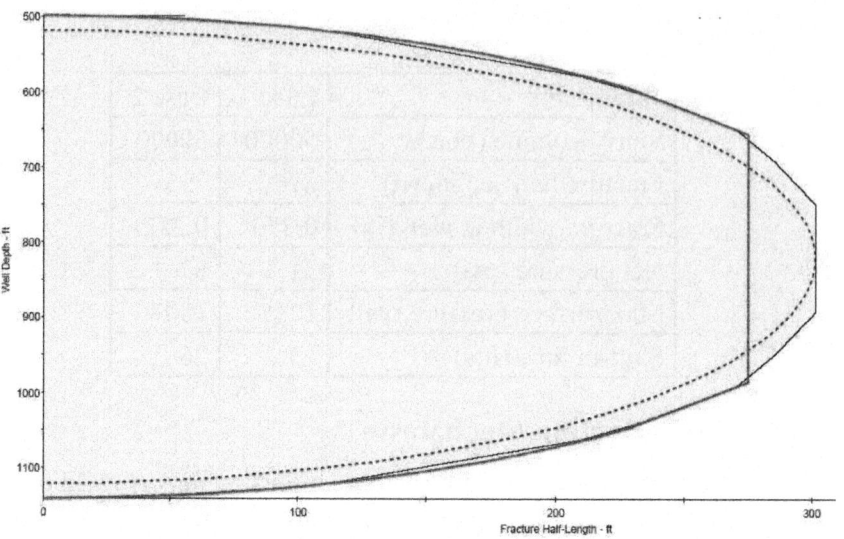

Figure 12. Fracture geometry history- annular injection well (Radial model).

DISCUSSION AND CONCLUSIONS

Assessment of environmental impacts of drilling operations and searching for the methodologies to protect nature and resources against negative impacts has become an interesting topic during the last thirty years in upstream petroleum industry. The necessity of environmental management in drilling operation, lessons learned, and a brief list of mitigation options from wastes generated by drilling operations in a southern Iranian oilfield were documented previously (Shadizadeh and Zoveidavianpoor, 2008, 2010). Most of the drilling wastes sources in the oilfields are OMBs and oily cuttings associated with them. Unfortunately, lack of demanding regulations regarding drilling waste discharge leaves room for drilling companies to leave the waste in the nature without treating them (Shadizadeh and Zoveidavianpoor, 2008, 2010). This chapter tried to study the possibilities of waste prevention and zero discharge by utilization of serviceable methods in drilling well sites. So, the feseability study of DCRI at Ahwaz oilfield was initiated and conducted to fulfill the needs of growing upstream petroleum industry in Iran. This article focuses on the design aspect of the technology. Design guidelines are given to include data required for project planning, injection scheme (annulus versus dedicated well) selection, injection well and disposal formation identification, subsurface fracturing simulation, and waste containment. Operational procedures such as slurry rheology were the area of investigation in this study; however, it was determined as input data for simulation that has conformity with the nature of selected drilling fluids and cuttings lithology of the Ahwaz oilfield. Well design requirements and estimation of disposal capacity in each of the injection schemes was performed. This study shows that the DCRI study at Ahwaz oilfield is practical by considering some potential risks involved in any DCRI job. It was determined that by using HF technology, drilling wastes could be reinjected to the Mishan formation or even a shallow formation such as the Aghajari formation without propagation of the fractures to the surface or near wellbores. The thickness of the Aghajari formation provides an appropriate barrier to upward growth of DCRI at the Mishan formation through a dedicated injection well. A dedicated injection well is more typical of longer-term, permanent injection operations and is more common onshore (Keck, 2002). It is simulated that a large amount of drilling waste can be safely injected to Mishan formation. The maximum surface pressure required to inject the slurry is in a range of 1,500 to 2,000 psi, which is completely reasonable with the current surface facilities. The propagation of the fracture to the surface showed to be efficient and safe in the two cases performed in the dedicated well injection scheme. The simulation results confirm that the drilling wastes produced from each wellbore could be injected through annulus of the same wellbore while drilling. The selected annulus for annular reinjection in Ahwaz oilfield is not very favorable because the injection point is close to the surface. As described before, other annuli are not suitable due to abnormal pressure or hydrocarbon bearing. The annular reinjection at Ahwaz oilfield has many serious risks that need a careful job planning. However, the amount of drilling wastes from a typical wellbore is not high and the simulations confirm that 15,000 bbl wastes from a typical wellbore can be injected without serious danger. Advantages and disadvantages of annular and dedicated well injectors are presented in Abou-Sayed and Guo (2001).

It should be noted that the simulations represent upper-bound predictions of the fracture geometry because low leak-off and high Young's modulus is assumed in different formations. In reality, even a very limited change in the amount of fluid leak-off, coupled with intermittent batch injection of slurry, would result in a significantly reduced fracture area. The analyses confirm the integrity and suitability of the injection operations and ensure safe application of this technology at Ahwaz oilfield.

Acknowledgement

The authors of this chapter would like to express thier gratitude to University Teknologi Malaysia due to the valuable supports during this study.

REFERENCES

1. A. S. Abou-Sayed, G. Quanxin, J. D. Mc Lennan, J. T. Hagan, 2000Case studies of waste disposal through hydraulic fracturing. Presented at the Workshop on Three Dimensional and Advanced Hydraulic Fracture Modeling, held in conjunction with the Fourth North American Rock Mechanics Symposium, July 29, 2000, Seattle.

2. A. S. Abou-Sayed, Q. Guo, 2001Design considerations in drill cuttings re-injection through downhole fracturing. Paper IADC/SPE 72308 presented at the IADC/SPE Middle East Drilling Technology, Bahrain. October 2224

3. G. T. Ankley, V. R. Mattson, E. N. Leonard, C. W. West, J. L. Bennett, 1993Predicting the acute toxicity of copper in freshwater sediments: Evaluating the role of acid-volatile sulfide. Environ. Toxic Chem. 12315320

4. Arthur, J.D., "A Comparative Analysis of Hydraulic Fracturing and Underground Injection", Presented at the GWPC Water/Energy Symposium, Pittsburgh, Pennsylvania.September 25292010

5. K. M. Bansal, Sugiarto, 1999Exploration and Production Operations-Waste Management A Comparative Overview: U.S. and Indonesia Cases. Paper SPE 54345 presented at the SPE Asia Pacific Oil and Gas Conference, Jakarta, Indonesia, April 2022

6. R. Bleier, A. J. J. Leuterman, C. Stark, 1993Drilling fluids making peace with the environment. J. Petrol. Tech. 45610

7. M. Bruno, A. Reed, S. Olmstead, 2000Environmental Management, Cost Management, and Asset Management for High-Volume Oil Field Waste Injection Projects. Paper SPE 51119 presented at the IADC/SPE Drilling Conference held in New Orleans, Louisiana, 2325February 2000.

8. A. Daneshy, 2010Hydraulic Fracturing to Improve Production. The Way Ahead: Tech 101. 6(3), 14-17.

9. M. J. Economides, 2010Design Flaws in Hydraulic Fracturing. Paper SPE 127870 presented at the SPE International Symposium and Exhibition on Formation Damage Control. 1012February. Lafayette, Louisiana, USA.

10. R. G. Keck, 2002Drill cuttings injection: A review of major operations and technical issues. Paper SPE 77553 presented at the SPE Annual Technical Conference and Exhibition, San Antonio, TX, September 29October 2.

11. S. J. Mc Millen, N. R. Gray, 1994Biotreatment of Exploration and Production Wastes. Paper SPE 27135 presented at the Second International

Conference on Health, Safety & Environment in Oil & Gas Exploration and Production, Jakarta, Indonesia, January 2527

12. S. J. Mc Millen, R. Smart, R. Bernier, 2004Biotreating E&P Wastes: Lessons Learned from 19922003Paper SPE 86794 presented at the 7th SPE International Conference on Health, Safety and Environment in Oil and Gas Exploration and Production, Calgary, Alberta, Canada, 29-31 March 2004.

13. J. M. Neff, 2002Bioaccumulation in Marine Organisms. Effects of Contaminants from Oil Well Produced Water. Amsterdam: Elsevier Science Publishers. 978-0-08043-716-3pages

14. J. E. Paulse, J. Getliff, R. Sørheim, 2004Vermicomposting and Best Available Technique for Oily Drilling Waste Management in Environmentally Sensitive Areas. Paper SPE 86730 presented at the 7th SPE International Conference on Health, Safety and Environment in Oil and Gas Exploration and Production, Calgary, Alberta, Canada, 2931March 2004.

15. J. Reddoch, 2008Why cuttings reinjection doesn't work everywhere-or does it? World Oil, January 20086970

16. S. R. Shadizadeh, M. Zoveidavianpoor, 2010A Drilling Reserve Mud Pit Assessment in Iran: Environmental Impacts and Awareness', Petroleum Science and Technology, 28: 14, 1513-1526. 1091-646610916466print/1532-2459 online

17. S. R. Shadizadeh, M. Zoveidavianpoor, 2008Environmental Impact Assessment of Onshore Drilling Operation in Iran, Abadan, Iran: Petroleum University of Technology.

18. S. R. Shadizadeh, S. Majidaie, M. Zoveidavianpoor, 2011Investigation of Drill Cuttings Reinjection: Environmental Management in Iranian Ahwaz Oilfield', Petroleum Science and Technology, 29: 11, 1093-1103. 1091-646610916466print/1532-2459 online.

19. R. W. Veatch Jr, Z. A. Moschovidis, 1989An Overview of Recent Advances in Hydraulic Fracturing Technology. Paper SPE 14085 presented at the International Meeting on Petroleum Engineering. 1720March. Beijing, China.

20. P. K. Zimmerman, J. D. Robert, 1991Oil-Based Drill Cuttings Treated by Landfarming. Oil & Gas Journal, 89328184

Chapter 2

URBAN TREE SPECIES SHOW THE SAME HYDRAULIC RESPONSE TO VAPOR PRESSURE DEFICIT ACROSS VARYING TREE SIZE AND ENVIRONMENTAL CONDITIONS

Lixin Chen[1], Zhiqiang Zhang[1] and Brent E. Ewers[2]

[1] Key Laboratory Soil and Water Conservation and Desertification Combating, Ministry of Education, College of Soil and Water Conservation, Beijing Forestry University, Beijing, People's Republic of China

[2] Program in Ecology, Department of Botany, University of Wyoming, Laramie, Wyoming, United States of America

ABSTRACT

Background

The functional convergence of tree transpiration has rarely been tested for tree species growing under urban conditions even though it is of significance to elucidate the relationship between functional convergence and species differences of urban trees for establishing sustainable urban forests in the context of forest water relations.

Methodology/Principal Findings

We measured sap flux of four urban tree species including *Cedrus deodara*, *Zelkova schneideriana*, *Euonymus bungeanus* and *Metasequoia glyptostroboides* in an urban park by using thermal dissipation probes (TDP). The concurrent microclimate conditions and soil moisture content were also measured. Our objectives were to examine 1) the influence of tree species and size on transpiration, and 2) the hydraulic control of urban trees under different environmental conditions over the transpiration in response to VPD as represented by canopy conductance. The results showed that the functional convergence between tree diameter at breast height (DBH) and tree canopy transpiration amount (E_c) was not reliable to predict stand transpiration and there were species differences within same DBH class. Species differed in transpiration patterns to seasonal weather progression and soil water stress as a result of varied sensitivity to water availability. Species differences were also found in their potential maximum transpiration rate and reaction to light. However, a same theoretical hydraulic relationship between G_c at VPD=1 kPa (G_{cref}) and the G_c sensitivity to VPD $(-dG_c/dlnVPD)$ across studied species as well as under contrasting soil water and R_s conditions in the urban area.

Conclusions/Significance

We concluded that urban trees show the same hydraulic regulation over response to VPD across varying tree size and environmental conditions and thus tree transpiration could be predicted with appropriate assessment of G_{cref}.

INTRODUCTION

Establishing urban forest/trees is widely accepted as one of the critical approaches to combat the rapid urbanization associated environmental, ecological, and human health problems[1]–[3]. The potential impact of urban forest/tree development on water resources availability is questioned due to the projected more sever water stress for urban area under global climate change [4], [5] and world-wide reorganization of reduced water yield at watershed scale from increased forest coverage due to land conversion, industry plantation, and forest ecological restoration programs [6], [7]. Our current understanding on forest/tree and water relations under urban environment is very limited [8]. Therefore, it is vital for assessing accurately the role vegetation may play in affecting urban water budget by understanding the eco-physiological response and environmental control of tree water use in urban environments [9].

Practically, tree species selection is of significance for the urban forest development from the water resources management viewpoint as observed species differences in controlling transpiration rate [10]–[12] add variability to the water flux leaving ecosystems and consequently to the hydrological cycle under natural environment [13], [14]. These differences can lead to a substantial spatial heterogeneity of canopy transpiration [15] and the amount of stand transpiration [16], [17]. For example, *Fraxinus excelsior* showed a greater tendency in enhancing transpiration towards the edge than *Quercus robur* did in natural mixed deciduous woodland stands [18]. Annual stand-scale transpiration from bamboo forest is higher than that from the coniferous forests in western Japan [19]. Dynamically, stand transpiration can be significantly enhanced during succession as species shift from woodland oaks and elms to Savanna oaks [20]. Therefore, it is highly probable that species differences in transpiration would influence local hydrology [6], [20]–[24].

Despite the convincing evidence of species effects on transpiration rates, analysis of the relationship between transpiration and tree size or hydrological control [25]–[27] have revealed the convergence in functioning across plant species. Meinzer et al. found that variation in diameter at breast height (DBH) accounted for 91% of the variation in total daily sap flux in the outermost 2 cm of sapwood by comparing 24 different species in a Panamanian rain forest[28]. Thus, the examination of species influence on transpiration is complicated by structural factors such as root, leaf and sapwood area. Regulation of canopy conductance (G_c) exerts a major control over plants transpiration. Stomatal closure is a well known mechanism to regulate plant water status and avoid fatal xylem cavitation under decreasing humidity [29]. Despite the fact that vulnerability to cavitation is highly species specific [30], a synthesis showed that the stomatal response to VPD can be described by a proportionate relationship between the G_c sensitivity to VPD ($-dG_c/d\ln VPD$ or $-m$) and G_c at VPD=1 kPa (G_{cref}) across 40 species and this relationship was theoretically verified [27]. The theoretical underpinnings suggests that for isohydric species that regulate minimum leaf water potential to prevent excessive cavitation, hydraulic control over transpiration and the ratio between G_c sensitivity to VPD and G_{cref} should be the same despite varying G_{cref} between species.

Urban areas tend to exhibit higher air temperature (e.g. "heat island" phenomena), more complex wind turbulence, and increased evaporative demand than the adjacent rural areas due to dense buildings and road system pavement. In addition, urban trees are usually planted in isolation or in rows. These characteristics may induce varied energy partitioning and thus microclimatic difference from natural forest stands. A natural question to ask is "Are there any differences for urban trees from that of natural forest in transpiration control?" as the tree hydraulic control tends to be related to the originating habitat [27]. Therefore, our study was conducted to explore the functional convergence across urban tree. Specifically, our objectives are to examine 1) the influence of tree species and size on transpiration, and 2) the hydraulic control of urban trees over the transpiration in response to VPD as represented by canopy conductance under different environmental conditions.

MATERIALS AND METHODS

Site Description and Tree Selection

The study was conducted in Laodong Park (38°54′N, 121°37′E), Dalian City, Liaoning Province, China. The temperate maritime climate is characterized by mean annual temperature ranging from 8°C–10°C and mean annual precipitation 550–800 mm with 60%–70% in summer.

The study plot is a man-made tree patch located in the north side of the park consisting of *Cedrus deodara* (Roxb) Loud., *Zelkova schneideriana* Hend.-Mazz., *Euonymus bungeanus* Maxim, and *Metasequoia glyptostroboides* Hu et cheng. The trees were originally planted with varying sizes to meet the aesthetic demand. Therefore, the DBH distribution is not the result of growth competition. Due to the restriction from park management, three trees were selected for each species with different DBHs. Trunks of sampled trees were bored to measure the sapwood. It was easy to distinguish the sapwood from heartwood by color difference contrasted by water content (Table 1, also see [31]).

Table 1. Characteristics of all sampled trees for sap flow measurement.

Species	DBH(cm)		Projected Crown area (m²)		Tree Height (m)		Sapwood area (cm²)	
	Mean	SE	Mean	SE	Mean	SE	Mean	SE
C. deodara	17.27	3.02	16.56	5.19	7.03	0.93	33.99	8.19
Z. schneideriana	13.93	3.11	18.77	4.32	5.30	0.49	18.99	5.84
M. glyptostroboides	19.33	4.10	4.80	2.10	11.60	1.06	36.63	14.46
E. bungeanus	13.50	2.53	34.96	10.17	5.63	0.48	51.44	12.31

coi:10.1371/journal.pone.0047882.t001

Sap flow Measurement and Canopy Conductance

Sap flux (J_s) of individual trees was measured continuously from June 25th to October 17th in 2009 using thermal dissipation probes (Dynamax, USA) [32]. A square of 5 cm*5 cm bark at a height of 1.3 m was removed to expose the cambium and the probes were installed. Thirty mm probes were used in all trees except two smaller *Z. schneideriana* trees whose shallow sapwood required the use of 20 mm probes. After the installation, the probes were sealed with silicon foam to prevent rain water infiltration and shielded with aluminum foil to insulate external thermal influences. The output from the probes was recorded as half-hourly average from measurements made at 10 s intervals and stored in a CR1000 data logger (Campbell Scientific, Inc., Logan, UT, USA). The sensors, the heaters and the data loggers were all powered by a 700 mA storage cell. Sap flux can be calculated according to standard calibration for the

TDP method based on temperature differences between the two probes [32]:

$$J_s = 0.0119 * [(\Delta T_m - \Delta T)/\Delta T]^{1.231} \tag{1}$$

where J_s is sap flux density (g cm^{-2} s^{-1}), ΔT_m is the maximum temperature difference between sensors during a day (°C), ΔT, Temperature difference between sensors at any given time (°C). This equation works well for non-porous species [33], and we acknowledge the existence of limitations when using it for ring-porous species due to non-uniform sap flow [34], [35]. However, given the small sapwood width for our studied trees, we assume the radial velocity gradient is not steep. Moreover, the contact between the probes and non-conducting xylem will give rise to greater errors [34],therefore, the sap flux density of only one depth was examined. Studies show good agreement between sap flux calculated by this equation and other independent transpiration measurements for ring-porous trees [36].

Daily canopy transpiration (E_c, mm d^{-1}) then can be calculated as

$$E_c = (\sum_{i=1}^{24} J_{si} * A_s)/A_c \tag{2}$$

Where A_s is the sapwood area (cm^2) and A_c is the crown area (m^2), i stands for the sequence of daily hours..

Canopy conductance (G_c) was calculated by using measured canopy transpiration and Penman-Monteith equation [37]:

$$\lambda E = \frac{\Delta R_n + 3600 \rho C_p VPD G_a}{\Delta + [\gamma(1 + G_a/G_c)]} \tag{3}$$

Where λ is the latent heat of vaporization of water (2.39 MJ kg^{-1}), Δ is the ratio of the saturated vapor pressure to temperature (kPa °C^{-1}), R_n is the net radiation (MJ m^{-2} h^{-1}), estimated from the regression equation $R_n = 0.7965 * R_s$-57.64 (recommended by Zeppel et al. [37], R_s is the total radiation, MJ m^{-2} h^{-1}), ρ is air density (kg m^{-3}), C_p is the specific heat of air (1.013 MJ kg^{-1} °C^{-1}), VPD is the vapor pressure deficit (kPa), γ is the psychrometer constant (0.066 kPa °C^{-1}). The constant 3600 is time conversion factor from second to hour. The unit of E here is m^3 m^2 h^{-1}.

Meteorology and Soil Moisture Content Measurement

Meteorological data were collected using an automatic weather system Weather-Hawk (Campbell Scientific, Logan, UT, USA) mounted at the height of 15 m and 5 m away from the sampled trees. Volumetric soil water content (θ, m^3 m^{-3}) was measured by two sets of ECH2O (Decagon Devices Inc., Pullman, WA, USA). The measurement was made at 25 cm, 50 cm, 75 cm, and 100 cm of soil profile. Relative extractable water (REW, unitless) [38]–[41] was calculated by using averaged θ across layers as

$$REW = \frac{\theta - \theta_{min}}{\theta_{max} - \theta_{min}} \qquad (4)$$

where θ_{max} and θ_{min} are the maximum and the minimum measured θ during the observation period, respectively. The cumulative soil water depletion for each drying cycle was calculated as the difference of profile averaged soil moisture content (θ) between the two consecutive rainfall events.

Model Description

A modified Jarvis-type model (i.e. multiplicative environmental drivers) was used to simulate measured tree transpiration as the parameters in this model are very effective in capturing species differences in responding to the environmental variables [42]. The two most responsible atmospheric factors, solar radiation and vapor pressure deficit (VPD), are used in the model to describe the sap flux variation. The modified Jarvis model is

$$E_{c\text{-}H} = a * \frac{R_s}{R_s + b} * \frac{1}{1 + e^{(\frac{c - VPD}{d})}} \qquad (5)$$

where $E_{c\text{-}H}$ is hourly canopy transpiration (mm.h^{-1}), R_s total radiation (W m^{-2}), and VPD vapor pressure deficit (kPa). Parameter a is maximum modeled canopy transpiration under ideal environmental condition. b describes light saturation level. c and d reflect plants response to VPD with c being interpreted as VPD level when half of maximum $E_{c\text{-}H}$ is reached and d as the slope between $E_{c\text{-}H}$ and VPD.

To estimate the average canopy stomatal sensitivity to VPD, we employed the simplified formula used by Oren et al. [27] :

$$G_c = -m * LnVPD + G_{cref} \qquad (6)$$

where G_c is canopy conductance, an estimate of average stomatal conductance over the canopy (mm s^{-1}) [43], $-m$ is the slope of G_c versus LnVPD (i.e. $-d\, G_c/dLnVPD$), which quantifies the sensitivity of average canopy stomatal conductance to VPD. G_{cref} is reference canopy conductance when VPD=1 kPa and can be used as surrogate for G_{cmax} [43].

Statistical Analysis

All statistical analysis was performed using SPSS (Version 16.0, Chicago, IL). Curve fitting was run through SigmaPlot (version 10.0, Systat Software,

California, USA) and parameters of individual trees were analyzed among species or environmental condition ranks via one-way ANOVA. ANOVA analysis was employed to test the existence of significant differences among groups, and the multiple comparisons of the results were performed by LSD post-hoc test.

In this study, we first tested the assumption that the relationship between −m and G_{cref} defined by Eq. (6) follows 0.6 slope [27] across the studied species, and then quantified G_c responses to θ using boundary line analysis between G_{cref} and θ as described by Schäfer et al. [44].

RESULTS

Micrometeorology and Soil Water Condition

Statistics of the averaged daily VPD, daily total solar radiation, and daily air temperature for July, August, and September, 2009 is shown in Table 2. There were significant differences in daily total solar radiation between the three months ($P<0.001$, n=92, One-way ANOVA) with September lower than the other two. Averaged daily VPD was significantly higher in August than in July ($P=0.000$, LSD post-hoc test) and September ($P=0.024$, LSD post-hoc test). Total rainfall was 432.5 mm from July 1st to September 31st with several intensive rainfall events. Soil moisture content in the upper layer depleted more quickly than deeper layers below 50 cm which maintained stable during several dry spells between the rain events (Fig. 1A).

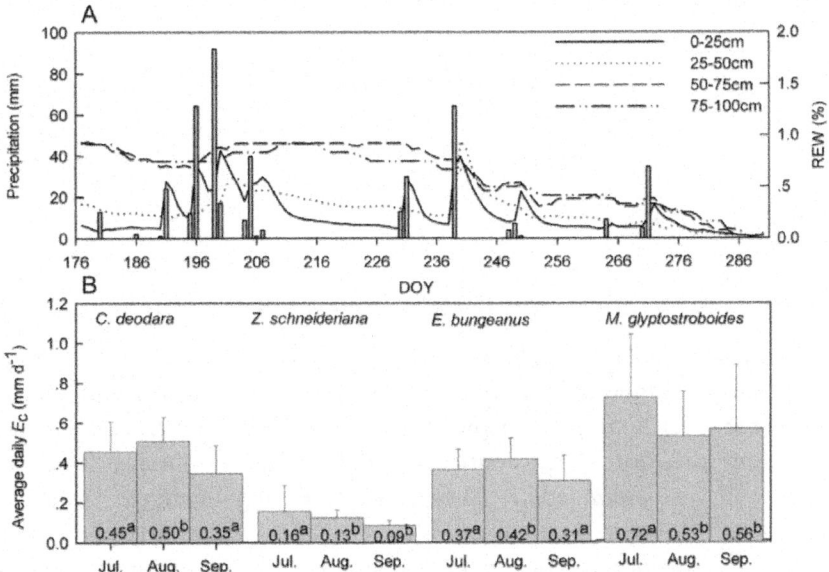

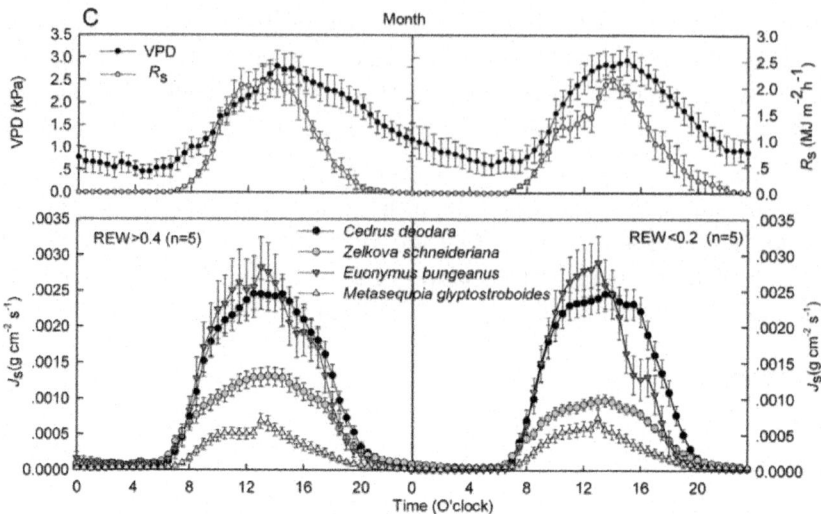

Figure 1. Water supply and transpiration. A: Rainfall and soil water condition during the studied period. B: Comparison of average daily E_c among different months and species. Significant differences among months within the same species are indicated by upper different lower case letters. Vertical bars stand for S.E.. C: Progression of sap flux density (J_s) under different soil water conditions (REW>0.4 and REW<0.2) in comparison with contemporary VPD and R_s.

doi:10.1371/journal.pone.0047882.g001

Table 2. Statistics of atmosphere variables.

Month	Averaged daily VPD (kPa)	Daily total Solar radiation (MJ.m⁻²)	Averaged daily Temperature (°C)	Monthly Rainfall(mm)
	Mean(S.E.)	Mean(S.E.)	Mean(S.E.)	Sum
July	0.85ᵃ(0.07)	13.54ᵃ (0.99)	23.94ᵃ (0.25)	265.25
August	1.33ᵇ (0.1)	14.59ᵃ (0.77)	25.47ᵇ (0.44)	106.75
September	1.05ᵃ(0.08)	9.67ᵇ (0.67)	20.85ᶜ (0.29)	60.5

Significant differences (P<0.05) among months are indicated by different upper lower case letters.
doi:10.1371/journal.pone.0047882.t002

Canopy Transpiration and Its Relationship with DBH

Transpiration varied considerably among the individuals both across and within species. The averaged daily total canopy transpiration was different across the species (Fig. 1B). Canopy transpiration of *C. deodara* and *E. bungeanus* were similar and did not show continuous decline as summer passed by. By contrast, *Z. schneideriana* showed decreasing canopy transpiration which was lower than the other species. Similarly, transpiration of *M. glyptostroboides* declined throughout the observation period but recovered a little in September from transpiration decrease in August (Fig. 1B). On a daily scale, J_s of different species followed similar pattern but varied in magnitude under different soil water conditions of similar R_s and VPD (Fig. 1C).

Individually, average daily transpiration rate ranged from 0.07 mm d⁻¹ by the smallest *E.bungeanus* tree to 0.20 mm d⁻¹ by the largest *C. deodara* tree. It could be partially ascribed to the dependence of water use on tree size (Fig. 2). However, there were considerable species-specific differences in canopy transpiration as indicated by different species within same DBH class. For instance, water transpired by *E. bungeanus* (0.195 mm. d⁻¹) was nearly 5-fold than that by *Z. schneideriana* (0.03 mm d⁻¹) as measured for ~14 cm DBH trees of these two species. When considering the statistical relationship between DBH and E_c, the biggest *M.glyptostroboides* triggered an exponential rise ($R^2=0.72$).

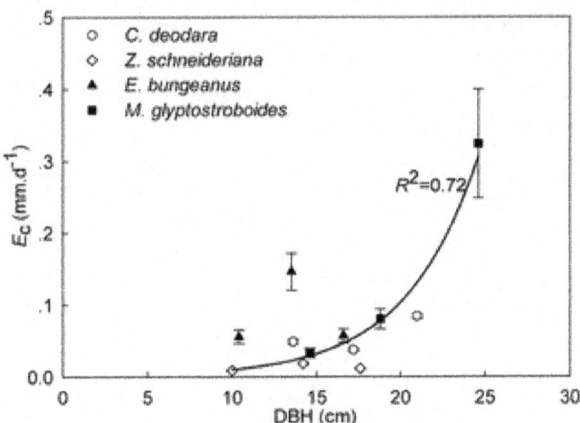

Figure 2. Tree size and transpiration. Relationship between daily canopy transpiration and DBH of individual trees of all investigated species.

doi:10.1371/journal.pone.0047882.g002

E_c Response to Environmental Drivers

Daily E_c was closely related to R_s and VPD ($P<0.05$) and the introduction of other variables did not further normalize the residual distribution. Curve fitting results showed that E_c saturated at higher VPD and total radiation (R_s) (Fig. 3, 4A). It was also observed that higher R_s would significantly enhance E_c under same VPD condition (Fig. 3). However, E_c under same R_s condition failed to show significant differences among VPD ranks ($P>0.05$). REW ranks neither affected the relationship between daily E_c and R_s nor E_c and VPD ($P>0.05$).

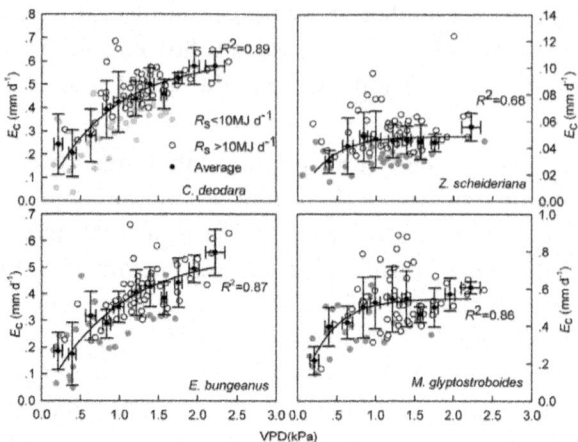

Figure 3. VPD and transpiration. Relationship between daily canopy transpiration in relation to VPD under different solar radiation (R_s) ranks with data on rainy days removed. Black dots represent average E_c values within every 0.2 kPa rank. If a rank has less than 3 data, this rank is not considered to reduce potential bias from lack of representativeness.

doi:10.1371/journal.pone.0047882.g003

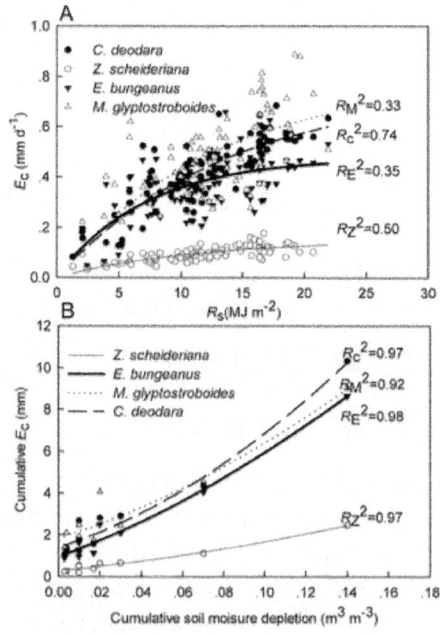

Figure 4. Influences of solar radiation and soil water over transpiration. Daily E_c in

response to R_s (A) and the relationship between sums of daily transpiration and cumulative soil water depletion for each drying cycle between two consecutive rainfall events (B). The lines are the fitting curves for different species. (A) and (B) share the same legends. R^2 for each species is denoted with subscripted first letter of the Latin name.

doi:10.1371/journal.pone.0047882.g004

Daily water transpired by trees did not show a significant correlation with REW ($P>0.40$) for plants did not react to concurrent soil water status. However, cumulative daily transpiration was significantly related to the cumulative soil water depletion during each drying cycle (Fig. 4B). When compared according to different REW ranks under similar VPD conditions ($P=0.024$), diurnal course of transpiration exhibited contrasting patterns (Fig. 1C). The J_s decline of *E.bungeanus* in afternoon was steeper under soil water stress and the J_s of *Z. schneideriana* was suppressed during the whole day. By contrast, no significant reduction of J_s was observed for *C. deodara* and *M. glyptostroboides*.

G_c Sensitivity

Logarithmic decrease of midday canopy conductance against VPD suggested progressive stomata closure to prevent excessive caviation as the air turned drier across species [31]. There were significant species difference in the G_c sensitivity to VPD ($-dG_c/d\ln VPD$ or $-m$) and the G_{cref} value within each REW and R_s rank (Table 3). No significant difference was observed for parameters of the same species under contrasting soil water and R_s ranges, and paired $-dG_c/dVPD$ and G_{cref} of different species still followed a strong linear relationship of ~0.6 slope (Fig. 5). Similar results were also obtained through boundary line analysis of the relationship between G_{cref} and REW (data not shown). The lack of significant deviation from a 0.6 slope indicates that all species were isohydric under all environmental conditions.

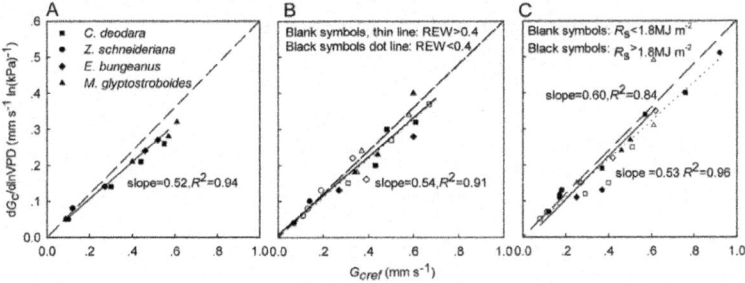

Figure 5. The sensitivity of canopy conductance. Relationship between sensitivity of

canopy conductance to VPD ($-dG_c/d\ln VPD$) and canopy conductance at VPD=1 kPa (G_{cref}) under (A) all environmental conditions, (B) contrasting REW ranks and (C) different solar radiation (R_s) levels. Data dots were from data in Table 3.

doi:10.1371/journal.pone.0047882.g005

Table 3. Parameters and significance for $G_c = -m * \mathrm{LnVPD} + G_{cref}$ under different soil moisture and radiation conditions.

doi:10.1371/journal.pone.0047882.t003

	General condition[1]		REW<0.4		REW>0.4		R_s<1.8 MJ. m^{-2}		R_s>1.8 MJ. m^{-2}	
	$-m$	G_{cref}	$-m$	G_{cref}	$-m$	G_{cref}	$-m$	G_{cref}	$-m$	G_{cref}
C. deodara	0.20[a](0.06)	0.43[a](0.12)	0.27[a](0.03)	0.51[a] (0.05)	0.26[a](0.06)	0.49[a](0.18)	0.17[a](0.07)	0.40[a](0.11)	0.31[a](0.11)	0.57[a](0.19)
Z. schneideriana	0.12[b](0.02)	0.10[b](0.02)	0.07[b](0.02)	0.11[b](0.03)	0.09[b](0.04)	0.14[b](0.02)	0.12[b](0.01)	0.10[b](0.02)	0.23[b](0.03)	0.16[b](0.03)
E. bungeanus	0.22[c](0.07)	0.42[c](0.13)	0.21[c](0.11)	0.44[c](0.13)	0.19[c](0.02)	0.36[c](0.02)	0.24[c](0.10)	0.44[c](0.17)	0.25[c](0.22)	0.51[c](0.35)
M.glyptostroboides	0.28[d](0.06)	0.50[d](0.09)	0.27[d](0.06)	0.46[d](0.07)	0.25[d](0.05)	0.43[d](0.07)	0.32[d](0.17)	0.49[d](0.21)	0.21[d](0.08)	0.38[d](0.18)

Data were given as mean of all sampled trees of the same species and S.E. in parenthesis.
R_s: Solar radiation.
[1]: Fitting curves run through G_c and VPD under entire REW and R_s range.
a,b,c,d: Significant difference of the parameter among species.
doi:10.1371/journal.pone.0047882.t003

Model Simulation

Average $E_{c\text{-}H}$ of all three trees within the same species from 30 days was included for model calibration and 6 days were randomly chosen for validation. Model performances were satisfactory for both calibration (see R^2_{adj}>0.89 in Table 4) and validation as the simulated half-hourly $E_{c\text{-}H}$ agreed well with observed values (Fig. 6). Model parameters showed significant differences in maximum transpiration (parameter a) ($P<0.001$, one-way ANOVA) and in light saturation level (parameter b, $P=0.004$) among species. For species response to VPD, significant differences were found for VPD corresponding to the level at half maximum transpiration (parameter c, $P=0.024$, one-way ANOVA) and the slope between $E_{c\text{-}H}$ and VPD (parameter d, $P=0.035$, one-way ANOVA) across species. Moreover, the level of response to VPD remained stable across days as c and d did not show significant differences among the replicate days ($P=0.552$ for c and 0.621 for d) for each species

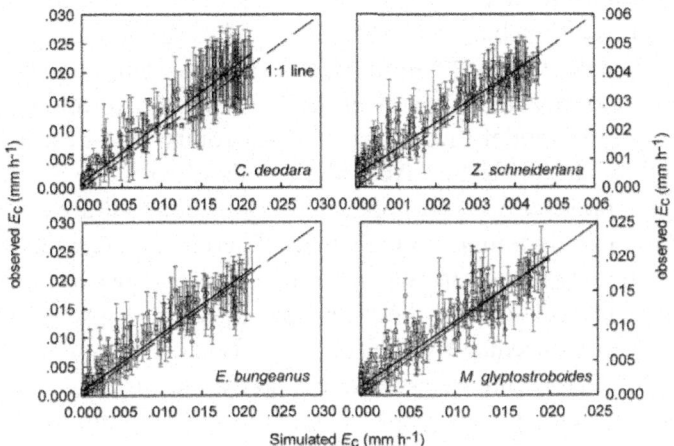

Figure 6. Model simulation. Simulated and observed hourly canopy conductance (E_{c-H}) for model $E_{c-H} = a * \dfrac{R_s}{R_s+b} * \dfrac{1}{1+e^{(\frac{c-VPD}{d})}}$ validation using six sampled days in August and September. Observed data were presented as mean of the trees within the same species±SD.

doi:10.1371/journal.pone.0047882.g006

Table 4. Calibrated parameters for $E_{c-H} = a * \dfrac{R_s}{R_s+b} * \dfrac{1}{1+e^{(\frac{c-VPD}{d})}}$ for species.

doi:10.1371/journal.pone.0047882.t004

Species	a (mm.h^{-1})		b (MJ.m^{-2}.h^{-1})		c (kPa)		d (kPa)		R^2_{adj}	
	Mean	SE	Mean	SE	Mean	SE	Mean	SE	Mean	SE
Cedrus deodara	0.035[a]	0.001	1.00[a]	0.20	0.26[a]	0.06	1.65[a]	0.1	0.96	0
Zelkova schneideriana	0.009[b]	0.004	1.88[b]	0.02	0.41[b]	0.11	1.45[b]	0.07	0.92	0.03
Euonymus bungeanus	0.065[c]	0.001	3.09[c]	0.03	1.41[c]	0.12	1.57[c]	0.09	0.88	0.11
Metasequoia glyptostroboides	0.050[d]	0.001	1.76[d]	0.03	1.72[d]	0.32	1.78[d]	0.16	0.92	0.04

Significant differences across species ($P<0.05$) are indicated with upper lower case letters.
doi:10.1371/journal.pone.0047882.t004

DISCUSSION

Influences from tree size and Species on Transpiration

Comparable to general conclusion drawn from previous studies that tree size plays an overwhelming role in influencing stand water use [9], [25], [26], [45], [46], our study observed a positive relationship between transpiration and DBH independent of species (Fig. 2). The significant relationship between transpiration and DBH is attributable to the positive relationship between DBH and sapwood area which directly transports sap flow [47]. A sigmoidal increase portrayed the relation between water use and DBH among 18 angiosperm species in tropical forest [45]. Similarly, tree size has been found in some studies to be more important than species in determining transpiration [28] which is explained by pipe model theory [48]. It is a robust non-destructive method to estimate biomass [49]–[52]. According to the theory, the amount of leaves on a tree is supported by a proportionate cross sectional area of a bundle of pipes with equal hydraulic conductance. Therefore, the greater the living biomass the tree has the more sapwood area it has. However, the theory also indicates that the relationship between leaf area and sapwood cross area is species-specific. Consequently, the species effects on transpiration also can be strong even with changing tree size. In our study, the variation of average daily transpiration by four species ranged from 0.07 to 0.20 mm. d^{-1}. This can also be ascribed to species specific canopy conductance, such as *E. bungeanus* in comparison with other species within ~14 cm DBH class. Such appreciable species-specific differences in transpiration suggested potential large influence of species composition at scales larger than stands [22].

Tree size can be a determinant factor on transpiration of highly structured stands [53], but the relationship may be more depend on the stand DBH composition. For example, Pataki et al. did not observe significant relationship between tree size and transpiration in urban areas and the author attributed it to limited range of DBH distribution which was common in the urban green spaces [8]. In our study the exponential increase between E_c and DBH was triggered by one big *M. glyptostroboides* without which the relationship changed to linear and the R^2 decreased significantly (Fig. 2). It shows the influence of DBH composition, especially the large ones, on the relationship with transpiration. In urban green space, the tree size is usually unbalanced to represent a continuous DBH range, so it is not reliable to predict water use of urban trees using its relationship with tree size.

Response to Environmental Variables and Seasonal Transpiration Patterns

All four species showed saturation of tree transpiration to high VPD and R_s and this phenomenon has been widely reported [54]–[57]. When the canopy is well-coupled with the atmosphere [31], plants exert effective stomatal control over transpiration to regulate the minimum water potential according to VPD and it was also predictable from the 0.6 relationship between G_{cref} and $-dG_c/$ dlnVPD. Therefore, E_c plateaus with increasing VPD as exhibited in this study. The E_c saturation to increasing R_s can be ascribed to the fact that stomata are fully opened at certain level of irradiance [58] and energy is not limiting to transpiration for these trees, and therefore do not react to higher radiation levels.

The influence of soil moisture on transpiration has not been consistent among studies [59]–[62]. In this study, daily E_c did not show significant correlation with soil moisture content. At the sub-daily time scales, mesic tree species show a greater response to light and VPD with soil moisture playing a limited role except under conditions where root to soil moisture resistance is high [63]. However, soil moisture content is an important environmental driver for tree transpiration for longer time scales (Fig. 4B) [64], [65].

Varied pattern of species transpiration in response to seasonal weather progression (Fig. 1B) was also observed. Due to J_s sensitivity to soil water stress (Fig. 1C), the transpiration of *Z.schneideriana* decreased along the progression of rain fall reduction through the months. However, the declining trend of transpiration of *M. glyptostroboides* should be ascribed to variation of VPD because of the sensitive G_c response to VPD by this species and high G_{cref}(Fig. 5A). Therefore, transpiration by this species declined as VPD increased in August and recovered in September as VPD declined. In addition to the low rainfall, the high transpiration of *C. deodara* and *E. bungeanus* in August may contribute to the build up cavitation in the xylem and subsequently reduce transpiration thereafter [30].

Sensitivity to VPD

On a plot basis, trees of different species observed same hydraulic control (Fig. 5) and species differences in G_{cref} led to varied magnitude of canopy transpiration across species (Fig. 1B, 3). In our study, even though $-dG_c/d\ln$VPD and G_{cref} showed significant difference among species, their ratios for all these four species converged to ~0.6 as reported [27], [66]. The changes of G_{cref} and $-dG_c/d\ln$VPD were along the 0.6 slope under varying REW and solar radiation conditions and no significant bias existed in relation to tree size similar to what was found by Ewers et al [43] for one tree species exposed to manipulated soil moisture and nutrients. The interpretation is that isohydric regulation of water loss through G_c response to VPD [27], [67] is the same among the four species regardless of varying tree sizes and environmental conditions, while the G_{cref} is species-specific. This relationship also means that G_c sensitivity to VPD [67] and the transpiration of urban trees can be reliably predicted from G_{cref} [68], [69].

Since environmental conditions did not influence the relationship between $-dG_c/d\ln$VPD and G_{cref}, G_{cref} can serve as an efficient indicator of species differences. The constant relationship between $-dG_c/d\ln$VPD and G_{cref} reflects an isohydric stomatal control which protects the xylem from developing runaway cavitation by guaranteeing safe water potential [63]. Species difference in G_{cref} may be related to xylem characteristics which tolerate cavitation. In addition to the partitioning of xylem resistance as well as aquaporin activity [70], [71], xylem anatomic differences were most commonly related to species› cavitation tolerances [30], [72]–[74]. Species difference in xylem tolerance to cavitation was explicitly revealed through the daily J_s rate under contrasting REW conditions (Fig. 1C). With larger diameter xylem vessels, ring-porous trees were less tolerant to negative water potential and xylem cavitation than diffuse-porous species due to intrinsic vulnerability of large diameter conduits to cavitation [30], [75], such as *Z. schneideriana* versus *E. bungeanus* in our case. This is further reflected by the decreased *Js* of *Z. schneideriana* under water stress (Fig. 1C). However *E. bungeanus* could afford the same magnitude of transpiration under water stress as under no water stress during first half of the day. By contrast, *C. deodara* and *M. glyptostroboides* were able to maintain normal transpiration rate. That is probably because as gymnosperm species, they have tracheids of small diameter and strong cell wall, features that are resistant to cavitation.

Sensitivity of canopy conductance to VPD has great implications for the survival of trees in urban landscape. Trees of high canopy conductance at low VPD shows higher sensitivity to VPD [27], [76], such as trees with higher G_{cref} in this study. Active control over G_c enables isohydric species to maximize carbon assimilation under low VPD and avoid the risk of runaway cavitation under atmospheric or soil drought [77]. Therefore, these species are suitable for urban environment where unpredictable local change is widespread [78]. Artificial activity produces extra thermal energy, which could be complicated temporally and spatially due to air turbulence caused by building arrangement [79]. Another factor comes from irrigation scope and intensity. Without irrigation, soil water condition is more likely to be stressed because of prevented rainfall percolation from pavements and enhanced soil evaporation in unpaved area. Although effective in G_c control that guarantees tree survival, species with low maximum canopy conductance through the entire VPD range have less advantage in competitive situation, such as multi-species urban forest in summer, because lower G_c does not favor carbon assimilation for growth in competition with high G_c species. Low carbon assimilation might be potentially harmful to tree survival through a major drought event from the perspective of water safety [80]. As a result, species with different water use strategies will be differently affected by shifts of the frequency, duration, and intensity of drought [81].

Implication of Species Differences on Urban Transpiration

Species differences tended to be revealed through the relationship between transpiration and environmental drivers. In our study, the modified Jarvis-type model described the environmental control over plants transpiration well (Fig. 6) and model parameters captured species differences in responding to the environment influences (Table 4). Parameter differences, i.e. species differences in responding to environmental factors, were also found among species by studies in tropical forest [42] and in Philippines [53]. Our study also showed increased expression of species differences under water stress (Fig. 1C).

Under urban environment, species differences in transpiration and response to the environment may cause spatial difference in hydraulic redistribution among green space even within small scope of area. Also, it may influence the effect of micro-meteorology modulation by vegetation. Even with such appreciable species differences, it is possible to predict urban canopy transpiration using their shared hydraulic control character. Unlike natural setting where transpiration can be assessed using the relationship with DBH, the influence of DBH on canopy transpiration is undermined because of the limited DBH distribution in the city. Our results recommend using the ~0.6 relationship between $-dG_c/dLnVPD$ and G_{cref}. As G_c under 1 kPa can be accurately tested via proper measurements including tree-based or remote sensed techniques, G_c, and thus canopy transpiration, can be assessed at any scale through concurrent VPD.

Conclusion

Species differences were found in an urban environment in the response of transpiration to environmental drivers including light, soil moisture and vapor pressure deficit and in the control of tree size on transpiration. Despite significant species differences, all species showed declining G_c against increasing VPD, and the theoretical $-dG_c/dlnVPD$ to G_{cref} ratio of ~0.6 was observed across studied species and under contrasting soil water and R_s conditions in the urban area. We, therefore, concluded that urban trees show isohydric regulation of minimum leaf water potential as reflected in their response to VPD and that transpiration can be predicted with appropriate assessment of G_c at VPD=1 kPa.

Acknowledgments

The authors would like to extend our sincere gratitude to Mr. Xiaofang Zhang and Mr. Keyu Dong for their help to collect data and providing logistic support during the study. We also thank the Dalian Forestry Administration and Dalian Landscape Architecture Administration, Liaoning Province, China for their valuable assistance on our field experiment and permitting our access to the park.

Author Contributions

Conceived and designed the experiments: ZZ LC. Performed the experiments: LC. Analyzed the data: LC ZZ. Contributed reagents/materials/analysis tools: ZZ. Wrote the paper: LC ZZ BEE.

REFERENCES

1.   Murphy DJ, Hall MH, Hall CAS, Heisler GM, Stehman SV, et al. (2011) The relationship between land cover and the urban heat island in northeastern Puerto Rico. Int J Climatol 31: 1222–1239

2.   Hedblom M, So¨derstro¨m B (2010) Landscape effects on birds in urban woodlands: an analysis of 34 Swedish cities. J Biogeogr 37: 1302–1316

3.   Heisler GM, Grant RH (2000) Ultraviolet radiation in urban ecosystems with consideration of effects on human health. Urban Ecosyst 4: 193–229

4.   Walther G-R, Post E, Convey P, Menzel A, Parmesan C, et al. (2002) Ecological responses to recent climate change. Nature 416: 389–395

5.   Allen CD, Macalady AK, Chenchouni H, Bachelet D, McDowell N, et al. (2010) A global overview of drought and heat-induced tree mortality reveals emerging climate change risks for forests. For Ecol Manage 259: 660–684

6.   Jackson RB, Jobbagy EG, Avissar R, Roy SB, Barrett DJ, et al. (2005) Trading water for carbon with biological sequestration. Science 310: 1944–1947

7.   Zhang L, Dawes WR, Walker GR (2001) Response of mean annual evapotranspiration to vegetation changes at catchment scale. Water Resour Res 37: 701–708

8.   Pataki DE, McCarthy HR, Litvak E, Pincetl S (2010) Transpiration of urban forests in the Los Angeles metropolitan area. Ecol Appl 21: 661–677

9.   Meinzer FC, Andrade JL, Goldstein G, Holbrook NM, Cavelier J, et al. (1999) Partitioning of soil water among canopy trees in a seasonally dry tropical forest. Oecologia 121: 293–301

10.   Franks PJ (2004) Stomatal control and hydraulic conductance, with special reference to tall trees. Tree Physiol 24: 865–878

11. Abril M, Hanano R (1998) Ecophysiological responses of three evergreen woody Mediterranean species to water stress. Acta Oecologica 19: 377–387

12. Benyon RG, Marcar NE, Theiveyanathan S, Tunningley WM, Nicholson AT (2001) Species differences in transpiration on a saline discharge site. Agric Water Manage 50: 65–81

13. Ford CR, Hubbard RM, Kloeppel BD, Vose JM (2007) A comparison of sap flux-based evapotranspiration estimates with catchment-scale water balance. Agric For Meteorol 145: 176–185

14. Oren R, Pataki D (2001) Transpiration in response to variation in microclimate and soil moisture in southeastern deciduous forests. Oecologia 127: 549–559

15. Ho"lscher D, Koch O, Korn S, Leuschner C (2005) Sap flux of five co-occurring tree species in a temperate broad-leaved forest during seasonal soil drought. Trees-Struct Funct 19: 628–637

16. Kaufmann MR (1985) Annual transpiration in subalpine forests: Large differences among four tree species. For Ecol Manage 13: 235–246

17. Licata JA, Pypker TG, Weigandt M, Unsworth MH, Gyenge JE, et al. (2011) Decreased rainfall interception balances increased transpiration in exotic ponderosa pine plantations compared with native cypress stands in Patagonia, Argentina. Ecohydrol 4: 83–93

18. Herbst M, Roberts JM, Rosier PTW, Taylor ME, Gowing DJ (2007) Edge effects and forest water use: A field study in a mixed deciduous woodland. For Ecol Manage 250: 176–186

19. Komatsu H, Onozawa Y, Kume T, Tsuruta K, Kumagai To, et al. (2010) Stand-scale transpiration estimates in a Moso bamboo forest: II. Comparison with coniferous forests. For Ecol Manage 260: 1295–1302

20. Asbjornsen H, Tomer MD, Gomez-Cardenas M, Brudvig LA, Greenan CM, et al. (2007) Tree and stand transpiration in a Midwestern bur oak savanna after elm encroachment and restoration thinning. For Ecol Manage 247: 209–219

21. van Dijk AIJM, Keenan RJ (2007) Planted forests and water in perspective. For Ecol Manage 251: 1–9

22. Mackay DS, Ahl DE, Ewers BE, Gower ST, Burrows SN, et al. (2002) Effects of aggregated classifications of forest composition on estimates of evapotranspiration in a northern Wisconsin forest. Global Change Biol 8: 1253–1265

23. Iida Si, Tanaka T, Sugita M (2005) Change of interception process due

to the succession from Japanese red pine to evergreen oak. J Hydrol 315: 154–166

24. Bittner S, Talkner U, Kra"mer I, Beese F, Ho"lscher D, et al. (2010) Modeling stand water budgets of mixed temperate broad-leaved forest stands by considering variations in species specific drought response. Agric For Meteorol 150: 1347–1357

25. Cienciala E, Kucera J, Malmer A (2000) Tree sap flow and stand transpiration of two Acacia mangium plantations in Sabah, Borneo. J Hydrol 236: 109–120

26. Meinzer FC, James SA, Goldstein G (2004) Dynamics of transpiration, sap flow and use of stored water in tropical forest canopy trees. Tree Physiol 24: 901–909

27. Oren R, Sperry JS, Katul GG, Pataki DE, Ewers BE, et al. (1999) Survey and synthesis of intra- and interspecific variation in stomatal sensitivity to vapour pressure deficit. Plant Cell Environ 22: 1515–1526

28. Meinzer FC, Goldstein G, Andrade JL (2001) Regulation of water flux through tropical forest canopy trees: do universal rules apply? Tree Physiol 21: 19–26

29. Sperry JS (2000) Hydraulic constraints on plant gas exchange. Agric For Meteorol 104: 13–23

30. Bush SE, Pataki DE, Hultine KR, West AG, Sperry JS, et al. (2008) Wood anatomy constrains stomatal responses to atmospheric vapor pressure deficit in irrigated, urban trees. Oecologia 156: 13–20

31. Chen L, Zhang Z, Li Z, Tang J, Caldwell P, et al. (2011) Biophysical control of whole tree transpiration under an urban environment in Northern China. J Hydrol 402: 388–400

32. Granier A (1987) Evaluation of transpiration in a Douglas-fir stand by means of sap flow measurements. Tree Physiol 3: 309–320

33. Bush SE, Hultine KR, Sperry JS, Ehleringer JR (2010) Calibration of thermal dissipation sap flow probes for ring- and diffuse-porous trees. Tree Physiol 30: 1545–1554

34. Clearwater MJ, Meinzer FC, Andrade JL, Goldstein G, Holbrook NM (1999) Potential errors in measurement of nonuniform sap flow using heat dissipation probes. Tree Physiol 19: 681–687

35. Steppe K, De Pauw DJW, Doody TM, Teskey RO (2010) A comparison of sap flux density using thermal dissipation, heat pulse velocity and heat field deformation methods. Agric For Meteorol 150: 1046–1056

36. Granier A, Anfodillo T, Sabatti M, Cochard H, Dreyer E, et al. (1994)

Axial and radial water flow in the trunks of oak trees: a quantitative and qualitative analysis. Tree Physiol 14: 1383–1396

37. Zeppel MJB, Macinnis-Ng CMO, Yunusa IAM, Whitley RJ, Eamus D (2008) Long term trends of stand transpiration in a remnant forest during wet and dry years. J Hydrol 349: 200–213

38. Gartner K, Nadezhdina N, Englisch M, Cermak J, Leitgeb E (2009) Sap flow of birch and Norway spruce during the European heat and drought in summer 2003. For Ecol Manage 258: 590–599

39. Kumagai T, Saitoh TM, Sato Y, Morooka T, Manfroi OJ, et al. (2004) Transpiration, canopy conductance and the decoupling coefficient of a lowland mixed dipterocarp forest in Sarawak, Borneo: dry spell effects. J Hydrol 287: 237–251

40. Palomo MJ, Moreno F, Ferna´ndez JE, Di´az-Espejo A, Giro´n IF (2002) Determining water consumption in olive orchards using the water balance approach. Agric Water Manage 55: 15–35

41. Tognetti R, Giovannelli A, Lavini A, Morelli G, Fragnito F, et al. (2009) Assessing environmental controls over conductances through the soil-plantatmosphere continuum in an experimental olive tree plantation of southern Italy. Agric For Meteorol 149: 1229–1243

42. O'Brien JJ, Oberbauer SF, Clark DB (2004) Whole tree xylem sap flow responses to multiple environmental variables in a wet tropical forest. Plant Cell Environ 27: 551–567

43. Ewers BE, Oren R, Johnsen KH, Landsberg JJ (2001) Estimating maximum mean canopy stomatal conductance for use in models. Can J For Res 31: 198– 207

44. Scha¨fer KVR, Oren R, Tenhunen JD (2000) The effect of tree height on crown level stomatal conductance. Plant Cell Environ 23: 365–375

45. Meinzer FC, Bond BJ, Warren JM, Woodruff DR (2005) Does water transport scale universally with tree size? Funct Ecol 19: 558–565

46. Wullschleger SD, Hanson PJ, Todd DE (2001) Transpiration from a multispecies deciduous forest as estimated by xylem sap flow techniques. For Ecol Manage 143: 205–213

47. Macfarlane C, Bond C, White DA, Grigg AH, Ogden GN, et al. (2010) Transpiration and hydraulic traits of old and regrowth eucalypt forest in southwestern Australia. For Ecol Manage 260: 96–105

48. Shinozaki K, Yoda K, Hozumi K, Kira T (1964) A quantitative theory of plant form-the pipe model theory,I.Basic analysis. Jpn J Ecol 14: 97–104

49. Valentine HT, Ma¨kela¨ A (2012) Modeling forest stand dynamics from

optimal balances of carbon and nitrogen. New Phytol 194: 961–971

50. Berninger F, Nikinmaa E, Sieva¨nen R, Nygren P (2000) Modelling of reserve carbohydrate dynamics, regrowth and nodulation in a N2-fixing tree managed by periodic prunings. Plant Cell Environ 23: 1025–1040

51. Yoneda T, Tamin R, Ogino K (1990) Dynamics of aboveground big woody organs in a foothill dipterocarp forest, West Sumatra, Indonesia. Ecol Res 5: 111–130

52. Ogawa K, Adu-Bredu S, Yokota T, Hagihara A (2010) Leaf biomass changes with stand development in hinoki cypress (Chamaecyparis obtusa [Sieb. et Zucc.] Endl.). Plant Ecolog 211: 79–88

53. Dierick D, Ho¨lscher D (2009) Species-specific tree water use characteristics in reforestation stands in the Philippines. Agric For Meteorol 149: 1317–1326

54. Granier A, Huc R, Barigah ST (1996) Transpiration of natural rain forest and its dependence on climatic factors. Agric For Meteorol 78: 19–29

55. Zhang H, Morison J, Simmonds L (1999) Transpiration and water relations of poplar trees growing close to the water table. Tree Physiol 19: 563–573

56. Hogg EH, Black TA, den Hartog G, Neumann HH, Zimmermann R, et al. (1997) A comparison of sap flow and eddy fluxes of water vapor from a boreal deciduous forest. J Geophys Res 102: 28929–28937

57. Hogg EH, Hurdle PA (1997) Sap flow in trembling aspen: implications for stomatal responses to vapor pressure deficit. Tree Physiol. 17: 501–509

58. Yunusa IAM, Aumann CD, Rab MA, Merrick N, Fisher PD, et al. (2010) Topographical and seasonal trends in transpiration by two co-occurring Eucalyptus species during two contrasting years in a low rainfall environment. Agric For Meteorol 150: 1234–1244

59. Lagergren F, Lindroth A (2002) Transpiration response to soil moisture in pine and spruce trees in Sweden. Agric For Meteorol 112: 67–85

60. Lundblad M, Lindroth A (2002) Stand transpiration and sapflow density in relation to weather, soil moisture and stand characteristics. Basic Appl. Ecol. 3: 229–243

61. Wallace J, McJannet D (2010) Processes controlling transpiration in the rainforests of north Queensland, Australia. J Hydrol 384: 107–117

62. Huang Y, Li X, Zhang Z, He C, Zhao P, et al. (2011) Seasonal changes in Cyclobalanopsis glauca transpiration and canopy stomatal conductance and their dependence on subterranean water and climatic factors in rocky

karst terrain. J Hydrol 402: 135–143

63. Sperry JS, Adler FR, Campbell GS, Comstock JP (1998) Limitation of plant water use by rhizosphere and xylem conductance: results from a model. Plant Cell Environ 21: 347–359

64. Llorens P, Poyatos R, Latron J, Delgado J, Oliveras I, et al. (2010) A multi-year study of rainfall and soil water controls on Scots pine transpiration under Mediterranean mountain conditions. Hydrol Processes 24: 3053–3064

65. Fisher RA, Williams M, Da Costa AL, Malhi Y, Da Costa RF, et al. (2007) The response of an Eastern Amazonian rain forest to drought stress: results and modelling analyses from a throughfall exclusion experiment. Global Change Biol 13: 2361–2378

66. Morison JI, Gifford RM (1983) Stomatal sensitivity to carbon dioxide and humidity: a comparison of two C3 and two C4 grass species. Plant Physiol 71: 789–796

67. Ewers BE, Gower ST, Bond-Lamberty B, Wang CK (2005) Effects of stand age and tree species on canopy transpiration and average stomatal conductance of boreal forests. Plant Cell Environ 28: 660–678

68. Novick K, Oren R, Stoy P, Juang J-Y, Siqueira M, et al. (2009) The relationship between reference canopy conductance and simplified hydraulic architecture. Adv Water Resour 32: 809–819

69. Mackay DS, Samanta S, Nemani RR, Band LE (2003) Multi-objective parameter estimation for simulating canopy transpiration in forested watersheds. J Hydrol 277: 230–247

70. Nardini A, Salleo S, Raimondo F (2003) Changes in leaf hydraulic conductance correlate with leaf vein embolism in Cercis siliquastrum L. Trees-Struct Funct 17: 529–534

71. Zwieniecki MA, Brodribb TJ, Holbrook NM (2007) Hydraulic design of leaves: insights from rehydration kinetics. Plant Cell Environ 30: 910–921

72. Markesteijn L, Poorter L, Bongers F, Paz H, Sack L (2011) Hydraulics and life history of tropical dry forest tree species: coordination of species' drought and shade tolerance. New Phytol 191: 480–495

73. Linton MJ, Sperry JS, Williams DG (1998) Limits to water transport in Juniperus osteosperma and Pinus edulis: implications for drought tolerance and regulation of transpiration. Funct Ecol 12: 906–911

74. Pratt RB, Jacobsen AL, Ewers FW, Davis SD (2007) Relationships among xylem transport, biomechanics and storage in stems and roots of

nine Rhamnaceae species of the California chaparral. New Phytol 174: 787–798

75. Li YY, Sperry JS, Taneda H, Bush SE, Hacke UG (2008) Evaluation of centrifugal methods for measuring xylem cavitation in conifers, diffuse- and ringporous angiosperms. New Phytol 177: 558–568

76. Yong JWH, Wong SC, Farquhar GD (1997) Stomatal responses to changes in vapour pressure difference between leaf and air. Plant Cell Environ 20: 1213– 1216

77. Katul GG, Palmroth S, Oren RAM (2009) Leaf stomatal responses to vapour pressure deficit under current and CO2-enriched atmosphere explained by the economics of gas exchange. Plant Cell Environ 32: 968–979

78. Liu W, You H, Dou J (2009) Urban-rural humidity and temperature differences in the Beijing area. Theor Appl Climatol 96: 201–207

79. Rizk AA, Henze GP (2010) Improved airflow around multiple rows of buildings in hot arid climates. Energy Build 42: 1711–1718

80. McDowell NG (2011) Mechanisms linking drought, hydraulics, carbon metabolism, and vegetation mortality. Plant Physiol 155: 1051–1059

81. West AG, Hultine KR, Sperry JS, Bush SE, Ehleringer JR (2008) Transpiration and hydraulic strategies in a pinon-juniper woodland. Ecol Appl 18: 911–927

Chapter 3

APPLICABILITY OF DIFFERENT HYDRAULIC PARAMETERS TO DESCRIBE SOIL DETACHMENT IN ERODING RILLS

Stefan Wirtz[1], Manuel Seeger[1,2], Andreas Zell[3], Christian Wagner[3], Jean-Frank Wagner[4], Johannes B. Ries[1]

[1]Department of Physical Geography, Trier University, Trier, Germany

[2]Department of Land Degradation and Development, Wageningen University, Wageningen, The Netherlands

[3]Department 7.3- Technical Physics, Saarland University, Saarbru¨cken, Germany

[4]Department of Geology, Trier University, Trier, Germany

ABSTRACT

This study presents the comparison of experimental results with assumptions used in numerical models. The aim of the field experiments is to test the linear relationship between different hydraulic parameters and soil detachment. For example correlations between shear stress, unit length shear force, stream power, unit stream power and effective stream power and the detachment rate does not reveal a single parameter which consistently displays the best correlation. More importantly, the best fit does not only vary from one experiment to another, but even between distinct measurement points. Different processes in rill erosion are responsible for the changing correlations. However, not all these procedures are considered in soil erosion models. Hence, hydraulic parameters alone are not sufficient to predict detachment rates. They predict the fluvial incising in the rill's bottom, but the main sediment sources are not considered sufficiently in its equations. The results of this study show that there is still a lack of understanding of the physical processes underlying soil erosion. Exerted forces, soil stability and its expression, the abstraction of the detachment and transport processes in shallow flowing water remain still subject of unclear description and dependence.

INTRODUCTION

Soil erosion models use different composite factors to describe and predict soil detachment and transport capacity. The most frequently used factors are average shear stress [1]–[4], unit length shear force [5], stream power [4], [6]–[9], unit stream power [10], [11] and effective stream power [12], [13].

In most cases, a linear equation describes the relation between the hydraulic parameters mentioned above and the detachment rate. By exceeding a certain threshold, erosion by concentrated flow begins and detachment rate increases. This threshold has a positive x-axis intercept, which means that there is no detachment below this point.

Another option is to consider concentrated flow erosion as a nonlinear threshold phenomenon or as a two-part linear threshold phenomenon: below the threshold soil detachment takes place (first linear relationship) but after exceeding the threshold, detachment rate increases much faster (second linear relationship) [14]. But it is unclear if this linear relationship is really suitable.

Knapen et al. [14] calculated the correlation between shear stress, unit length shear force, stream power and Reynolds number and the detachment rate from several WEPP datasets. The best average correlation was determined for stream power with $R^2=0.59$. The WEPP-used shear stress is a variable that reaches only low R^2 values for all of the tested data sets. Knapen et al. [14] describes the shear stress as follows (p. 80 f.): "Although the use of flow shear stress as soil detachment predictor can be contested, critical shear stress (τ_{cr}) and concentrated flow erodibility KC (…) have been selected as the most universal parameters to describe soil erosion resistance to concentrated flow." The correlations between these factors and the soil detachment rate show very varying results. There is not a single parameter that always reveals the best correlation. These considerations lead to two main questions:

- Are soil erosion, detachment and transport, directly dependent on water flow characteristics?
- Are these concepts, as implemented in soil erosion models, suitable to describe rill erosion?

These questions have been tackled by many research groups that have been searching for the equation that suits their observations best [1]–[13], [15]–[42]. However, taking into consideration the numerous and variable results, a deeper insight into the rill erosion processes on hillslopes is essential. To get this insight, different strategies can be applied [43]: (1) Modelling, (2) laboratory experiments (3) field observations and (4) field experiments. Each of these methods shows different advantages and disadvantages.

Due to difficulties to measure certain parameters, models have to be calibrated. During this process, the phenomenon of equifinality can appear: different parameter sets show the same result. Another weakness of rill erosion models is that the model parameters are often adapted from river hydrodynamics equations. Govers and his colleagues [13], [44] showed that these equations are not suitable for rill erosion processes. Therefore, there is often a mismatch between model results and observed or measured "reality" [43]. Additionally, models only project the concepts of the designer, not necessarily the reality.

In laboratory experiments, the initial and boundary conditions are well controlled. Soil parameters are well known and rill forms and slope can be adapted to the specific question. Thus, physical laws can be tested in a well-defined environment. However, Giménez and Govers [5] showed that parameters determined under laboratory conditions are not easily transformable to natural environments. One disadvantage of former laboratory experiments or field observations is the fact that in most cases only total runoff and sediment output are measured while the relative contribution of the individual processes is not considered [45].

Field data currently reflect the reality as close as possible. Nevertheless, observations as well as experiments show certain disadvantages: (1) Measurement techniques may disturb the observed processes, (2) time scale of human observations is shorter than that of the process under study, (3) some processes cannot be measured directly or indirectly and (4) some processes are chaotic and the spatial and temporal variations are difficult to specify [43].

The relationship between soil detachment and hydraulic parameters used in soil erosion models is in most cases deduced from laboratory experiments but the transferability of these results to natural rills is not generally given. Our setup in natural rills enables to measure the input parameters for calculating hydraulic parameters combining the advantages of laboratory experiments with the advantages of testing natural rills.

The main purpose of the field experiments was to quantify in a detailed temporal and spatial resolution the soil erosion dynamics in natural rills under concentrated flow for comparison of the measured sediment dynamics with those calculated by means of the most common detachment and transport equations.

Specifically, this study›s objectives are:

• elucidating the relationship between hydraulic parameters such as shear stress, unit length shear force, unit stream power, stream power, effective stream power and the Reynolds number and soil detachment in natural rills,

- providing an explanation why physically-based soil erosion models do not capture rill erosion processes and
- addressing the question whether current modelling approaches are generally suited to describe rill erosion processes.

The overall aim of this study is to have a critical view on concepts for modelling rill erosion based on experiments performed in naturally developed rills.

MATERIALS AND METHODS

Ethics Statement

No specific permits were required for the described field studies. The mayors of the towns next to the study sites or the owners of the fields were informed about the intended activities and were asked for permission. The test sites Freila, Negratin and Salada are abandoned fields which are sporadically used as pasture for goats or sheep and in Belerda the experiment was accomplished on an almond field. The locations Freila, Negratin end Salada are not privately-owned and permission was granted from the owner of the study site Belerda. None of the study sites are protected in any way and the field studies did not involve endangered or protected species.

Study Areas

The four study areas in Andalusia are located at Negratin, Freila, Salada and Belerda. UTM coordinates of the tested rills are given in Table 1.

Table 1. Description table of the experiments: Temperature and precipitation with the nearest meteorological station (INM).

Experiment	Meteorological station	Average annual temperature	Annual precipitation	Northing of the rill	Easting of the rill
Freila 1+3	Baza	14.2°C	368 mm	4154368	509860
Freila 2	Baza	14.2°C	368 mm	4154398	509826
Negratin	Baza	14.2°C	368 mm	4156324	505710
Salada	Embalse Valdeinfierno	13.4°C	311 mm	4187266	595761
Belerda	Granada	15.6°C	473 mm	4133440	478070

UTM 30 coordinates of the five tested rills are presented.
Freila 1 and Freila 3 are two experiments in the same rill.
doi:10.1371/journal.pone.0064861.t001

Negratin and Freila

The areas are located within the Hoya de Baza sedimentary basin and composed of marls, in which calcareous Regosols have developed. The climate is semi-arid and vegetation is dominated by low shrubs and *Stipa tenacissima* grass tussocks. The land cover at the south side of the Negratin-dam is dominated by abandoned cereal fields, which are extensively grazed by sheep and agricultural land comprised mainly of cereal dry-farming and almond grooves [46].

Salada

Located at the SE-margin of the Betic range (SE-Spain), inside the penibetic complex. The area is composed of conglomerates with a clayey to loamy matrix, in which Regosols as well as to fairly developed (Calcic) Cambisols have developed. Vegetation is similar to that found in the Freila and Negratin-area. The climate is semi-arid too, but less accentuated than in the previously mentioned area [46]. Here the land use consists of rain fed agricultural areas (where cereals, olives, and almonds are cultivated), and abandoned or uncultivated areas.

Belerda

This test area is located in the Guadix basin. The parent material consists of tertiary and quaternary conglomerates, sands, silts and clays. The soil texture class following the FAO [47]is a silty clay loam. The land use is separated into cultivated areas, with almond and olive groves, and abandoned agricultural fields [48]. The climate is, though still semi-arid, characterised by higher average annual temperatures and precipitations in comparison with the other test zones. The climatic parameters of the test fields are summarized in Table 1.

Tested Rills

The main descriptors of the rills are summarized in Table 2. In this table, grain size class limits are from [49], texture class is determined following [47]. Photographies of the rills are presented in Figure 1.

Figure 1. Photographies of the tested rills. Informations about the rills are presented in Table 2. doi:10.1371/journal.pone.0064861.g001

Table 2. Rill parameters: Grain size class limits are from [49], texture class is determined following [47].

		Freila 1	Freila 2	Freila 3	Negratin	Salada	Belerda
	Ø Slope [°]	9.4	7.7	9.4	5.6	25.6	16.9
	Max. Slope [°]	15.2	14.1	15.2	12.9	7.3	12.5
	Tested flow length [m]	16	21	16	30	17	23
	Texture class	SL	SL	SL	SiL	SiCL	L
Gravel	>2000 µm [%]	30	30	30	1	1	13
Sand	2000-630 µm [%]	14	14	14	1	2	10
	630-200 µm [%]	14	14	14	5	2	10
	200-125 µm [%]	13	13	13	6	1	8
	125-63 µm [%]	16	16	16	11	7	17
Silt	63-20 µm [%]	13	13	13	11	17	13
	20-6.3 µm [%]	10	10	10	20	17	13
	6.3-2 µm [%]	11	11	11	24	24	14
Clay	<2 µm [%]	9	9	9	21	29	15
	Starting soilmoisture [% w/w]	3.1	3.5	3.1	3.1	5.8	2.4
	K_t [s^2 $m^{0.5}$ $kg^{-0.5}$]	0.0090	0.0090	0.0090	0.0095	0.0096	0.0093
	Location WEPP dataset	Academy	Academy	Academy	Frederick	Mexico	Caribou
	Maximum width [m]	~0.4	~2.2	~0.4	~0.4	~0.5	~0.3
	Maximum depth [m]	~0.05	~0.7	~0.05	~0.2	~0.25	~0.15
	Vegetation cover [%]	~40	~40	~40	~0	~15	~5
	Rock fragment cover [%]	~80	~80	~80	~5	~20	~50
	Grain density [g cm^{-3}]	2.69	2.69	2.69	2.65	2.66	2.61
	Dry bulk density [g cm^{-3}]	1.44	1.55	1.44	1.57	1.52	1.68
	Org. material [%]	1.29	1.29	1.29	1.75	2.97	1.34
	Critical shear stress [Pa]	1.97	2.07	1.97	2.93	3.20	2.77
	Land use	rangeland	rangeland	rangeland	rangeland	rangeland	cropland

K_t is a transport coefficient, which has been adopted from the WEPP dataset. The WEPP-location is given. Measured values are starting soil moisture, maximum width, maximum depth, grain density, dry bulk density, org. material; parameters estimated in the field are vegetation cover and rock fragment cover; critical shear stress is calculated following WEPP.
doi:10.1371/journal.pone.0064861.t002

The tested rills in Freila have developed on a sandy loam with high gravel content. Sand content is 57% with a relatively homogeneous contribution between coarse, medium, fine and very fine sand. The same is true in the silt fraction, the 34% are homogeneously contributed in the complete silt fraction between 63 and 2 µm. The rills show all a dense rock fragment cover and the highest vegetation cover of the four test sides.

In Negratin, the soil material is nearly gravel free, coarse, medium and fine sand also show low amounts, most of the fine material is in the grain size class <20 µm. The rock fragment cover in the rill is higher than the gravel content of the soil material thus it is possible that residual rock fragment accumulation has occurred.

In Salada the grain size distribution is similar to Negratin. The highest account of the fine soil material is in the class <63 µm. The residual rock fragment accumulation is formed even more clearly as in Negratin; the vegetation cover is relatively high compared to the other test sites.

The rill in Salada is the only rill that has developed in a field being used for agriculture. The soil material is composed by a mixture of all particle size classes from gravel to clay. The rock fragment cover is high compared to the other test sites and the vegetation cover comparatively low. This test site shows the highest dry bulk density which can be declared by the actual agricultural use.

Rill Experiment (RE)

The rill experiments consist of two runs: first the rill is tested under field conditions (run a); in a second run (run b), approximately 15 minutes later, the same rill is tested under almost saturated soil conditions. A constant discharge of 250 L (or 330 L, respectively) is maintained during 4 minutes (or 3 minutes, respectively), using a motor-driven pump, resulting in a total water inflow of 1000 L. Mobilisation of material at the inflow has been avoided.

The flow velocity within the rill is characterized by the travel time of the waterfront and of two colour tracers (started at 1 and 2 minutes of the experiment), measured for every meter using a chronograph. By means of this procedure, three velocity curves are recorded and changes in flow dynamics can be detected. As colour tracers, food colourings (E 124 (red) and E 13 (blue)) are used for reasons of safety.

The rill's slope is characterized by measuring with a spring bow of 1 m range and a digital spirit level. It must be considered that slope measuring provides only average slopes for 1 meter. A step or a knick-point in the rill is not accounted, but its position and height are recorded.

Four water samples are taken at three different measuring points (MP1–MP3). The first sample is taken as soon as the waterfront has reaching the sampling point, the second 30 seconds later, the third 90 second later, and the fourth 150 seconds later.

The (suspended) sediment concentration SSC is determined by filtration of the samples in laboratory [50].

At each measuring point, rill cross section is measured. With a laser rangefinder, the distance between sensor and rill bottom is measured in 0.002 m steps. This allows an accurate calculation of the rills cross section area and an estimation of the rills volume.

Water level is continuously measured by ultrasonic sensors at each measuring point.

Descriptors for Soil Detachment

Soil detachment can be described by shear stress τ, unit length shear force Γ, stream power ω, unit stream power ω_U and effective stream power ω_{eff}

$$\tau = \rho * g * R * S \quad [Pa] \tag{1}$$

$$\Gamma = \rho * g * A * S = \tau * W_P \quad [N\ m^{-1}] \tag{2}$$

$$\omega = \rho * g * R * S * v = \tau * v \quad [W\ m^{-2}] \tag{3}$$

$$\omega_U = S * v \quad [m\ s^{-1}] \tag{4}$$

$$\varpi_{eff} = \frac{(\tau * v)^{1.5}}{d^{\frac{2}{3}}} = \frac{\omega^{1.5}}{d^{\frac{2}{3}}} \quad [W\ m^{-1}] \tag{5}$$

with ρ=liquid density [kg m^{-3}], g the gravitational acceleration (9.81 m s^{-2}), R the hydraulic radius [m], A the flow cross section area [m^2], S the effective slope (sin(slope angle)), W_p the wetted perimeter [m], v the flow velocity [m s^{-1}] and d the water depth [m]; abbreviations of the units are Pa=Pascal, N=Newton, W=Watt.

Reynolds number describes the balance between the inertial flow forces represented by the product in the numerator and the viscous forces as described by the dynamic viscosity in the denominator. It is a criterion for stability of a flowing medium. When Reynolds number is small, viscous forces dominate the motion and inertial ones can be ignored whereas at high Reynolds numbers inertial forces dominate and it is often possible to ignore viscosity [51]. Reynolds Number Re is calculated as follows:

$$Re = \frac{\rho * v * R}{\eta} \tag{6}$$

with ρ=liquid density [kg m^{-3}], v=flow velocity [m s^{-1}], R=hydraulic radius [m] and η=dynamic viscosity [Pa s].

Liquid density is calculated using sediment concentration and grain density. The use of water's density is not practicable due to sediment concentrations of more than 400 g L^{-1}. Grain density was measured by a capillary pycnometer following DIN 18124 [52]. Flow velocity for each sample is interpolated between three measured velocities (arrival of the waterfront and arrival of the two colour tracers). Hydraulic radius and wetted cross section area can be calculated by measuring water level and the rill profile. The viscosity of the sediment suspensions was measured with a shear rate controlled rheometer

(Haake MARS from Thermo Fisher Scientific, Karlsruhe, Germany) and a cone-plate geometry with an angle of 2° and a diameter of 60 mm [53]. The shear rate γ is defined as:

$$\gamma = \frac{dv}{dy} \tag{7}$$

with v=fluid velocity and y=the gap between the cone and base plate. The rheomter controls the shear rate and measures the shear stress τ, from which the viscosity η is calculated via

$$\eta = \frac{\tau}{\gamma} \tag{8}$$

The sample volume is always 2.0 ml and the cell is tempered to 20°C+/−0.01°C. Data points are taken at shear rates between 150 s^{-1} and 1500 s^{-1}. The viscosity does not depend on the shear rate. This is according to theoretical considerations. For a suspension of monodisperse particles one expects a linear relation [54], [55] for volume concentrations up to approximately 10%.

Detachment rate D_R [kg s^{-1} m^{-2}] is calculated from the measured sediment concentrations and different hydraulic parameters:

$$D_R = \frac{SSC * v * A}{L * W_P} \tag{9}$$

with SSC=sediment concentration [g L^{-1}=kg m^{-3}] and L=flow length [m].

For the calculation of the critical shear stress, the equations from the WEPP model [34] is used. The authors separate between "cropland with sand content >30%" and "rangeland".

$$\tau_{cr}(cropland) = 2.67 + 0.065 * (\%clay)$$
$$- 0.058 * (\%very\ fine\ sand) \tag{10}$$

$$\tau_{cr}(rangeland) = 3.23 - 0.056 * (\%sand) - 0.244 * (\%org.\ mat.)$$

$$+ 0.9 * (dry\ bulk\ density) \tag{11}$$

For quantification of the different processes in the rill, the transport rate T_R [kg s^{-1}] and the transport capacity T_C [kg s^{-1}] are calculated:

$$T_R = SSC * v * A \tag{12}$$

$$T_C = R * K_t * \tau^{1.5} \tag{13}$$

Kt [s^2 $m^{0.5}$ $kg^{-0.5}$] is a transport coefficient depending on soil substrate. The Kt value of the WEPP substrate which was most similar to the given test site conditions is used.

Quantification of Different Erosion Processes

Following shear stress based model concepts, the transport rate cannot exceed the transport capacity [56]. Shear stress of the flowing water controls also the detachment. Therefore the transport rate up to the transport capacity is considered here as shear stress dependent uptake. The transport rate exceeding the transport capacity is considered as shear stress independent erosion caused by processes such as bank failure and headcut retreat. The resulting quantities are set into relation and given in percent of total transport rate.

RESULTS

Initial Data

The used parameters show a wide range of data. In most cases (12 of 19), the standard deviation is higher than the mean values, the highest standard deviation – mean - percentage reaches the transport capacity (224%), the effective stream power (188.9%), the sediment concentration (168.3%) and detachment and transport rate (both 150%). The lowest percentage is calculated for sample density (0.5%). All initial data are presented in supporting information Tables S1, S2, S3, S4, S5, S6, S7, S8, S9, S10, S11, S12, S13, S14, S15, S16,S17, S 18 and the statistical values of the data in Table 3.

Table 3. Descriptive statistics of the initial data.

variable	Maximum	Minimum	Mean	Standard Deviation	Percentage from Mean
SSC [g L^{-1}]	422.30	0.001	52.15	87.78	168.3
D_R [kg s^{-1} m^{-2}]	0.96	0.001	0.10	0.15	150.0
T_R[kg s^{-1}]	2.06	0.001	0.16	0.24	150.0
p [g cm^{-3}]	1.26	1.00	1.03	0.005	0.5
Slope [°]	24.50	1.70	9.73	6.90	70.9
T_C [kg s^{-1}]	3.38	0.001	0.25	0.56	224.0
v [m s^{-1}]	2.94	0.04	0.79	0.49	62.0
η [kg s^{-1} m^{-1}]	0.00311	0.00100	0.00126	0.00044	34.9
Water depth [cm]	21.00	0.20	3.99	4.23	106.0
A [cm^2]	877.69	0.80	149.21	195.84	131.3
W_P [cm]	107.58	4.85	38.21	24.16	63.2
R [cm]	9.65	0.10	2.92	2.12	72.6
τ [Pa]	246.70	0.96	52.38	55.18	105.3
Γ [N m^{-1}]	172.58	0.10	23.99	35.10	146.3
ω [W m^{-2}]	365.28	0.31	41.54	55.91	134.6
$ω_u$ [m s^{-1}]	0.88	0.001	0.14	0.17	121.4
$ω_{eff}$ [W m^{-1}]	37864.55	5.81	3807.14	7192.32	188.9
Re [ ]	86918.88	237.00	19053.94	16226.56	85.2
τ - $τ_{cr}$ [Pa]	244.73	−1.46	49.89	55.11	110.5

SSC = sediment concentration, D_R = detachment rate, T_R = transport rate, p = sample density, T_C = transport capacity, v = flow velocity, η = dynamic viscosity, A = flow cross section, W_P = wetted perimeter, R = hydraulic radius, τ = shear stress, Γ = unit length shear force, ω = stream power, $ω_u$ = unit stream power, $ω_{eff}$ = effective stream power, Re = Reynolds-Number, $τ_{cr}$ = critical shear stress.
doi:10.1371/journal.pone.0064861.t003

Dynamic Viscosity

The dynamic viscosity of the liquid shows a clear positive correlation with sediment concentration, i.e. dynamic viscosity increases with sediment concentration (see Figure 2). However, clear deviations from the trend line were observed for samples with low sediment concentrations, which were often rich in transported organic material. The small branchlets with low weight imply a low sediment concentration, but in rheometer measurements, they tilt and a high shear stress is erroneously measured. The trend line equation has been calculated for samples from different test sites, the R^2-value of 0.92 indicates that this equation can be used for further experiments.

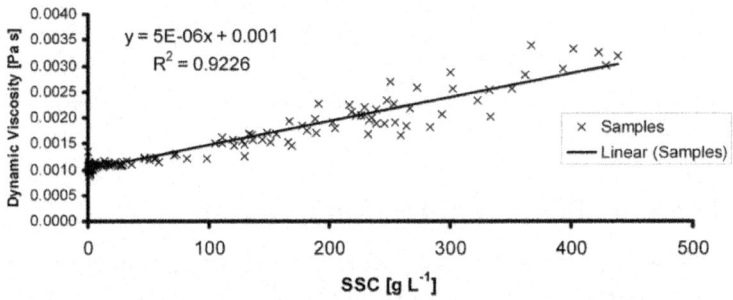

Figure 2. Correlation between sediment concentration of each sample and the measured dynamic viscosity. The linear correlation function and the R^2 value is presented.

doi:10.1371/journal.pone.0064861.g002

Correlations between Detachment Rate and Hydraulic Parameters

The R^2 values of the correlations between the detachment rate and different hydraulic parameters show the complete possible range from $R^2=0$ up to $R^2=0.99$ (see Table S19). Trend lines are increasing, decreasing and almost constant and thus it is not possible to find any clear dependency. Notably, only 40 of 252 correlations (about 16%) show an increasing trend line with an R^2 value$\geq$0.7. Table 4 shows that the highest average R^2-value is calculated for the $(\tau-\tau_{cr})$ – detachment rate - relationship if all R^2 values are used (0.53), if only the R^2- values with increasing trend line are considered in calculation, the τ – detachment rate relationship shows the highest average R^2 (0.55). Separating the experiments into two groups, Freila 1–3 with low sediment concentrations (LSSC) and Negratin, Salada, Belerda with high sediment concentrations (HSSC), the highest R^2-values of the LSSC-experiments reach

τ, Γ and the $(\tau\text{-}\tau_{cr})$ – detachment rate - relationship (0.65) if all values are used respectively the $(\tau\text{-}\tau_{cr})$ – detachment rate - relationship (0.39) if only the R^2 values$\geq$0.7 with increasing trend lines are used. In the HSSC-experiments, the Γ reaches the highest value (0.70) if all values are included and ω_{eff} (0.52) if only the R^2 values$\geq$0.7 with increasing trend lines are used, respectively.

Table 4. R^2 **- correlation values between different hydraulic parameters** and the detachment rate.

	τ	Γ	ω	ω_U	ω_{eff}	**Re**	$\tau - \tau_{cr}$
all values	0.52	0.50	0.37	0.43	0.40	0.39	0.53
only values with increasing trendline	0.55	0.52	0.40	0.46	0.45	0.39	0.53
all Freila experiments	0.65	0.65	0.50	0.53	0.48	0.53	0.65
Negratin, Salada, Belerda all values	0.69	0.70	0.43	0.56	0.56	0.49	0.64
Freila only values with increasing trend line	0.38	0.36	0.25	0.33	0.32	0.26	0.39
Negratin, Salada, Belerda only values with increasing trend lines	0.44	0.41	0.39	0.43	0.52	0.35	0.45

τ = shear stress, Γ = unit length shear force, ω = stream power, ω_U = unit stream power, ω_{eff} = effective stream power, Re = Reynolds number, τ_{cr} = critical shear stress. The complete dataset is presented in table S19.
doi:10.1371/journal.pone.0064861.t004

Quantification of Different Erosion Processes

Figure 3 shows the relationships between the measured transport rates and the predicted transport capacities. From 144 samples, in 82 cases the transport rate exceeds the capacity, corresponding to approximately 57% of all cases. Tables S20 and S21 present the differences between transport rates and transport capacities (S20) and the percentage of transport rate exceeding the capacity (S21) and hence the percentage of processes which are not controlled by the influence of shear stress. The percentage of material which is transported by processes independent of shear stress is on average 41.5% (see Table 5). Remarkably, the distribution is uneven, i.e. in the three Freila-experiments, the mean is 24.3% while in Negratin, Salada and Belerda, the average value is as high as 58.7% (see Table 5). The second group shows clearly higher sediment concentrations, meaning that the processes independent of shear stress provide higher sediment concentrations than the shear stress-based processes. This indicates that the influence of hydraulic parameters is higher for low sediment concentrations, or, in other words that high sediment concentrations are not caused by hydraulic parameters.

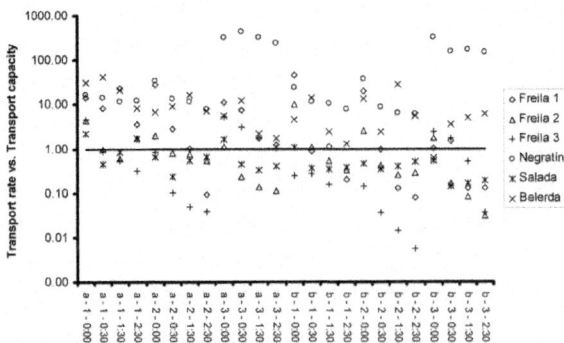

Figure 3. Transport rate vs. Transport capacity for each sample. The different experiments are represented by different symbols. On the x-axis, the following parameters are presented: run a or run b – measuring point 1–3 – sampling time at measuring point. The horizontal line marks the 1:1-relation between transport rate and transport capacity.

doi:10.1371/journal.pone.0064861.g003

Table 5. Percentage of exceedance: Share of transport rate exceeding transport capacity.

doi:10.1371/journal.pone.0064861.t005

Experiment	Value
Freila1	41.4%
Freila 2	16.0%
Freila 3	15.7%
Negratin	94.0%
Salada	6.0%
Belerda	76.1%
Average Freila 1–3	24.3%
Average Negratin, Salada, Belerda	58.7%
Average of all experiments	41.5%

In the first six rows, the exceedance percentage of each experiment is presented, in the next two rows the average percentage of the Freila experiments and of the Negratin-, Salada- and Belerda experiment and in the last row the average of all experiments. The complete dataset is presented in table S20.
doi:10.1371/journal.pone.0064861.t005

DISCUSSION

A comparison with results of other research groups shows that the measured values are in a realistic range. Ghebreiyessus [3] measured shear stress values up to 40 Pa and in the experiments of Nearing et al. [4], Reynolds numbers of up to 100000 and unit stream power values of up to 10 m s^{-1} were reached. Giménez & Govers [5] found unit stream power values of up to 0.4 m s^{-1} and unit length shear force values of up to 6 N m^{-1}. In a study of Zhang et al. [9], shear stress values of up to 30 Pa and unit stream power values of up to 0.5 m s^{-1} were reported. Govers [13] measured shear stress values of up to 100 Pa and effective stream power values of up to 10000 W m^{-1}. While the measurements presented here are in the same order of magnitude compared to the previously published research, there are no clear linear correlations between hydraulic parameters and erosion parameters in the results of the field experiments. Therefore, these outcomes indicate that linear models may generally not be sufficient in order to describe the complex processes in natural rills.

Four possible improvements may help to improve this important concept which has been studied already for over thirty years, (1) including a clear description of the employed parameters, (2) including the turbulence, (3) considering the impact of processes that do not depend on the shear stress and likewise (4) consider the high spatial and temporal variability observed in natural rills. These potential improvements will be discussed in more detail below.

For instance, the flow shear stress, a hydraulic parameter, and the critical shear stress, a soil parameter (similar to soil strength), must be differentiated. In particular, the flow shear stress must exceed the critical shear stress for erosion to occur. A number of hydraulic parameters, such as the flow velocity or the fluid density, water depth or width and roughness are used for the computation of the flow shear stress. The actual version of the shear stress equation calculates the average shear stress by depth averaging of momentum equation for steady uniform flow per area and time. Some factors used in shear stress calculation have been developed from empirical studies [15]–[26]. In most cases, the theoretical basis of the equations is however not clear. The formula applied by Chisci et al. [27] is derived from Landau and Lifchitz [57]. Other versions of the Landau-Lifchitz equation can be found in the literature[2]–[5].

The critical shear stress is the force needed to detach a soil particle. So it corresponds to a soil parameter and therefore, input for calculations should also depend on soil characteristics. However, this is the case for the WEPP model [34] only, where the critical shear stress is calculated using soil parameters such as texture, organic matter content and dry bulk density. In other cases, both hydraulic and soil parameters are used [34]. The discrepancies in the methods of computation of the shear stress may be due to the conditions under which the equations are deduced, as these equations are based on empirical observations. The empirical nature of the development of the different expressions is clearly highlighted in previous work[30]–[32]. That means the equations are not deduced from physical laws but from empirical studies. In many studies [12], [35], [37]–[41], neither critical shear stress nor shear stress are used for the calculation of the transport capacity at all. In other studies shear stress is used to calculate transport capacity and detachment capacity [36] or transport rate [42] and critical shear stress to calculate the detachment capacity [36]. In both cases it is clear that shear stress and critical shear stress operate against each other, the important parameter is the difference between these two variables.

A summary of these equations can be found in Reid and Dunne [58], on the EPA-homepage[59] and in Hessel and Jetten [60].

The second reason for the low R^2- values in the correlations between hydraulic parameter and soil detachment can be the lack of turbulence parameters in the equations.

In the study of Knapen et al. [14] the Reynolds number shows very different correlations to the detachment rate, and this holds as well for the results of this study. The reason could be that the turbulence, described by the Reynolds number, does not directly operate on substrate, it influences the acting shear stress, that means the calculated shear stress is much lower than the operating shear stress, a relation which has been confirmed in several studies: Nearing et al. [61] found that turbulence can increase the active shear stress by a factor of several thousands. They measured flow shear stresses ranging from 0.5 to 2 Pa, and tensile strengths ranging from 1 to 2 kPa. Despite the fact that the tensile strengths are 1000 times larger than the flow shear stresses, the authors also measured detachment rates in the order of 300 g $m^{-2}s^{-1}$. Such large detachment rates were attributed to turbulent burst events. Another study about the influence of turbulence on detachment rates was published by Nearing & Parker [62]. They showed that under turbulent flow conditions the same shear stress value caused a clearly higher detachment rate. In their flume experiments the difference between detachment rate caused by turbulent and laminar flow increased with increasing shear stress value, i.e., if given hydraulic conditions lead to a high shear stress value, the influence of turbulence on soil erosion is higher than in low shear stress value ranges.

The shear stress equation, as well as the equations describing other hydraulic parameters, assumes that drag forces are dominant for controlling erosion. But rill erosion is the result of the combination of different processes including headcut erosion, sidewall sloughing, tunnelling, micro-piping, slaking piping and sapping [14], [45], [63]–[67]. This is the third possible improvement for the problems of the model equations. The percentage of headcutting in the different studies ranges between "four times higher than the contribution of bed scours" [67] to "60% of total rill erosion" [68]. Stefanovic and Bryan [69] showed that concentrated flow causes sediment production primarily from knickpoints, chutes, meanders and bank failure. Govers [45] distinguished between hydraulic erosion, mass wasting processes on rill sidewalls, gullying and piping. Hydraulic rill erosion mostly occurred during three extreme runoff events. Mass wasting processes caused 37% of total erosion in rills. Gullying, the retreat erosion at knickpoints and headcuts caused about 12% of rill erosion rates. In the experiments presented here, the main mechanisms causing rill erosion were mass wasting and gullying processes, hence the correlations between hydraulic parameters and detachment rate are generally low. However, the hydraulic rill erosion only occurs in extreme runoff events, in most cases, the runoff values are too low to cause hydraulic rill erosion. The percentage of material which is transported independent of shear stress is very high on the water front samples. Here the transport of loose material is probably more important than in the other samples meaning that this process is mainly independent of shear stress. In these cases of transport rate vs. transport capacity <1 the independence of shear stress cannot be excluded, in the other cases the processes controlled by shear stress can occur. Thus, it can be deduced that, in the case of $T_R > T_C$, not only shear stress controlled processes provide the material; at least the difference between T_R and T_C is caused by processes independent of shear stress.

The experiments presented here show that the correlation between hydraulic parameters and detachment rate does neither change from one experiment to another, nor from one run to another, but from one measuring point and run to another. Thus, sediment producing processes have a high spatial and temporal variability. This is the fourth possible improvement for models. It is very difficult to propose a single factor that always describes the soil detachment satisfactory. The high variability of erosion processes, even under controlled experimental conditions, has been highlighted in different studies. Measured variability shows a wide range between 3.4% and 173.2% [70]–[75]. This is partially the result of non-homogeneous parameters concerning soil characteristics and rainfall. On experimental plots, infiltration rates and soil aggregate stability can be highly variable [76] and rainfall also shows a high spatial and temporal variability [77]. Therefore, the input parameters to the different measurements reflected in the mentioned studies were not really comparable. Nevertheless, the results also make clear that modelling soil erosion has to include uncertainty in model input, as well as in the data used for model calibration and validation.

In field experiments, the spatial and temporal variability of soil conditions cannot be avoided, and is, furthermore, part of the investigations. Thus, additional input parameters as rainfall or flow should be maintained constant in the experiments to generate reproducible data. The high variability in soil erosion processes cannot be represented by a single factor like shear stress.

The results show that there is not a simple linear correlation between a certain hydraulic parameter and soil detachment rate. Depending on model purpose and scale, the factors can be used to predict the magnitude of rill detachment but they are not applicable for the simulation of rill erosion with high-resolution spatial and temporal change in processes.

A newer approach is the use of probability density functions to predict soil detachment [78],[79]. Sidorchuk gives two sources of stochasticity in erosion modelling: (1) the necessity of spatial and temporal averaging when determining deterministic equations, which describe concentrated flow erosion and (2) the fact that the main erosion factors, if these can be determined anyway, can only be measured with limited accuracy. This is not the first attempt to model erosion by relating the probability of soil detachment with the excess of erosion driving forces over soil erosion resistance forces, other articles using a stochastic approach to describe soil erosion were published by Nearing [80], Wilson [81] and Sidorchuk [82]–[87]. Notably, one of the earliest articles about stochastic in erosion processes has been published by Einstein[88]. These stochastic models reduce the number of empirical components. Applying these models to the experiments presented here is beyond the scope of the current study.

CONCLUSIONS

The results show that a linear correlation between hydraulic parameter and soil detachment is not sufficient to describe processes in natural rills. The reason for this behaviour is the combination of various processes that can cause different amounts of soil erosion. The shear stress, for instance, only describes one process, while the results clearly show that there is not one fixed parameter that always predicts soil detachment best. Applicability of one certain hydraulic parameter to predict the sediment concentration changes at a certain point in time within a few minutes, because the temporal and spatial distribution of the different erosion processes is highly randomly determined. Therefore, it might be more useful to formulate results in probabilistic terms, an approach which has already been implemented by previous researchers, but is beyond the current work.

Supporting Information

Table S1. Freila 1 erosion data. doi:10.1371/journal.pone.0064861.s001 (DOC)

Run - MP - flow length [m]- sampling time [min:sec]	Sediment Concentration [g L⁻¹]	Detachment rate [kg s⁻¹ m⁻²]	Transport rate [kg s⁻¹]	Sample density [g cm⁻³]	Slope [°]	Transport capacity [kg s⁻¹]
a-1-3.4-0:00	10.6	0.0156	0.017750523	1.01	2.4	0.00125
a-1-3.4-0:30	2.1	0.0092	0.010500304	1.00	2.4	0.00128
a-1-3.4-1:30	3.3	0.0257	0.029404312	1.00	2.4	0.00129
a-1-3.4-2:30	0.4	0.0040	0.004617795	1.00	2.4	0.00126
a-2-8.6-0:00	14.9	0.0008	0.000726251	1.01	7.4	0.00003
a-2-8.6-0:30	2.4	0.0004	0.000914653	1.00	7.4	0.00032
a-2-8.6-1:30	1.4	0.0005	0.001633948	1.00	7.4	0.00156
a-2-8.6-2:30	0.6	0.0010	0.007309575	1.00	7.4	0.07782
a-3-13.1-0:00	26.6	0.0407	0.240254717	1.02	6	0.02132
a-3-13.1-0:30	8.0	0.0079	0.041958790	1.01	6	0.00555
a-3-13.1-1:30	3.1	0.0049	0.028349123	1.00	6	0.01600
a-3-13.1-2:30	2.5	0.0042	0.024596630	1.00	6	0.01957
b-1-3.4-0:00	35.1	0.0524	0.060019839	1.02	2.4	0.00132
b-1-3.4-0:30	0.5	0.0010	0.001109306	1.00	2.4	0.00126
b-1-3.4-1:30	0.6	0.0012	0.001397038	1.00	2.4	0.00126
b-1-3.4-2:30	0.1	0.0002	0.000302770	1.00	2.4	0.00144
b-2-8.6-0:00	4.8	0.0001	0.000126626	1.00	7.4	0.00001
b-2-8.6-0:30	0.6	0.0001	0.000260038	1.00	7.4	0.00027
b-2-8.6-1:30	0.7	0.0013	0.010225583	1.00	7.4	0.07834
b-2-8.6-2:30	0.1	0.0000	0.000132974	1.00	7.4	0.00161
b-3-13.1-0:00	9.9	0.0336	0.247153306	1.01	6	0.24591
b-3-13.1-0:30	2.6	0.0021	0.012025713	1.00	6	0.00812
b-3-13.1-1:30	1.0	0.0042	0.030456868	1.00	6	0.22937
b-3-13.1-2:30	0.8	0.0047	0.035191499	1.00	6	0.26525

Table S2: Freila 1 runoff data. doi:10.1371/journal.pone.0064861.s002

Run - MP - flow length [m]- sampling time [min:sec]	Flow velocity [m s⁻¹]	Dynamic viscosity [kg s⁻¹ m⁻¹]	Water depth [cm]	Flow cross section [cm²]	Wetted Perimeter [cm]	Hydraulic radius [cm]
a-1-3.4-0:00	0.41	0.001053	0.20	40.9	33.5	1.2

a-1-3.4-0:30	1.20	0.001011	0.30	41.7	33.7	1.2
a-1-3.4-1:30	2.12	0.001017	0.30	41.7	33.7	1.2
a-1-3.4-2:30	2.61	0.001002	0.40	42.0	34.2	1.2
a-2-8.6-0:00	0.34	0.001075	0.30	1.4	10.9	0.1
a-2-8.6-0:30	0.39	0.001012	1.30	10.0	27.6	0.4
a-2-8.6-1:30	0.41	0.001007	2.50	28.0	41.1	0.7
a-2-8.6-2:30	0.41	0.001003	10.30	289.7	88.9	3.3
a-3-13.1-0:00	0.92	0.001133	3.60	98.3	45.1	2.2
a-3-13.1-0:30	1.01	0.001040	2.40	51.8	40.5	1.3
a-3-13.1-1:30	1.05	0.001016	3.10	86.9	44.3	2.0
a-3-13.1-2:30	1.06	0.001012	3.30	94.9	44.6	2.1
b-1-3.4-0:00	0.41	0.001176	0.30	41.7	33.7	1.2
b-1-3.4-0:30	0.57	0.001002	0.40	42.0	34.2	1.2
b-1-3.4-1:30	0.59	0.001003	0.40	42.0	34.2	1.2
b-1-3.4-2:30	0.91	0.001000	0.50	47.2	36.3	1.3
b-2-8.6-0:00	0.32	0.001024	0.50	0.8	10.9	0.1
b-2-8.6-0:30	0.49	0.001003	1.30	9.4	28.0	0.3
b-2-8.6-1:30	0.53	0.001003	10.40	290.7	89.0	3.3
b-2-8.6-2:30	0.46	0.001001	2.40	26.1	37.7	0.7
b-3-13.1-0:00	0.76	0.001050	9.20	327.5	56.2	5.8
b-3-13.1-0:30	0.72	0.001013	2.80	64.0	42.8	1.5
b-3-13.1-1:30	0.98	0.001005	9.20	318.0	55.9	5.7
b-3-13.1-2:30	1.32	0.001004	9.30	346.0	57.4	6.0

Table S3: Freila 1 hydraulic data. doi:10.1371/journal.pone.0064861.s003

Run - MP - flow length [m]- sampling time [min:sec]	τ [Pa]	Γ [N m^{-1}]	ω [W m^{-2}]	ω_U [m s^{-1}]	ω_{eff} [W m^{-1}]	Re [ ]	$\tau - \tau_{cr}$ [Pa]
a-1-3.4-0:00	5.05	1.69	2.07	0.02	187.91	4791	3.08
a-1-3.4-0:30	5.09	1.72	6.10	0.05	725.17	14707	3.12
a-1-3.4-1:30	5.09	1.72	10.79	0.09	1703.78	25834	3.12
a-1-3.4-2:30	5.05	1.73	13.18	0.11	1898.77	32015	3.08
a-2-8.6-0:00	1.68	0.18	0.57	0.04	20.68	420	-0.30
a-2-8.6-0:30	4.58	1.26	1.77	0.05	42.55	1384	2.61
a-2-8.6-1:30	8.62	3.54	3.53	0.05	77.65	2776	6.64
a-2-8.6-2:30	41.19	36.61	16.89	0.05	315.87	13326	39.22
a-3-13.1-0:00	22.72	10.25	20.90	0.10	876.60	17996	20.75

a-3-13.1-0:30	13.20	5.34	13.31	0.11	583.29	12474	11.23
a-3-13.1-1:30	20.14	8.93	21.11	0.11	982.55	20267	18.17
a-3-13.1-2:30	21.83	9.74	23.08	0.11	1077.88	22237	19.86
b-1-3.4-0:00	5.20	1.75	2.13	0.02	149.50	4412	3.23
b-1-3.4-0:30	5.05	1.73	2.90	0.02	196.40	7053	3.08
b-1-3.4-1:30	5.05	1.73	2.98	0.02	204.64	7245	3.08
b-1-3.4-2:30	5.33	1.94	4.84	0.04	363.78	11770	3.36
b-2-8.6-0:00	0.96	0.10	0.31	0.04	5.81	237	-1.01
b-2-8.6-0:30	4.25	1.19	2.08	0.06	54.38	1644	2.28
b-2-8.6-1:30	41.30	36.75	21.83	0.07	461.19	17220	39.33
b-2-8.6-2:30	8.73	3.29	4.00	0.06	96.25	3166	6.76
b-3-13.1-0:00	60.18	33.79	45.74	0.08	1517.71	42492	58.21
b-3-13.1-0:30	15.35	6.58	11.01	0.08	396.41	10602	13.38
b-3-13.1-1:30	58.39	32.63	56.94	0.10	2108.18	55255	56.42
b-3-13.1-2:30	61.89	35.49	81.83	0.14	3606.26	79497	59.92

Table S4: Freila 2 erosion data. doi:10.1371/journal.pone.0064861.s004

Run - MP - flow length [m]- sampling time [min:sec]	Sediment Concentration [g L⁻¹]	Detachment rate [kg s⁻¹ m⁻²]	Transport rate [kg s⁻¹]	Sample density [g cm⁻³]	Slope [°]	Transport capacity [kg s⁻¹]
a-1-4-0:00	16.8	0.0668	0.267274644	1.01	5.7	0.05891
a-1-4-0:30	2.4	0.0078	0.023207900	1.00	5.7	0.02325
a-1-4-1:30	1.4	0.0058	0.014046683	1.00	5.7	0.02362
a-1-4-2:30	2.3	0.0086	0.019756646	1.00	5.7	0.01134
a-2-8.5-0:00	11.3	0.0079	0.033779926	1.01	3.4	0.01692
a-2-8.5-0:30	3.9	0.0033	0.014486969	1.00	3.4	0.01837
a-2-8.5-1:30	2.8	0.0028	0.011787666	1.00	3.4	0.01570
a-2-8.5-2:30	1.8	0.0023	0.010364921	1.00	3.4	0.01946
a-3-13.3-0:00	19.2	0.0239	0.287705065	1.01	7.4	0.25306
a-3-13.3-0:30	6.1	0.0127	0.181365395	1.00	7.4	0.77566
a-3-13.3-1:30	3.1	0.0070	0.098724460	1.00	7.4	0.71743
a-3-13.3-2:30	2.3	0.0057	0.079795433	1.00	7.4	0.71687
b-1-4-0:00	12.4	0.0193	0.037281867	1.01	5.7	0.00362
b-1-4-0:30	2.0	0.0062	0.014347225	1.00	5.7	0.01351
b-1-4-1:30	0.8	0.0027	0.006088968	1.00	5.7	0.01132
b-1-4-2:30	0.4	0.0017	0.003861665	1.00	5.7	0.01132

b-2-8.5-0:00	12.0	0.0103	0.044903275	1.01	3.4	0.01704
b-2-8.5-0:30	1.6	0.0018	0.008409765	1.00	3.4	0.01955
b-2-8.5-1:30	0.8	0.0010	0.005012019	1.00	3.4	0.01954
b-2-8.5-2:30	0.7	0.0012	0.005702665	1.00	3.4	0.01954
b-3-13.3-0:00	38.0	0.0371	0.445794531	1.02	7.4	0.25751
b-3-13.3-0:30	2.9	0.0092	0.132228769	1.00	7.4	0.77333
b-3-13.3-1:30	1.6	0.0042	0.059490319	1.00	7.4	0.71636
b-3-13.3-2:30	1.4	0.0016	0.022511209	1.00	7.4	0.71624

Table S5: Freila 2 runoff data. doi:10.1371/journal.pone.0064861.s005

Run - MP - flow length [m]- sampling time [min:sec]	Flow velocity [m s^{-1}]	Dynamic viscosity [kg s^{-1} m^{-1}]	Water depth [cm]	Flow cross section [cm^2]	Wetted Perimeter [cm]	Hydraulic radius [cm]
a-1-4-0:00	0.47	0.001084	4.7	338.62	99.98	3.39
a-1-4-0:30	0.54	0.001012	2.8	174.04	74.13	2.35
a-1-4-1:30	0.69	0.001007	2.4	142.28	60.20	2.36
a-1-4-2:30	0.84	0.001012	1.6	100.62	57.12	1.76
a-2-8.5-0:00	0.21	0.001057	0.3	142.06	50.60	2.81
a-2-8.5-0:30	0.24	0.001020	0.5	152.10	52.27	2.91
a-2-8.5-1:30	0.31	0.001014	0.2	134.99	49.38	2.73
a-2-8.5-2:30	0.38	0.001009	0.5	152.10	52.27	2.98
a-3-13.3-0:00	0.32	0.001096	4.2	469.04	90.43	5.19
a-3-13.3-0:30	0.34	0.001031	8.2	877.69	107.58	8.16
a-3-13.3-1:30	0.37	0.001016	8	838.70	105.94	7.92
a-3-13.3-2:30	0.41	0.001012	8	838.70	105.94	7.92
b-1-4-0:00	0.56	0.001062	0.8	53.72	48.34	1.11
b-1-4-0:30	0.67	0.001010	1.8	109.04	57.70	1.89
b-1-4-1:30	0.80	0.001004	1.6	100.62	57.12	1.76
b-1-4-2:30	0.87	0.001002	1.6	100.62	57.12	1.76
b-2-8.5-0:00	0.26	0.001060	0.4	143.66	51.04	2.81
b-2-8.5-0:30	0.31	0.001008	0.8	168.31	56.37	2.99
b-2-8.5-1:30	0.39	0.001004	0.8	168.31	56.37	2.99
b-2-8.5-2:30	0.48	0.001004	0.8	168.31	56.37	2.99
b-3-13.3-0:00	0.25	0.001190	4.2	469.04	90.43	5.19
b-3-13.3-0:30	0.52	0.001015	8.2	877.69	107.58	8.16
b-3-13.3-1:30	0.45	0.001008	8	838.70	105.94	7.92
b-3-13.3-2:30	0.19	0.001007	8	838.70	105.94	7.92

Table S6: Freila 2 hydraulic data. doi:10.1371/journal.pone.0064861.s006

Run - MP - flow length [m]- sampling time [min:sec]	τ [Pa]	Γ [N m^{-1}]	ω [W m^{-2}]	ω_U [m s^{-1}]	ω_{eff} [W m^{-1}]	Re [ ]	$\tau - \tau_{cr}$ [Pa]
a-1-4-0:00	33.35	33.34	15.67	0.05	476.44	14840.13	31.28
a-1-4-0:30	22.91	16.98	12.47	0.05	477.57	12643.85	20.84
a-1-4-1:30	23.05	13.87	15.97	0.07	767.17	16276.69	20.98
a-1-4-2:30	17.19	9.82	14.47	0.08	866.60	14676.93	15.12
a-2-8.5-0:00	16.45	8.32	3.45	0.01	308.68	5619.61	14.38
a-2-8.5-0:30	16.97	8.87	4.14	0.01	287.89	6976.46	14.90
a-2-8.5-1:30	15.93	7.87	4.96	0.02	696.51	8412.52	13.86
a-2-8.5-2:30	17.36	9.07	6.58	0.02	577.49	11212.34	15.29
a-3-13.3-0:00	66.33	59.98	21.22	0.04	809.25	15329.03	64.26
a-3-13.3-0:30	103.48	111.32	34.97	0.04	1095.81	26859.14	101.41
a-3-13.3-1:30	100.22	106.18	37.48	0.05	1235.77	29202.70	98.15
a-3-13.3-2:30	100.17	106.12	41.06	0.05	1417.16	32125.42	98.10
b-1-4-0:00	10.91	5.27	6.11	0.06	377.59	5905.24	8.84
b-1-4-0:30	18.44	10.64	12.30	0.07	628.24	12503.04	16.37
b-1-4-1:30	17.17	9.81	13.73	0.08	801.27	14039.07	15.10
b-1-4-2:30	17.17	9.81	14.92	0.09	908.07	15284.39	15.10
b-2-8.5-0:00	16.50	8.42	4.29	0.02	352.57	6954.96	14.43
b-2-8.5-0:30	17.39	9.80	5.34	0.02	308.90	9112.39	15.32
b-2-8.5-1:30	17.38	9.80	6.84	0.02	447.03	11708.97	15.31
b-2-8.5-2:30	17.38	9.80	8.25	0.03	592.94	14138.38	15.31
b-3-13.3-0:00	67.10	60.68	16.78	0.03	568.65	11156.44	65.03
b-3-13.3-0:30	103.27	111.10	53.49	0.07	2072.79	41729.93	101.20
b-3-13.3-1:30	100.12	106.07	45.47	0.06	1651.29	35706.89	98.05
b-3-13.3-2:30	100.11	106.06	19.44	0.03	461.72	15281.78	98.04

Table S7: Freila 3 erosion data. doi:10.1371/journal.pone.0064861.s007

Run - MP - flow length [m]- sampling time [min:sec]	Sediment Concentration [g L^{-1}]	Detachment rate [kg s^{-1} m^{-2}]	Transport rate [kg s^{-1}]	Sample density [g cm^{-3}]	Slope [°]	Transport capacity [kg s^{-1}]
a-1-2.9-0:00	45.0	0.2958	0.379845863	1.03	4.7	0.08011
a-1-2.9-0:30	3.0	0.0174	0.017769517	1.00	4.7	0.02040

a-1-2.9-1:30	1.6	0.0135	0.014059196	1.00	4.7	0.02558
a-1-2.9-2:30	0.8	0.0081	0.008537001	1.00	4.7	0.02670
a-2-11-0:00	43.2	0.1264	0.653849085	1.03	15.1	0.77604
a-2-11-0:30	7.8	0.0298	0.189057846	1.00	15.1	1.85097
a-2-11-1:30	4.3	0.0173	0.117480752	1.00	15.1	2.37381
a-2-11-2:30	3.7	0.0161	0.120276643	1.00	15.1	3.15647
a-3-13.8-0:00	56.3	0.0619	0.414999121	1.04	4.4	0.07706
a-3-13.8-0:30	11.1	0.0144	0.087141739	1.01	4.4	0.02722
a-3-13.8-1:30	5.5	0.0080	0.048446063	1.00	4.4	0.02707
a-3-13.8-2:30	3.9	0.0068	0.043053290	1.00	4.4	0.04278
b-1-2.9-0:00	3.5	0.0257	0.036744778	1.00	4.7	0.14627
b-1-2.9-0:30	1.4	0.0178	0.023345189	1.00	4.7	0.08610
b-1-2.9-1:30	0.4	0.0054	0.006318732	1.00	4.7	0.03960
b-1-2.9-2:30	0.0	0.0000	0.000000000	1.00	4.7	0.03846
b-2-11-0:00	10.2	0.0422	0.267669423	1.01	15.1	1.85510
b-2-11-0:30	3.3	0.0122	0.080111940	1.00	15.1	2.12315
b-2-11-1:30	1.9	0.0055	0.038627457	1.00	15.1	2.57786
b-2-11-2:30	1.2	0.0026	0.019920464	1.00	15.1	3.37596
b-3-13.8-0:00	26.1	0.0421	0.302333459	1.02	4.4	0.12407
b-3-13.8-0:30	6.1	0.0121	0.076629194	1.00	4.4	0.04286
b-3-13.8-1:30	2.5	0.0034	0.021755547	1.00	4.4	0.04272
b-3-13.8-2:30	2.2	0.0002	0.001559311	1.00	4.4	0.04271

Table S8: Freila 3 runoff data. doi:10.1371/journal.pone.0064861.s008

Run - MP - flow length [m]- sampling time [min:sec]	Flow velocity [m s⁻¹]	Dynamic viscosity [kg s⁻¹ m⁻¹]	Water depth [cm]	Flow cross section [cm²]	Wetted Perimeter [cm]	Hydraulic radius [cm]
a-1-2.9-0:00	0.44	0.001225	7.6	191.68	45.06	4.25
a-1-2.9-0:30	0.65	0.001015	4.3	89.43	35.77	2.50
a-1-2.9-1:30	0.90	0.001008	4.8	100.07	36.54	2.74
a-1-2.9-2:30	1.00	0.001004	4.9	103.52	37.15	2.79
a-2-11-0:00	0.61	0.001216	10.00	248.00	47.02	5.27
a-2-11-0:30	0.55	0.001039	15.00	436.45	57.68	7.57
a-2-11-1:30	0.52	0.001022	17.00	517.84	61.88	8.37
a-2-11-2:30	0.51	0.001018	20.00	638.72	68.08	9.38
a-3-13.8-0:00	0.35	0.001281	7.00	210.67	48.55	4.34
a-3-13.8-0:30	0.62	0.001055	5.00	128.02	44.00	2.91

a-3-13.8-1:30	0.69	0.001027	5.00	128.02	44.00	2.91
a-3-13.8-2:30	0.68	0.001020	6.00	160.05	45.78	3.50
b-1-2.9-0:00	0.38	0.001018	9.7	276.10	50.23	5.50
b-1-2.9-0:30	0.82	0.001007	5.9	204.80	46.02	4.45
b-1-2.9-1:30	1.33	0.001002	5.7	134.37	41.18	3.26
b-1-2.9-2:30	1.55	0.001000	5.6	123.89	38.41	3.23
b-2-11-0:00	0.6	0.001051	15.00	436.45	57.68	7.57
b-2-11-0:30	0.51	0.001016	16.00	479.77	59.92	8.01
b-2-11-1:30	0.38	0.001009	18.00	550.13	63.54	8.66
b-2-11-2:30	0.25	0.001006	21.00	674.83	69.96	9.65
b-3-13.8-0:00	0.42	0.001130	9.00	276.30	52.05	5.31
b-3-13.8-0:30	0.78	0.001031	6.00	160.05	45.78	3.50
b-3-13.8-1:30	0.54	0.001013	6.00	160.05	45.78	3.50
b-3-13.8-2:30	0.04	0.001011	6.00	160.05	45.78	3.50

Table S9: Freila 3 hydraulic data. doi:10.1371/journal.pone.0064861.s009

Run - MP - flow length [m]- sampling time [min:sec]	τ [Pa]	Γ [N m^{-1}]	ω [W m^{-2}]	ω_U [m s^{-1}]	ω_{eff} [W m^{-1}]	Re [ ]	$\tau - \tau_{cr}$ [Pa]
a-1-2.9-0:00	35.16	15.84	15.47	0.04	339.16	15709.50	33.19
a-1-2.9-0:30	20.13	7.20	13.13	0.05	387.49	16085.57	18.16
a-1-2.9-1:30	22.03	8.05	19.74	0.07	664.13	24368.99	20.06
a-1-2.9-2:30	22.41	8.33	22.33	0.08	788.02	27665.64	20.44
a-2-11-0:00	138.45	65.10	84.45	0.16	3602.41	27174.42	136.48
a-2-11-0:30	194.31	112.09	107.33	0.14	3938.82	40414.91	192.34
a-2-11-1:30	214.45	132.70	111.95	0.14	3859.87	42875.13	212.48
a-2-11-2:30	240.30	163.61	123.07	0.13	3992.08	47287.94	238.33
a-3-13.8-0:00	33.81	16.42	11.83	0.03	239.70	12271.29	31.84
a-3-13.8-0:30	22.05	9.70	13.56	0.05	368.01	17075.36	20.08
a-3-13.8-1:30	21.97	9.67	15.20	0.05	436.48	19653.60	20.00
a-3-13.8-2:30	26.38	12.08	17.97	0.05	497.22	23420.27	24.41
b-1-2.9-0:00	44.28	22.24	16.83	0.03	326.93	20572.31	42.31
b-1-2.9-0:30	35.80	16.48	29.31	0.07	1046.85	36207.56	33.83
b-1-2.9-1:30	26.23	10.80	34.84	0.11	1388.43	43265.86	24.26
b-1-2.9-2:30	25.93	9.96	40.30	0.13	1748.03	50139.66	23.96
b-2-11-0:00	194.60	112.25	116.76	0.16	4468.98	43466.94	192.63
b-2-11-0:30	205.03	122.86	105.19	0.13	3660.45	40501.41	203.06

b-2-11-1:30	221.51	140.75	83.15	0.10	2378.54	32237.17	219.54
b-2-11-2:30	246.70	172.58	61.60	0.07	1368.60	23964.67	244.73
b-3-13.8-0:00	40.61	21.14	17.05	0.03	350.70	20049.07	38.64
b-3-13.8-0:30	26.41	12.09	20.69	0.06	614.06	26675.66	24.44
b-3-13.8-1:30	26.35	12.06	14.15	0.04	347.22	18563.41	24.38
b-3-13.8-2:30	26.35	12.06	1.17	0.00	8.22	1533.53	24.38

Table S10: Negratin erosion data. doi:10.1371/journal.pone.0064861.s010

Run - MP - flow length [m]- sampling time [min:sec]	Sediment Concentration [g L^{-1}]	Detachment rate [kg s^{-1} m^{-2}]	Transport rate [kg s^{-1}]	Sample density [g cm^{-3}]	Slope [°]	Transport capacity [kg s^{-1}]
a-1-3.2-0:00	67.3	0.3509	0.378972891	1.04	3.2	0.02284
a-1-3.2-0:30	55.7	0.3594	0.417851939	1.03	3.2	0.02926
a-1-3.2-1:30	37.0	0.2853	0.331751054	1.02	3.2	0.02877
a-1-3.2-2:30	35.2	0.3353	0.411477439	1.02	3.2	0.03395
a-2-5.1-0:00	128.8	0.1191	0.083361100	1.08	7.7	0.00238
a-2-5.1-0:30	102.6	0.2553	0.258820005	1.06	7.7	0.01888
a-2-5.1-1:30	60.6	0.1840	0.183346450	1.04	7.7	0.01584
a-2-5.1-2:30	41.3	0.1941	0.219213489	1.03	7.7	0.02768
a-3-11.5-0:00	170.4	0.0370	0.039566213	1.11	1.7	0.00012
a-3-11.5-0:30	155.3	0.0391	0.039947665	1.10	1.7	0.00009
a-3-11.5-1:30	95.0	0.0333	0.035583905	1.06	1.7	0.00011
a-3-11.5-2:30	81.0	0.0364	0.040214682	1.05	1.7	0.00016
b-1-3.2-0:00	82.1	0.5253	0.567410676	1.05	3.2	0.02314
b-1-3.2-0:30	35.9	0.2925	0.340134613	1.02	3.2	0.02874
b-1-3.2-1:30	24.7	0.2555	0.297103809	1.02	3.2	0.02845
b-1-3.2-2:30	12.4	0.2164	0.265524812	1.01	3.2	0.03324
b-2-5.1-0:00	128.2	0.1551	0.112506225	1.08	7.7	0.00302
b-2-5.1-0:30	48.3	0.1641	0.171502072	1.03	7.7	0.01942
b-2-5.1-1:30	28.8	0.1376	0.147576853	1.02	7.7	0.02288
b-2-5.1-2:30	28.3	0.1823	0.219256736	1.02	7.7	0.03540
b-3-11.5-0:00	168.5	0.0358	0.038265903	1.10	1.7	0.00012
b-3-11.5-0:30	79.8	0.0234	0.025837099	1.05	1.7	0.00016
b-3-11.5-1:30	52.8	0.0137	0.013972868	1.03	1.7	0.00008
b-3-11.5-2:30	51.4	0.0153	0.016371852	1.03	1.7	0.00011

Table S11: Negratin runoff data. doi:10.1371/journal.pone.0064861.s011

Run - MP - flow length [m]- sampling time [min:sec]	Flow velocity [m s⁻¹]	Dynamic viscosity [kg s⁻¹ m⁻¹]	Water depth [cm]	Flow cross section [cm²]	Wetted Perim- eter [cm]	Hydraulic radius [cm]
a-1-3.2-0:00	0.53	0.001336	0.5	106.28	33.75	3.15
a-1-3.2-0:30	0.59	0.001278	2	126.85	36.33	3.49
a-1-3.2-1:30	0.71	0.001185	2	126.85	36.33	3.49
a-1-3.2-2:30	0.82	0.001176	3	143.13	38.35	3.73
a-2-5.1-0:00	0.64	0.001644	0.4	10.11	13.72	0.74
a-2-5.1-0:30	0.75	0.001513	2.1	33.86	19.88	1.70
a-2-5.1-1:30	0.96	0.001303	2	31.49	19.53	1.61
a-2-5.1-2:30	1.18	0.001207	2.6	44.94	22.14	2.03
a-3-11.5-0:00	0.46	0.001852	0.40	5.05	9.30	0.54
a-3-11.5-0:30	0.59	0.001776	0.30	4.33	8.89	0.49
a-3-11.5-1:30	0.74	0.001475	0.40	5.05	9.30	0.54
a-3-11.5-2:30	0.81	0.001405	0.50	6.12	9.61	0.64
b-1-3.2-0:00	0.65	0.001411	0.5	106.28	33.75	3.15
b-1-3.2-0:30	0.75	0.001180	2	126.85	36.33	3.49
b-1-3.2-1:30	0.95	0.001123	2	126.85	36.33	3.49
b-1-3.2-2:30	1.50	0.001062	3	143.13	38.35	3.73
b-2-5.1-0:00	0.76	0.001641	0.5	11.54	14.22	0.81
b-2-5.1-0:30	0.99	0.001242	2.3	36.00	20.49	1.76
b-2-5.1-1:30	1.29	0.001144	2.5	39.73	21.03	1.89
b-2-5.1-2:30	1.46	0.001141	3	53.06	23.58	2.25
b-3-11.5-0:00	0.45	0.001843	0.40	5.05	9.30	0.54
b-3-11.5-0:30	0.53	0.001399	0.50	6.12	9.61	0.64
b-3-11.5-1:30	0.61	0.001264	0.30	4.33	8.89	0.49
b-3-11.5-2:30	0.63	0.001257	0.40	5.05	9.30	0.54

Table S12: Negratin hydraulic data. doi:10.1371/journal.pone.0064861.s012

Run - MP - flow length [m]- sampling time [min:sec]	τ [Pa]	Γ [N m⁻¹]	ω [W m⁻²]	ω_U [m s⁻¹]	ω_{eff} [W m⁻¹]	Re [ ]	$\tau - \tau_{cr}$ [Pa]
a-1-3.2-0:00	17.96	6.06	9.52	0.03	1004.67	13009.48	15.04
a-1-3.2-0:30	19.78	7.19	11.70	0.03	543.29	16715.29	16.85
a-1-3.2-1:30	19.56	7.11	13.84	0.04	698.89	21333.97	16.63

a-1-3.2-2:30	20.89	8.01	17.08	0.05	730.82	26518.86	17.96
a-2-5.1-0:00	10.46	1.44	6.70	0.09	687.70	3099.18	7.54
a-2-5.1-0:30	23.82	4.74	17.75	0.10	982.41	8925.34	20.90
a-2-5.1-1:30	21.99	4.30	21.12	0.13	1317.11	12329.08	19.06
a-2-5.1-2:30	27.36	6.06	32.28	0.16	2090.11	20354.33	24.44
a-3-11.5-0:00	1.75	0.16	0.80	0.01	28.58	1490.41	-1.18
a-3-11.5-0:30	1.56	0.14	0.92	0.02	42.67	1786.38	-1.37
a-3-11.5-1:30	1.67	0.16	1.24	0.02	54.86	2890.54	-1.26
a-3-11.5-2:30	1.95	0.19	1.58	0.02	67.92	3863.99	-0.98
b-1-3.2-0:00	18.12	6.12	11.78	0.04	1382.74	15249.04	15.20
b-1-3.2-0:30	19.55	7.10	14.58	0.04	755.90	22575.90	16.62
b-1-3.2-1:30	19.41	7.05	18.44	0.05	1074.74	29978.88	16.48
b-1-3.2-2:30	20.60	7.90	30.90	0.08	1778.67	53133.31	17.67
b-2-5.1-0:00	11.52	1.64	8.76	0.10	886.05	4058.91	8.59
b-2-5.1-0:30	23.79	4.87	23.45	0.13	1404.17	14369.29	20.86
b-2-5.1-1:30	25.28	5.32	32.62	0.17	2178.77	21693.14	22.35
b-2-5.1-2:30	30.10	7.10	44.00	0.20	3022.79	29329.14	27.17
b-3-11.5-0:00	1.74	0.16	0.79	0.01	27.61	1464.08	-1.18
b-3-11.5-0:30	1.94	0.19	1.03	0.02	35.72	2528.39	-0.98
b-3-11.5-1:30	1.47	0.13	0.90	0.02	40.73	2434.20	-1.46
b-3-11.5-2:30	1.63	0.15	1.03	0.02	41.37	2810.22	-1.30

Table S13: Salada erosion data. doi:10.1371/journal.pone.0064861.s013

Run - MP - flow length [m]- sampling time [min:sec]	Sediment Concentration [g L⁻¹]	Detachment rate [kg s⁻¹ m⁻²]	Transport rate [kg s⁻¹]	Sample density [g cm⁻³]	Slope [°]	Transport capacity [kg s⁻¹]
a-1-2.3-0:00	65.7	0.2744	0.163638470	1.04	14.9	0.07299
a-1-2.3-0:30	21.9	0.0840	0.053634448	1.01	14.9	0.12230
a-1-2.3-1:30	11.4	0.0767	0.044566797	1.01	14.9	0.05316
a-1-2.3-2:30	15.4	0.4565	0.309609576	1.01	14.9	0.18004
a-2-4.7-0:00	30.8	0.0812	0.094584461	1.02	24.5	0.14384
a-2-4.7-0:30	18.3	0.0529	0.070550578	1.01	24.5	0.29789
a-2-4.7-1:30	40.7	0.2666	0.437682105	1.03	24.5	0.79787
a-2-4.7-2:30	11.4	0.0980	0.117807544	1.01	24.5	0.17785
a-3-4.7-0:00	113.4	0.3238	0.315564275	1.07	24.5	0.18940
a-3-4.7-0:30	42.2	0.1193	0.126580737	1.03	24.5	0.28476

a-3-4.7-1:30	25.0	0.1295	0.149733833	1.02	24.5	0.43935
a-3-4.7-2:30	14.7	0.2187	0.282982246	1.01	24.5	0.70410
b-1-2.3-0:00	36.8	0.1280	0.076308115	1.02	14.9	0.07110
b-1-2.3-0:30	8.1	0.0812	0.091376454	1.01	14.9	0.25308
b-1-2.3-1:30	5.8	0.0783	0.081788592	1.00	14.9	0.23377
b-1-2.3-2:30	7.0	0.1136	0.143162068	1.00	14.9	0.36293
b-2-4.7-0:00	42.9	0.2433	0.404671007	1.03	24.5	0.84780
b-2-4.7-0:30	19.7	0.1253	0.189952323	1.01	24.5	0.54642
b-2-4.7-1:30	9.6	0.0684	0.084120916	1.01	24.5	0.20340
b-2-4.7-2:30	10.9	0.1511	0.215740177	1.01	24.5	0.41807
b-3-4.7-0:00	39.0	0.1429	0.151710680	1.02	24.5	0.28391
b-3-4.7-0:30	9.5	0.0551	0.063738761	1.01	24.5	0.43310
b-3-4.7-1:30	7.8	0.0905	0.117090916	1.00	24.5	0.69957
b-3-4.7-2:30	6.6	0.1257	0.174190550	1.00	24.5	0.90918

Table S14: Salada runoff data. doi:10.1371/journal.pone.0064861.s014

Run - MP - flow length [m]- sampling time [min:sec]	Flow velocity [m s^{-1}]	Dynamic viscosity [kg s^{-1} m^{-1}]	Water depth [cm]	Flow cross section [cm^2]	Wetted Perimeter [cm]	Hydraulic radius [cm]
a-1-2.3-0:00	0.48	0.001328	2.5	51.91	25.92	2.00
a-1-2.3-0:30	0.35	0.001110	3.2	69.41	27.75	2.50
a-1-2.3-1:30	0.86	0.001057	2.2	45.47	25.27	1.80
a-1-2.3-2:30	2.33	0.001077	4.1	86.29	29.49	2.93
a-2-4.7-0:00	0.62	0.001154	4.7	49.50	24.79	2.00
a-2-4.7-0:30	0.51	0.001091	3.7	76.10	28.35	2.68
a-2-4.7-1:30	0.78	0.001204	7.7	137.92	34.93	3.95
a-2-4.7-2:30	1.84	0.001057	4.1	56.01	25.58	2.19
a-3-4.7-0:00	0.62	0.001567	5	44.87	20.73	2.16
a-3-4.7-0:30	0.51	0.001211	6	59.01	22.58	2.61
a-3-4.7-1:30	0.78	0.001125	7	76.97	24.61	3.13
a-3-4.7-2:30	1.84	0.001073	8	104.38	27.53	3.79
b-1-2.3-0:00	0.40	0.001184	2.5	51.91	25.92	2.00
b-1-2.3-0:30	0.68	0.001041	6.3	164.46	48.91	3.36
b-1-2.3-1:30	0.95	0.001029	5.9	148.03	45.41	3.26
b-1-2.3-2:30	0.96	0.001035	7.5	212.97	54.81	3.89
b-2-4.7-0:00	0.66	0.001214	8.0	143.01	35.38	4.04
b-2-4.7-0:30	0.87	0.001098	6.6	110.25	32.24	3.42

b-2-4.7-1:30	1.45	0.001048	4.2	60.45	26.15	2.31
b-2-4.7-2:30	2.12	0.001054	5.7	93.66	30.39	3.08
b-3-4.7-0:00	0.66	0.001195	6	59.01	22.58	2.61
b-3-4.7-0:30	0.87	0.001047	7	76.97	24.61	3.13
b-3-4.7-1:30	1.45	0.001039	8	104.38	27.53	3.79
b-3-4.7-2:30	2.12	0.001033	9	124.17	29.48	4.21

Table S15: Salada hydraulic data. doi:10.1371/journal.pone.0064861.s015

Run - MP - flow length [m]- sampling time [min:sec]	τ [Pa]	Γ [N m^{-1}]	ω [W m^{-2}]	ω_U [m s^{-1}]	ω_{eff} [W m^{-1}]	Re [ ]	$\tau - \tau_{cr}$ [Pa]
a-1-2.3-0:00	52.57	13.63	25.24	0.12	1482.75	7531.25	49.37
a-1-2.3-0:30	63.95	17.75	22.53	0.09	1061.25	8050.56	60.75
a-1-2.3-1:30	45.71	11.55	39.35	0.22	3144.11	14760.25	42.50
a-1-2.3-2:30	74.52	21.98	173.86	0.60	19280.91	64004.64	71.32
a-2-4.7-0:00	82.79	20.52	51.33	0.26	2823.76	10932.84	79.59
a-2-4.7-0:30	110.44	31.31	56.09	0.21	3783.55	12635.30	107.23
a-2-4.7-1:30	164.68	57.53	128.40	0.32	8038.43	26224.38	161.48
a-2-4.7-2:30	89.70	22.95	165.43	0.76	17895.79	38472.08	86.50
a-3-4.7-0:00	94.27	19.54	58.44	0.26	3292.10	9166.87	91.06
a-3-4.7-0:30	109.10	24.64	55.42	0.21	2691.66	11247.11	105.90
a-3-4.7-1:30	129.22	31.80	100.75	0.32	5953.66	22018.10	126.02
a-3-4.7-2:30	155.65	42.85	287.09	0.76	26199.37	65737.82	152.45
b-1-2.3-0:00	51.66	13.39	20.67	0.10	1098.76	6920.70	48.46
b-1-2.3-0:30	85.25	41.70	58.17	0.18	2802.12	22158.58	82.04
b-1-2.3-1:30	82.54	37.48	78.04	0.24	4548.92	30059.69	79.33
b-1-2.3-2:30	98.44	53.96	94.08	0.25	5131.14	36030.09	95.24
b-2-4.7-0:00	168.82	59.74	111.42	0.27	6334.72	22554.07	165.62
b-2-4.7-0:30	140.82	45.40	123.21	0.36	8374.20	27572.20	137.61
b-2-4.7-1:30	94.60	24.74	136.77	0.60	13237.98	32076.24	91.40
b-2-4.7-2:30	126.24	38.36	267.99	0.88	29621.25	62486.08	123.04
b-3-4.7-0:00	108.88	24.59	71.86	0.27	3974.97	14785.43	105.68
b-3-4.7-0:30	127.99	31.50	111.99	0.36	6977.44	26284.46	124.79
b-3-4.7-1:30	154.99	42.67	224.07	0.60	18065.59	53022.45	151.78
b-3-4.7-2:30	172.07	50.72	365.28	0.88	34762.59	86918.88	168.87

Table S16: Belerda erosion data. doi:10.1371/journal.pone.0064861.s016

Run - MP - flow length [m]- sampling time [min:sec]	Sediment Concentration [g L^{-1}]	Detachment rate [kg s^{-1} m^{-2}]	Transport rate [kg s^{-1}]	Sample density [g cm^{-3}]	Slope [°]	Transport capacity [kg s^{-1}]
a-1-6-0:00	111.1	0.6584	0.463968090	1.07	11.7	0.01464
a-1-6-0:30	250.8	0.9576	0.674872393	1.15	11.7	0.01644
a-1-6-1:30	288.4	0.5173	0.364531283	1.18	11.7	0.01694
a-1-6-2:30	176.1	0.4415	0.387089673	1.11	11.7	0.04693
a-2-13-0:00	243.8	0.1124	0.242154919	1.15	15.1	0.03603
a-2-13-0:30	359.6	0.1026	0.084365950	1.22	15.1	0.00926
a-2-13-1:30	234.8	0.0617	0.038866537	1.14	15.1	0.00230
a-2-13-2:30	88.0	0.0533	0.041378347	1.05	15.1	0.00577
a-3-17-0:00	341.2	0.2734	0.876936459	1.21	15.4	0.15169
a-3-17-0:30	422.3	0.6314	2.058803483	1.26	15.4	0.16862
a-3-17-1:30	62.6	0.0887	0.297213079	1.04	15.4	0.13513
a-3-17-2:30	383.4	0.0896	0.296150064	1.24	15.4	0.16984
b-1-6-0:00	93.0	0.1522	0.118800584	1.06	11.7	0.02513
b-1-6-0:30	212.6	0.3207	0.226026124	1.13	11.7	0.01594
b-1-6-1:30	16.5	0.0211	0.012073263	1.01	11.7	0.00496
b-1-6-2:30	13.5	0.0389	0.030357874	1.01	11.7	0.02340
b-2-13-0:00	312.2	0.2399	0.516754731	1.19	15.1	0.03803
b-2-13-0:30	97.3	0.0387	0.038785872	1.06	15.1	0.01574
b-2-13-1:30	421.1	0.2325	0.171978951	1.26	15.1	0.00611
b-2-13-2:30	61.8	0.0860	0.086292771	1.04	15.1	0.01525
b-3-17-0:00	31.7	0.0240	0.078397448	1.02	15.4	0.12267
b-3-17-0:30	141.3	0.1565	0.524266143	1.09	15.4	0.14471
b-3-17-1:30	125.3	0.2185	0.722405305	1.08	15.4	0.13812
b-3-17-2:30	109.2	0.2682	0.909694568	1.07	15.4	0.14510

Table S17: Belerda runoff data. doi:10.1371/journal.pone.0064861.s017

Run - MP - flow length [m]- sampling time [min:sec]	Flow velocity [m s^{-1}]	Dynamic viscosity [kg s^{-1} m^{-1}]	Water depth [cm]	Flow cross section [cm^2]	Wetted Perimeter [cm]	Hydraulic radius [cm]
a-1-6-0:00	2.94	0.001556	1.0	14.20	11.75	1.21
a-1-6-0:30	1.89	0.002254	1.0	14.20	11.75	1.21
a-1-6-1:30	0.89	0.002442	1.0	14.20	11.75	1.21

a-1-6-2:30	0.80	0.001881	2.0	27.55	14.61	1.89
a-2-13-0:00	0.42	0.002219	3.0	23.65	16.57	1.43
a-2-13-0:30	0.46	0.002798	0.7	5.06	6.33	0.80
a-2-13-1:30	0.72	0.002174	0.2	2.31	4.85	0.48
a-2-13-2:30	1.09	0.001440	0.6	4.32	5.98	0.72
a-3-17-0:00	0.56	0.002706	0.2	45.89	18.87	2.43
a-3-17-0:30	1.03	0.003111	0.3	47.50	19.18	2.48
a-3-17-1:30	0.95	0.001313	0.5	50.15	19.70	2.55
a-3-17-2:30	0.16	0.002917	0.4	48.85	19.45	2.51
b-1-6-0:00	0.65	0.001465	1.5	19.65	13.01	1.51
b-1-6-0:30	0.75	0.002063	1.0	14.20	11.75	1.21
b-1-6-1:30	0.94	0.001083	0.2	7.72	9.52	0.81
b-1-6-2:30	1.14	0.001068	1.5	19.65	13.01	1.51
b-2-13-0:00	0.70	0.002561	3.0	23.65	16.57	1.43
b-2-13-0:30	0.48	0.001486	1.0	8.31	7.72	1.08
b-2-13-1:30	1.08	0.003105	0.5	3.78	5.69	0.66
b-2-13-2:30	1.68	0.001309	1.0	8.31	7.72	1.08
b-3-17-0:00	0.52	0.001159	0.3	47.50	19.18	2.48
b-3-17-0:30	0.74	0.001706	0.5	50.15	19.70	2.55
b-3-17-1:30	1.18	0.001627	0.4	48.85	19.45	2.51
b-3-17-2:30	1.62	0.001546	0.6	51.41	19.95	2.58

Table S18: Belerda hydraulic data. doi:10.1371/journal.pone.0064861.s018

Run - MP - flow length [m]- sampling time [min:sec]	τ [Pa]	Γ [N m^{-1}]	ω [W m^{-2}]	ω_U [m s^{-1}]	ω_{eff} [W m^{-1}]	Re [ ]	$\tau - \tau_{cr}$ [Pa]
a-1-6-0:00	25.71	3.02	75.58	0.60	14155.26	24423.50	22.94
a-1-6-0:30	27.78	3.26	52.62	0.38	8224.56	11735.83	25.01
a-1-6-1:30	28.34	3.33	25.22	0.18	2728.19	5190.96	25.57
a-1-6-2:30	41.58	6.08	33.17	0.16	2592.39	8864.96	38.81
a-2-13-0:00	41.96	6.95	17.62	0.11	766.31	3107.72	39.19
a-2-13-0:30	24.97	1.58	11.58	0.12	1076.71	1619.26	22.19
a-2-13-1:30	13.93	0.67	10.00	0.19	1991.25	1799.58	11.16
a-2-13-2:30	19.47	1.16	21.19	0.28	2954.62	5758.07	16.70
a-3-17-0:00	76.70	14.47	42.95	0.15	17731.56	6092.17	73.92
a-3-17-0:30	81.32	15.60	83.46	0.27	36654.75	10296.35	78.55
a-3-17-1:30	68.88	13.57	65.17	0.25	17991.78	19049.17	66.11

a-3-17-2:30	80.93	15.74	12.80	0.04	1816.39	1683.83	78.16
b-1-6-0:00	31.78	4.13	20.66	0.13	1543.50	7087.10	29.01
b-1-6-0:30	27.21	3.20	20.36	0.15	1979.84	4961.43	24.44
b-1-6-1:30	16.31	1.55	15.41	0.19	3811.70	7155.59	13.54
b-1-6-2:30	30.30	3.94	34.59	0.23	3345.39	16287.61	27.53
b-2-13-0:00	43.50	7.21	30.45	0.18	1740.33	4652.71	40.73
b-2-13-0:30	29.16	2.25	14.00	0.13	1128.08	3684.88	26.39
b-2-13-1:30	21.39	1.22	23.11	0.28	3798.42	2911.66	18.62
b-2-13-2:30	28.56	2.20	47.98	0.44	7159.25	14340.51	25.79
b-3-17-0:00	65.77	12.62	34.20	0.14	9616.37	11330.97	63.00
b-3-17-0:30	72.10	14.20	53.35	0.20	13326.36	12001.08	69.32
b-3-17-1:30	70.51	13.71	83.19	0.31	30113.84	19632.98	67.74
b-3-17-2:30	71.65	14.29	116.06	0.43	37864.55	28811.68	68.87

Table S19: R^2 - values between the detachment rate and different hydraulic parameters. doi:10.1371/journal.pone.0064861.s019

		τ		Γ		ω		ω_U		ω_{eff}		Re		$\tau - \tau_{cr}$	
Freila 1 run a	MP 1	0.29	0	0.12	0	0.02	-	0.02	-	0.00	0	0.02	-	0.29	0
	MP 2	0.49	0	0.55	0	0.49	+	0.01	0	0.48	+	0.49	+	0.49	0
	MP 3	0.18	+	0.18	+	0.01	+	0.94	-	0.01	-	0.02	-	0.18	+
Freila 1 run b	MP 1	0.03	0	0.06	0	0.41	-	0.47	-	0.33	-	0.51	-	0.03	0
	MP 2	0.95	0	0.98	0	0.95	+	0.30	0	0.94	+	0.95	+	0.95	0
	MP 3	0.16	+	0.15	+	0.00	0	0.15	-	0.02	-	0.00	0	0.16	+
Freila 2 run a	MP 1	0.81	+	0.91	+	0.17	+	0.45	-	0.32	-	0.01	+	0.81	+
	MP 2	0.08	-	0.08	-	0.58	-	0.61	-	0.37	-	0.61	-	0.08	0
	MP 3	0.8	-	0.79	-	0.97	-	0.78	-	0.94	-	0.98	-	0.80	-
Freila 2 run b	MP 1	0.84	-	0.85	-	0.99	-	0.84	-	0.92	-	0.99	-	0.84	-
	MP 2	0.99	-	0.99	-	0.58	-	0.53	-	0.19	-	0.62	-	0.99	-
	MP 3	0.92	-	0.92	-	0.22	-	0.08	-	0.13	-	0.27	-	0.92	-
Freila 3 run a	MP 1	0.97	+	0.98	+	0.14	-	0.69	-	0.43	-	0.36	-	0.97	-

	MP 2	0.88	-	0.82	-	0.89	-	0.92	+	0.89	-	0.94	-	0.88	-
	MP 3	0.80	+	0.81	+	0.62	-	0.99	-	0.86	-	0.78	-	0.80	+
Freila 3 run b	MP 1	0.96	+	0.96	+	0.93	-	0.99	-	0.94	-	0.94	-	0.96	+
	MP 2	0.64	-	0.60	-	0.68	+	0.72	+	0.74	+	0.79	+	0.64	-
	MP 3	0.93	+	0.93	+	0.26	+	0.03	+	0.09	+	0.19	+	0.93	+
Negra-tin run a	MP 1	0.03	-	0.03	-	0.16	-	0.23	-	0.01	+	0.20	-	0.03	-
	MP 2	0.60	+	0.53	+	0.22	+	0.05	+	0.06	+	0.14	+	0.60	+
	MP 3	0.31	-	0.09	+	0.19	+	0.96	+	0.15	0	0.23	0	0.05	0
Ne-gratin run b	MP 1	0.86	-	0.86	-	0.53	-	0.51	-	0.00	0	0.58	-	0.86	-
	MP 2	0.14	+	0.19	+	0.16	+	0.08	+	0.17	0	0.14	+	0.14	+
	MP 3	0.31	+	0.27	+	0.39	+	0.24	+	0.99	0	0.79	0	0.31	0
Salada run a	MP 1	0.44	+	0.45	+	0.70	+	0.64	+	0.70	+	0.67	+	0.44	+
	MP 2	0.78	+	0.81	+	0.22	+	0.00	0	0.02	+	0.12	+	0.78	+
	MP 3	0.09	-	0.07	-	0.00	0	0.02	0	0.01	0	0.00	0	0.09	-
Salada run b	MP 1	0.23	-	0.18	-	0.19	-	0.27	-	0.17	-	0.20	-	0.23	-
	MP 2	0.86	+	0.87	+	0.01	-	0.18	-	0.05	-	0.03	-	0.86	+
	MP 3	0.00	-	0.00	-	0.03	0	0.02	0	0.05	0	0.02	0	0.00	0
Belerda run a	MP 1	0.35	-	0.35	-	0.25	+	0.28	+	0.26	+	0.08	+	0.34	-
	MP 2	0.71	+	0.55	+	0.02	0	0.85	-	0.92	-	0.23	-	0.71	+
	MP 3	0.27	+	0.19	+	0.47	+	0.31	+	0.78	+	0.00	0	0.27	+
Belerda run b	MP 1	0.12	+	0.07	+	0.05	-	0.46	-	0.59	-	0.37	-	0.12	+
	MP 2	0.06	+	0.20	+	0.00	0	0.01	0	0.02	-	0.16	-	0.06	+
	MP 3	0.71	+	0.74	+	0.88	+	0.86	+	0.83	+	0.72	+	0.71	+

Table S20: Comparison of the transport rate with the transport capacity: Transport rate T_R [kg s^{-1}] - Transport capacity T_C [kg s^{-1}]. doi:10.1371/journal.pone.0064861.s020

run-MP-time	Freila1	Freila 2	Freila 3	Negratin	Salada	Belerda
a - 1 - 0:00	0.016496	0.208367	0.299736	0.356133	0.090648	0.449332
a - 1 - 0:30	0.009216	-0.000042	-0.002630	0.388589	-0.068663	0.658430
a - 1 - 1:30	0.028119	-0.009571	-0.011526	0.302982	-0.008596	0.347591
a - 1 - 2:30	0.003356	0.008419	-0.018163	0.377529	0.129568	0.340157
a - 2 - 0:00	0.000700	0.016861	-0.122188	0.080985	-0.049255	0.206128
a - 2 - 0:30	0.000594	-0.003886	-1.661911	0.239944	-0.227338	0.075105
a - 2 - 1:30	0.000078	-0.003910	-2.256325	0.167504	-0.360192	0.036567
a - 2 - 2:30	-0.070511	-0.009096	-3.036197	0.191535	-0.060043	0.035613
a - 3 - 0:00	0.218935	0.034645	0.337938	0.039447	0.126165	0.725246
a - 3 - 0:30	0.036412	-0.594298	0.059926	0.039858	-0.158184	1.890184
a - 3 - 1:30	0.012344	-0.618707	0.021373	0.035472	-0.289620	0.162081
a - 3 - 2:30	0.005022	-0.637078	0.000277	0.040050	-0.421114	0.126305
b - 1 - 0:00	0.058696	0.033665	-0.109527	0.544266	0.005208	0.093669
b - 1 - 0:30	-0.000152	0.000836	-0.062758	0.311392	-0.161699	0.210084
b - 1 - 1:30	0.000136	-0.005232	-0.033282	0.268657	-0.151983	0.007109
b - 1 - 2:30	-0.001141	-0.007456	-0.038459	0.232281	-0.219765	0.006954
b - 2 - 0:00	0.000120	0.027868	-1.587432	0.109482	-0.443133	0.478729
b - 2 - 0:30	-0.000007	-0.011142	-2.043042	0.152084	-0.356465	0.023048
b - 2 - 1:30	-0.068115	-0.014524	-2.539236	0.124697	-0.119275	0.165872
b - 2 - 2:30	-0.001477	-0.013833	-3.356041	0.183860	-0.202332	0.071039
b - 3 - 0:00	0.001240	0.188280	0.178267	0.038147	-0.132203	-0.044270
b - 3 - 0:30	0.003910	-0.641102	0.033766	0.025673	-0.369358	0.379557
b - 3 - 1:30	-0.198913	-0.656871	-0.020963	0.013890	-0.582478	0.584281
b - 3 - 2:30	-0.230061	-0.693729	-0.041146	0.016264	-0.734992	0.764599

Table S21: Comparison of the transport rate with the transport capacity: Percentage of T_R exceeding T_C.

doi:10.1371/journal.pone.0064861.s021

run-MP-time	Freila1	Freila 2	Freila 3	Negratin	Salada	Belerda
a - 1 - 0:00	92.9	78.0	78.9	94.0	55.4	96.8
a - 1 - 0:30	87.8	0	0	93.0	0	97.6
a - 1 - 1:30	95.6	0	0	91.3	0	95.4
a - 1 - 2:30	72.7	42.6	0	91.7	41.8	87.9
a - 2 - 0:00	96.5	49.9	0	97.1	0	85.1

a - 2 - 0:30	65.0	0	0	92.7	0	89.0
a - 2 - 1:30	4.8	0	0	91.4	0	94.1
a - 2 - 2:30	0	0	0	87.4	0	86.1
a - 3 - 0:00	91.1	12.0	81.4	99.7	40.0	82.7
a - 3 - 0:30	86.8	0	68.8	99.8	0	91.8
a - 3 - 1:30	43.5	0	44.1	99.7	0	54.5
a - 3 - 2:30	20.4	0	0.6	99.6	0	42.6
b - 1 - 0:00	97.8	90.3	0	95.9	6.8	78.8
b - 1 - 0:30	0	5.8	0	91.5	0	92.9
b - 1 - 1:30	9.7	0	0	90.4	0	58.9
b - 1 - 2:30	0	0	0	87.5	0	22.9
b - 2 - 0:00	94.9	62.1	0	97.3	0	92.6
b - 2 - 0:30	0	0	0	88.7	0	59.4
b - 2 - 1:30	0	0	0	84.5	0	96.4
b - 2 - 2:30	0	0	0	83.9	0	82.3
b - 3 - 0:00	0.5	42.2	59.0	99.7	0	0
b - 3 - 0:30	32.5	0	44.1	99.4	0	72.4
b - 3 - 1:30	0	0	0	99.4	0	80.9
b - 3 - 2:30	0	0	0	99.3	0	84.1

Acknowledgments

We thank all participants of the field trip to Andalusia in September 2009 who supported the performance of the experiments and Olli, Seta and Andreas for revising the whole manuscript.

Author Contributions

Conceived and designed the experiments: SW MS JBR. Performed the experiments: SW MS AZ CW JBR. Analyzed the data: SW MS JFW JBR. Contributed reagents/materials/analysis tools: SW MS AZ CW JFW JBR. Wrote the paper: SW MS AZ CW JFW JBR.

REFERENCES

1. Lyle WM, Smerdon ET (1965) Relation of compaction and other soil properties to erosion resistance of soils. Transactions of the ASAE 8: 419–422.

2. Torri D, Dfalanga M, Chisci G (1987) Threshold conditions for incipient rilling. Catena Supplement 8: 97–105.

3.   Ghebreiyessus YT, Gantzer CJ, Alberts EE, Lentz RW (1994) Soil erosion by concentrated flow: shear stress and bulk density. Transactions of the ASAE 37 (6): 1791–1797.

4.   Nearing MA, Norton LD, Bulgakov DA, Larionov GA, West LT, et al. (1997) Hydraulics and Erosion in Eroding Rills. Water Resources Research 33 (4): 865– 876.

5.   Gime'nez R, Govers G (2002) Flow Detachment by Concentrated Flow on Smooth and Irregular Beds. Soil Sci Soc Am J 66(5): 1475–1483.

6.   Bagnold RA (1977) Bed load transport by natural rivers. Water resources Research 13: 303–312.

7.   Hairsine PB, Rose CW (1992) Modeling water erosion due to overland flow using physical principles, 2. Rill flow. Water resources Research 28: 245–250.

8.   Elliot WJ, Laflen JM (1993) A Process-based rill erosion model. Transactions of the ASAE 36 (1): 35–72.

9.   Zhang G, Liu B, Liu G, He X, Nearing MA (2003): Detachment of undisturbed soil by shallow flow. Soil Science Society of America Journal 67: 713–719.

10. Yang CT (1972) Unit stream power and sediment transport. J Hydraulics Div Am Soc Civil Eng 98: 1805–1825.

11. Moore ID, Burch GJ (1986) Sediment transport capacity of sheet and rill flow: Application to unit stream power theory. Water Resour Res 22: 1350–1360.

12. Bagnold RA (1980): An empirical correlation of bedload transport rates in flumes and natural rivers. Proc. Royal Society Series A372: 453–473.

13. Govers G (1992) Evaluation of transport capacity formulae for overland flow. In: Parsons AJ, Abrahams AD editors. Overland flow: hydraulics and erosion mechanics. London: UCL Press. pp. 243–273.

14. Knapen A, Poesen J, Govers G, Gyssels G, Nachtergaele J (2007): Resistance of soils to concentrated flow erosion: A review. Earth-Science Reviews 80(1–2): 75– 109.

15. Foster GR (1982) Modelling the erosion process. In: Johnson HP, Brakensiek DL, editors. Hydraulic Modelling of Small Watersheds. St. Joseph, MI: ASAE monograph 5, Hann CT.

16. Shields A (1936) Anwendung der A¨ hnlichkeitsmechanik und der Turbulenzforschung auf die Geschiebebewegung. Berlin: Dissertation, TU Berlin.

17. Miller M, McCave IN, Komar PD (1977) Threshold sediment motion

under unidirectional currents. Sedimentology 24: 507–527.

18. Parker G, Klingeman PC, McLean DG (1982) Bedload and size distribution in paved gravel-bed streams. Journal of the Hydraulics Division, American Society of Civil Engineers 108: 544–571.

19. Diplas P (1987) Bedload transport in gravel-bed streams. J. Hydraul. Eng. ASCE 113: 277–292.

20. Parker G (1990) Surface-based bedload transport relation for gravel rivers. J Hydraul Res 28: 417–436.

21. Komar PD (1987 a) Selective grain entrainment by a current from a bed of mixed sizes: A reanalysis. J Sediment Petrol 57: 203–211.

22. Komar PD (1987 b) Selective gravel entrainment and the empirical evaluation of flow competence. Sedimentology 34: 1165–1176.

23. Andrews ED (1983) Entrainment of gravel from naturally sorted riverbed material. Geol Soc Am Bull 94: 1225–1231.

24. Ashworth PJ, Ferguson RI (1989 a) Size-selective entrainment of bed load in gravel bed streams. Water Resources Research 25: 627–634.

25. Ashworth PJ, Ferguson RI (1989 b) Quantifying gravel deposition on river bars using flexible netting. Journal of Sedimentary Research Vol. 59 No. 4: 623–624.

26. Komar PD, Carling PA (1991) Grain sorting in gravel-bed streams and the choice of particle sizes for flow-competence evaluations. Sedimentology 38: 489–502.

27. Chisci G, Sfalanga M, Torri D (1985) An experimental model for evaluating soil erosion on a single-rainstorm basis. In: El-Swaify SA, Moldenhauer WC, Lo A, editors. Soil Erosion and Conservation. Ankeny, Iowa: Soil conservation Society of America. pp. 558–565.

28. Ott WP, van Uchelen JC (1936) Application of similarity principles and turbulence research to bedload movement. Hydrodynamics Laboratory California Institute of Technology Publication No. 167. Soil conservation Service.

29. Graf WH (1971) Hydraulics of sediment transport. New York: McGraw-Hill.

30. Partheniades E, Paaswell RE (1970) Erodibility of channels with cohesive boundary. Journal of the Hydraulic Division, Proceedings of the American Society of civil Engineers: 755–771.

31. Andrews ED (1984) Bed-material entrainment and hydraulic geometry of gravelbed rivers in Colorado. Geological Society of America Bulletin 95(3): 371–378.

32. Andrews ED, Erman DC (1986) Persistence in the size distribution of superficial bed material during an extreme snowmelt flood. Water Resources Research 22: 191–197.

33. De Ploey J (1990) Threshold conditions for thalweg gullying with special reference to loess areas. Catena Supplement 17: 147–151.

34. Flanagan DC, Livingston SJ (1995) WEPP user summary, USDA-water erosion prediction project. NSERL Report No. 11. National soil erosion research Laboratory.

35. Yalin MS (1963) An expression for bedload transportation. J Hydr Eng Div ASCE 89: 221–250.

36. Foster GR, Flanagan DC, Nearing MA, Lane LJ, Risse LM, et al. (1995) Hillslope erosion component. In: USDA Water Erosion Prediction Project: Hillslope Profile and Watershed Model Documentation, Chapter 11, USDAARS NSERL Report 10.

37. Govers G (1991) Rill erosion on arable land in central Belgium: Rates, controls and predictability. Catena 18: 133–155.

38. Abrahams AD, Gao P, Aebly FA (2000) Relation of sediment transport capacity to stone cover and size in rain-impacted interrill flow. Earth Surf Proc Land 25: 497–504.

39. Low HS (1989) Effect of sediment density on bed-load transport. J Hydraul Eng ASAE 115: 124–138.

40. Rickenmann D (1991) Hyperconcentrated flow and sediment transport at steep slopes. J Hydraul Eng ASCE 117: 1419–1439.

41. Yang CT (1973) Incipient motion and sediment transport. J Hydr Eng Div ASCE 99: 1679–1703.

42. Parker G (1979) Hydraulic geometry of active gravel rivers. Journal of the Hydraulics Division, American Society of Civil Engineers 105: 1185–1201.

43. Kleinhans MG, Bierkens MFP, van der Perk M (2010) HESS opinions. On the use of laboratory experimentation: "Hydrologists, bring out shovel and garden hoses and hit the dirt". Hydrology and Earth System Sciences 14: 369–382.

44. Govers G, Gime´nez R, van Oost K (2007) Exploring the relationsship between experiments, modelling and field observations. Earth-Science Reviews 84 (3–4): 87–102.

45. Govers G (1987) Spatial and temporal variability in rill development processes at the Huldenberg experimental site. Catena Supplement 8: 17–34.

46. Seeger M (2007) Uncertainty of factors determining runoff and erosion processes as quantified by rainfall simulations. Catena 71 (1): 56–67.

47. FAO (2006) Guidelines for soil description. Rome: 4. ed. Food and Agriculture Organization of the United Nations.

48. Vandekerckhove L, Poesen J, Oostwoud Wijdenes D, de Figueiredo T (2003) Topographical thresholds for ephemeral gully initiation in intensively cultivated areas of the Mediterranean. Catena 33(3–4): 271–292.

49. Ad-hoc-Arbeitsgruppe Boden (2005) Bodenkundliche Kartieranleitung. Hannover: 5. Auflage, Bundesamt fu" r Geowissenschaften und Rohstoff in Zusammenarbeit mit den staatlichen Geologischen Diensten der Bundesrepublik Deutschland. 438 p.

50. Wirtz S, Seeger M, Ries JB (2010) The rill experiment as a method to approach a quantification of rill erosion process activity. Zeitschrift fu" r Geomorphologie Vol. 54,1: 47–64.

51. Allen JRL (1994) Fundamental properties of fluids and their relation to sediment transport processes. In: Pye K, editor. Sediment transport and depositional processes. Blackwell Scientific Publications. pp. 25–88.

52. DIN 18124 (1997) Baugrund, Untersuchung von Bodenproben - Bestimmung der Korndichte - Kapillarpyknometer, Weithalspyknometer. Deutsches Institut fu" r Normung e.V., Ausgabe : 1997–07, Deutsch.

53. Macosko CW (1994) Rheology: Principles, Measurements and Applications. New York: VCH: Wiley-VCH. New York.

54. Einstein A (1906) Eine neue Bestimmung der Moleku"ldimensionen. Analen der Physik 19: 289–306.

55. Einstein A (1911) Berichtigung zu meiner Arbeit: "Eine neue Bestimmung der Moleku"ldimensionen". Analen der Physik 34: 591–593.

56. Scherer U (2008) Prozessbasierte Modellierung der Bodenerosion in einer Lo"sslandschaft. Karlsruhe: Disserationsschrift, Fakulta"t fu" r Bauingenieur-, Geound Umweltwissenschaften, Universita"t Fridericiana zu Karlsruhe (TH). 248 p.

57. Landau L, Lifchitz E (1971) Me´caniques des fluids. Moscow: MIR

58. Reid DM, Dunne T (1996) Rapid evaluation of sediment budgets. Reiskirchen: Catena Verlag.

59. EPA-homepage (2009) Channel Processes: Bedload transport. United States Environmental protection agency. Available: http://water.epa.gov/scitech/ datait/tools/warsss/bedload.cfm Accessed 2010 September 14.

60. Hessel R, Jetten V (2007) Suitability of transport equations in modelling

soil erosion for a small Loess plateau catchment. Eng Geol 91: 56–71.

61. Nearing MA, Bradford JM, Parker SC (1991) Soil detachment by shallow flow at low slopes. Soil Science Society of America Journal 55 (2): 339–344.

62. Nearing MA, Parker SC (1994) Detachment of soil by flowing water under turbulent and laminar conditions. Soil Science Society of America Journal 58 (6): 1612–1614.

63. Bryan RB, Govers G, Poesen J (1989) The concept of soil erodibility and some problems of assessment and application. Catena 16 (4–5): 393–412.

64. Bryan RB (1990) Knickpoint evolution in rillwash. Catena 17: 111–132.

65. Owoputi LO, Stolte WJ (1995) Soil detachment in the physically based soil erosion process: a review. Transactions of the ASAE 38 (4): 1099–1110.

66. Rapp I. (1998) Effects of soil properties and experimental conditions on the rill erodibilities of selected soils. Pretoria: Ph. D. Thesis, Faculty of Biological and Agricultural Sciences, University of Pretoria, South Africa.

67. Zhu JC, Gantzer CJ, Peyton RL, Alberst EE, Anderson SH (1995) Simulated small-channel bed scour and head cut erosion rates compared. Soil Science Society of America Journal 59 (1): 211–218.

68. Kohl KD (1988) Mechanics of rill headcutting. Ames: Phd. Diss. Iowa State University.

69. Stefanovic JR, Bryan RB (2009) Flow energy and channel adjustments in rills developed in loamy sand and sandy loam soils. Earth Surface Processes and Landforms 34: 133–144.

70. Nearing MA (1998) Why soil erosion models over-predict small soil losses and under-predict large soil losses. Catena 32: 15–22.

71. Ruttimann M, Schaub D, Prasuhn V, Ruegg W (1995) Measurement of runoff and soil erosion on regulary cultivated fields in Switzerland – some critical considerations. Catena 25: 127–139.

72. Wendt RC, Alberts EE, Hjelmfelt AT (1986) Variability of runoff and soil loss from fallow experimental plots. Soil Sci Soc Am J 50: 730–736.

73. Risse LM, Nearing MA, Nicks AD, Laflen JM (1993) Assessment of error in the universal soil loss equation. Soil Sci Soc Am J 57: 825–833.

74. Zhang XC, Nearing MA, Risse LM, McGregor KC (1996) Evaluation of runoff and soil loss predictions using natural runoff plot data. Trans ASAE 39: 855– 863.

75. Liu BY, Nearing MA, Baffaut C, Ascough II JC (1996) The WEPP

watershed model: III. Comparisons to measured data from small watersheds. Transactions of the ASAE 40: 945–951.

76. Ajayi AE, Horta IDMF (2007) The effect of spatial variability of soil hydraulic properties on surface runoff processes. Anais XIII Simpo´ sio Brasileiro de Sensoriamento Remoto, Floriano´polis, Brasil, 21–26 abril 2007, p. 3243– 3248.

77. Dunkerley D (2008) Rain event properties in nature and in rainfall simulation experiments: a comparative review with recommendations for increasingly systematic study and reporting. Hydrological Processes 22: 4415– 4435.

78. Sidorchuk A (2005 b) Stochastic modelling of erosion and deposition in cohesive soils. Hydrological Processes 19: 1399–1417.

79. Sidorchuk A (2009 b) A third generation erosion model: The combination of probabilistic and deterministic components. Geomorphology 110 (1–2): 2–10.

80. Nearing MA (1991) A probabilistic model of soil detachment by shallow turbulent flow. Transactions of the ASAE 34 (1):

81. 81–85. 81. Wilson BN (1993) Development of a fundamentally-based detachment model. Transactions of the ASAE 36 (4): 1105–1114.

82. Sidorchuk A (2001) Calculation of the rate of erosion in soils and cohesive sediments. Eurasien Soil Science 34 (8): 893–900.

83. Sidorchuk A (2002) Stochastic Modelling of soil erosion and deposition. 12th ISCO Conference, Beijing 2002.

84. Sidorchuk A, Smith A, Nikora V (2004) Probability distribution function approach in stochastic modelling of soil erosion. Sediment transfer trough the fluvial system Vol. 288. IHAS Publ., pp. 345–353.

85. Sidorchuk A (2005 a) Stochastic components in the gully erosion modelling. Catena 63: 299–317.

86. Sidorchuk A, Schmidt J, Cooper G (2008) Variability of shallow overland flow velocity and soil aggregate transport observed with digital videography. Hydrological Processes 22: 4035–4048.

87. Sidorchuk A (2009 a) High-Frequency variability of aggregate transport under water erosion of well-structured soils. Eurasian Soil ScienceVol. 42, No. 5: 543–552.

88. Einstein HA (1936) Der Geschiebetrieb als Wahrscheinlichkeitsproblem. Zu¨ rich: Diss.-Druckerei A.-G. Gebr. Leemann & Co. 112 p.

Chapter 4

STOMATAL CONTROL AND LEAF THERMAL AND HYDRAULIC CAPACITANCES UNDER RAPID ENVIRONMENTAL FLUCTUATIONS

Stanislaus J. Schymanski[1], Dani Or[1] and Maciej Zwieniecki[2]

[1] Department of Environmental Systems Sciences, ETH Zurich, Zurich, Switzerland
[2] Department of Plant Sciences, University of California Davis, Davis, California, United States of America

ABSTRACT

Leaves within a canopy may experience rapid and extreme fluctuations in ambient conditions. A shaded leaf, for example, may become exposed to an order of magnitude increase in solar radiation within a few seconds, due to sunflecks or canopy motions. Considering typical time scales for stomatal adjustments, (2 to 60 minutes), the gap between these two time scales raised the question whether leaves rely on their hydraulic and thermal capacitances for passive protection from hydraulic failure or over-heating until stomata have adjusted. We employed a physically based model to systematically study effects of short-term fluctuations in irradiance on leaf temperatures and transpiration rates. Considering typical amplitudes and time scales of such fluctuations, the importance of leaf heat and water capacities for avoiding damaging leaf temperatures and hydraulic failure were investigated. The results suggest that common leaf heat capacities are not sufficient to protect a non-transpiring leaf from over-heating during sunflecks of several minutes duration whereas transpirative cooling provides effective protection. A comparison of the simulated time scales for heat damage in the absence of evaporative cooling with observed stomatal response times suggested that stomata must be already open before arrival of a sunfleck to avoid over-heating to critical leaf temperatures. This is consistent with measured stomatal conductances in shaded leaves and has implications for water use efficiency of deep canopy leaves and vulnerability to heat damage during drought. Our results also suggest that typical leaf water contents could sustain several minutes of evaporative cooling during a sunfleck without increasing the xylem water supply and thus risking embolism. We thus submit that shaded leaves rely on hydraulic capacitance and evaporative cooling to avoid over-heating and hydraulic failure during exposure to typical sunflecks, whereas thermal capacitance provides limited protection for very short sunflecks (tens of seconds).

INTRODUCTION

Leaves may be subjected to rapidly fluctuating irradiance due to motion of sunflecks and clouds that may span two orders of magnitude from light compensation points of shade-adapted leaves to almost full irradiance intensities [1]. Such environmental fluctuations occur at time scales ($<$ 1 min) much shorter than characteristic time scales for stomatal adjustments (2 to 60 min.) [2]. For leaves with slowly adjusting stomata, rapid fluctuations at shorter time scales could push leaf hydraulic and thermal status beyond operational limits resulting in xylem cavitation, overheating or wilting. Chazdon [1] pointed out that whereas intense sunflecks may lead to an increase in leaf temperatures by 18 K, heat damage due to such occurrences was rarely observed. Thenceforth, most analyses of stomatal adjustments to fluctuating irradiance in the canopy tended to focus on carbon gain and water stress, and much less on the need to avoid heat damage (e.g. [1]–[5]). On the other hand, Beerling et al. [6] simulated steady-state leaf temperatures of planar leaves with low and high stomatal numbers and concluded that high stomatal density is necessary to allow for sufficient evaporative cooling and avoid lethal leaf temperatures (assumed in the range of 45–55°C) under high irradiance.

Since evaporative cooling is essential to avoid heat damage in leaves exposed to full sunlight, and time scales of stomatal adjustments are longer than fluctuations in solar irradiance within a canopy, the question arises whether typical sunfleck intensities and durations could damage non-transpiring leaves. If this is the case, then adaptation for cooling would appear as a more imperative driver for stomatal adjustments than the potential increase in carbon gain, assumed in most studies on sunfleck effects to date.

The interlinked leaf thermal and hydraulic capacitances (embedded in leaf water content per leaf area) may provide passive protection and thus play a critical role in autonomous capacitive-based responses to rapid fluctuations in irradiance. For example, a variable leaf water content per unit leaf area can affect both thermal and hydraulic capacitances. When a leaf is exposed to a sunfleck, its temperature can rise by up to 20 K with an initial rate of 1–2 K min^{-1} for leaves with about 50–100 g m^{-2} water content [7]. Given the effect of leaf temperature on leaf-to-air vapour pressure gradient, transpiration rates are expected to rise accordingly. Increasing leaf water content (thicker leaves) can be an effective measure to increase capacitive buffering of such environmental fluctuations, until more robust but slower regulatory measures such as stomatal adjustments can take over and prevent detrimental effects.

An alternative protective measure may involve keeping stomata open even under low light conditions, in anticipation of autonomous evaporative cooling in response to a rapid increase in irradiation. The necessity to avoid damaging temperatures may thus impact water use efficiency in water-limited environments. Researchers have found that a number of shade tolerant species maintain open stomata and very low water use efficiencies in the shade, while others maintain lower stomatal conductances in the shade but are able to open their stomata faster in response to a sunfleck (e.g. [1], [3], [8]).

An important factor to consider is that a spike in transpiration flux due to rapidly changing environmental conditions (e.g. due to a sunfleck or wind gust), may trigger cavitation and failure of the water supply network to the leaf [9]. To mitigate such a scenario, stored water in leaf tissue could buffer the effect of such a spike in demand and thus reduce the risk of cavitation. For a range of living plant tissues including leaves, the water content can vary by up to 10% of its maximum value before turgor loss and irreversible plasmolysis sets in [10]–[14]. Consequently a leaf with a water storage of 0.2 mm (0.2 kg m^{-2}) could lose up to 0.02 mm of water (0.02 kg m^{-2}) before permanent damage occurs. In this context, turgor loss and passive stomatal closure can be seen as an autonomous measure to stop water loss before this critical stage is reached. Furthermore, it has been shown for a number of tree species that leaves are more vulnerable to xylem embolism than stems [15]–[18], suggesting that the hydraulic pathways in trees are organised in a way to protect the stem xylem from pressure drops emanating from the leaves [18].

Considering the disparity in time scale of environmental fluctuations relative to stomatal adjustment times, the primary objective of this study is to investigate the protective roles of leaf heat and water capacitances under fast environmental fluctuations (relative to stomatal response times).

We aim to answer the following questions:

- Do natural fluctuations in leaf irradiance necessitate stomatal regulation to avoid heat damage or hydraulic failure?

- What is the role of leaf heat and water capacities in negotiating the trade-off between cavitation and over-heating?

A physically-based leaf energy balance model was formulated to simulate leaf temperature and transpiration dynamics as a function of varying environmental conditions (irradiance, air temperature, vapour pressure, wind speed). The effect of rapid environmental fluctuations (e.g. irradiance due to moving sunflecks) on the heat and mass exchange of the leaf and resulting changes in leaf temperature and hydration status were simulated. In a first step, simulations were performed using an observed time series of irradiance and air temperatures in the understorey of a tropical rainforest [19], which allowed comparison of simulated leaf temperature dynamics with observations. In a second step, typical amplitudes and time scales of irradiance fluctuations were considered to investigate the importance of leaf heat and water capacities for avoiding damaging extremes in leaf temperatures and hydration status.

METHODS

All relevant symbols used in this section and their respective units are given in Table 1. All derivations and analyses were performed using the freely available software SAGE (version 5.0, http://sagemath.org). The steady-state temperature for given leaf dimensions, environmental conditions and stomatal conductance (g_{sv}) was obtained by numerical root finding of Eq. 1 (see below), whereas the dynamics were simulated using a finite time step discretisation.

Table 1. Symbols, standard values and units used in this paper.

Symbol	Description (standard value)	Units
z_a	Thermal diffusivity of air	$m^2\,s^{-1}$
λ_E	Latent heat of vaporisation (2.45×10^6)	$J\,kg^{-1}$
v_a	Kinematic viscosity of air	$m^2\,s^{-1}$
ρ_a	Density of dry air	$kg\,m^{-3}$
σ	Stefan-Boltzmann constant (5.67×10^{-8})	$W\,m^{-2}\,K^{-4}$
a_s	Fraction of transpiring leaf surface area (relative to 1-sided leaf area)	-
C_i	Conductive heat flux away from leaf subsection	$W\,m^{-2}$
C_{wa}	Concentration of water vapour in the free air	$mol\,m^{-3}$
C_{wl}	Concentration of water vapour inside the leaf	$mol\,m^{-3}$
c_{pa}	Specific heat of dry air (1010)	$J\,K^{-1}\,kg^{-1}$
c_{pl}	Leaf heat capacity at constant pressure	$J\,K^{-1}\,m^{-2}$
c_{pv}	Heat capacity of water at constant pressure	$J\,K^{-1}\,kg^{-1}$
D_{va}	Binary diffusion coefficient of water vapour in air	$m^2\,s^{-1}$
E_l	Latent heat flux away from leaf	$W\,m^{-2}$
E_{lm}	Transpiration rate	$kg\,m^{-2}\,s^{-1}$
E_{lm0}	Steady-state transpiration rate prior to arrival of sunfleck	$kg\,m^{-2}\,s^{-1}$
$E_{l,mol}$	Transpiration rate in molar units	$mol\,m^{-2}\,s^{-1}$
g_{bv}	Leaf boundary layer conductance to water vapour	$m\,s^{-1}$
g_{sv}	Stomatal conductance to water vapour	$m\,s^{-1}$
g_{lv}	Total leaf conductance to water vapour	$m\,s^{-1}$
$g_{lv,mol}$	Total leaf conductance to water vapour	$mol\,m^{-2}\,s^{-1}$
h_c	Average one-sided convective heat transport coefficient	$m\,s^{-1}$
h_{cl}	Convective heat transport coefficient for the lower leaf side	$m\,s^{-1}$
h_{cu}	Convective heat transport coefficient for the upper leaf side	$m\,s^{-1}$
H_l	Sensible heat flux emitted by the leaf	$W\,m^{-2}$
k_a	Thermal conductivity of air in leaf boundary layer	$W\,K^{-1}\,m^{-1}$

H_l	Sensible heat flux emitted by the leaf	W m^{-2}
k_a	Thermal conductivity of air in leaf boundary layer	W K^{-1} m^{-1}
L_l	Characteristic leaf length scale (0.05)	m
m_w	Leaf water content	kg m^{-2}
M_w	Molar mass of water (0.018)	kg mol^{-1}
n_a	Amount of matter	mol
N_{Le}	Lewis number	-
N_{Nu}	Nusselt number	-
N_{Nu_L}	Average Nusselt number for whole leaf	-
N_{Pr}	Prandtl number for air (0.71)	-
N_{Re}	Reynolds number	-
N_{Re_L}	Average Reynolds number for whole leaf	-
N_{Sh_L}	Average Sherwood number	-
P_{va}	Vapour pressure in free air	Pa
P_{vi}	Vapour pressure inside the leaf	Pa
R_{mol}	Molar gas constant (8.314472)	J K^{-1} mol^{-1}
R_s	Absorbed short wave radiation	W m^{-2}
R_{ll}	Net longwave radiation emission by a leaf	W m^{-2}
t	Time	s
T_a	Air temperature	K
T_b	Boundary layer temperature, $T_b = (T_l + T_a)/2$	K
t_{crit}	Critical time to heat damage or turgor loss	s
T_{crit}	Critical leaf temperature for the onset of heat damage (322)	K
T_l	Leaf temperature	K
v_w	Wind velocity	m s^{-1}
x	Distance from leading edge along a leaf	m

All area-related variables are expressed per unit leaf area.
doi:10.1371/journal.pone.0054231.t001

Leaf energy Balance Model

The leaf energy balance is determined by the dominant energy fluxes between the leaf and its surroundings, including radiative, sensible, and latent energy exchange (linked to mass exchange). The dominant energy fluxes considered here are illustrated in Fig. 1.

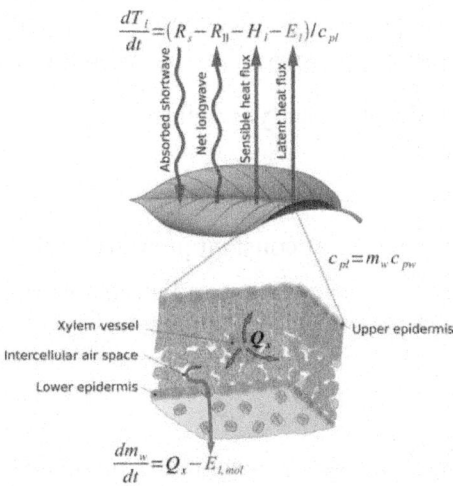

$$\frac{dT_l}{dt} = (R_s - R_{ll} - H_l - E_l)/c_{pl}$$

$$c_{pl} = m_w c_{pw}$$

$$\frac{dm_w}{dt} = Q_x - E_{l,mol}$$

Figure 1. Components of the leaf mass and energy balance and their conventional directions considered in this study. Arrows point in the direction of a positive flux. Both leaf temperature (T_l) and water content (m_w) depend on the transpiration rate (E_l and $E_{l,mol}$ in energetic and molar units respectively). The leaf water content (m_w) affects the leaf heat capacity (c_{pl}) and turgor pressure, which becomes critical when leaf water content declines below 90% of its maximum value (see text). Changes in leaf water content result from differences in the water supply rate from the xylem (Q_x) and evaporative losses ($E_{l,mol}$). doi:10.1371/journal.pone.0054231.g001

Neglecting heat transport through the petiole, the energy balace of a spatially homogeneous leaf can be written as:

$$\frac{dT_l}{dt} = (R_s - R_{ll} - H_l - E_l)/c_{pl}$$

(1)

where T_l is leaf temperature, R_s absorbed short wave radiation, R_{ll} is the net longwave balance, i.e. the emitted minus the absorbed, H_l is the sensible heat flux away from the leaf, E_l is the latent heat flux away from the leaf and c_{pl} is the leaf heat capacity at constant pressure. In the above, extensive variables are defined per unit leaf area.

The special case of a partly illuminated leaf would involve formulation of the energy balances for the illuminated and the shaded leaf areas separately and an additional term for the heat transport by conduction between these two leaf subsections (C_l):

$$\frac{dT_l}{dt} = (R_s - R_{ll} - H_l - E_l - C_l)/c_{pl}$$

(2)

where all terms refer to the sunlit part of the leaf. For simplicity, we will limit the present analysis to spatially homogeneous planar leaves, i.e. full illumination and a negligible temperature gradient between the two sides of the leaf.

Assuming that leaf heat capacity is mainly determined by its water content (m_w), c_{pl} is represented as:

$$c_{pl} = m_w c_{pw}$$

(3)

where c_{pw} is the heat capacity at constant pressure of liquid water.

Assuming further that the longwave radiation absorbed by the leaf is equal to its emission at air temperature (T_a), the net longwave emission is represented by the difference between blackbody radiation at T_l and that at T_a:

$$R_{ll} = 2\sigma(T_l^4 - T_a^4)$$

(4)

where σ is the Stefan-Boltzmann constant and the factor 2 represents the two sides of a broad leaf. Note that this formulation is a coarse approximation, but it represents a standard procedure (e.g. [20]). A more accurate account of the longwave radiation balance would have to involve longwave sky radiation as well as longwave radiation originating from the ground and neighbouring leaves in the canopy.

Sensible Heat Flux

The exchange of sensible and latent heat between the leaf and the free air is dominated by convective transport, which is generally formulated as the product of a convective transport coefficient and the temperature difference between the surface and the free air. Convective transport coefficients depend on leaf orientation, geometry, and surface properties (e.g. hairs), wind conditions and temperature (p. 168–172 in [20]). In this study, we neglect the effects of leaf surface properties, orientation and geometry by assuming that leaves behave like horizontal rectangular metal plates of width D_l (in wind direction).

The total convective heat transport away from the leaf is represented as:

$$H_l = (h_{cu} + h_{cl})(T_l - T_a) = 2h_c(T_l - T_a) \tag{5}$$

where h_{cu}, h_{cl} and h_c are the convective heat transport coefficients for the upper, the lower and the average of both leaf sides respectively.

Different textbooks propose different empirical equations to calculate heat transfer coefficients for flat plates. The differences may originate from different experimental data, different reference length scales or different boundary conditions. In order to avoid the risk of mismatch between empirical equations and applicable boundary conditions and for better traceability, we drew most of the below relations from a single textbook (Incropera et al., 2006 [21]).

Following Incropera et al. [21], different convective heat transfer coefficients were formulated for forced and free convection (presence and absence of significant wind), and laminar vs. turbulent conditions. The coefficients are generally formulated as a function of the dimensionless Nusselt number (N_{Nu_L}):

$$h_c = k_a \frac{N_{Nu_L}}{L_l} \tag{6}$$

where k_a is the thermal conductivity of the air in the boundary layer and L_l is a characteristic length scale of the leaf. In the absence of wind, buoyancy forces, driven by the density gradient between the air at the surface of the leaf and the free air dominate convective heat exchange (free or natural convection). The influence of vapour pressure gradients across the stomatal pores on the density gradient would add a significant level of complexity to the solution of the sensible and latent heat exchange equations. For simplicity, we will therefore limit this study to forced conditions, i.e. where wind velocity is greater than 0.5 m s^{-2} for the leaf properties and environmental conditions considered here.

Under strong enough wind, inertial forces drive the convective heat transport (forced convection) and the relevant dimensionless number is the Reynolds number (N_{Re_L}), which defines the balance between inertial and viscous forces:

$$N_{Re_L} = \frac{v_w L_l}{v_a}$$
(7)

where v_w is the wind velocity (m s^{-1}), L_l (m) the length of the leaf in wind direction and v_a is the kinematic viscosity of air.

The local Reynolds number changes from the leading edge downwind as ([21], Eq. 6.23):

$$N_{Re_x} = \rho_a v_a x / \mu_a = \frac{v_a x}{v_a}$$
(8)

where ρ_a is the air density, v_a is the wind velocity outside the boundary layer, x is the distance from the leading edge, μ_a is the dynamic viscosity and v_a is the kinematic viscosity of air ($\mu = v_a \rho_a$). All of the fluid properties are evaluated at the mean boundary layer temperature, defined as $T_b = \frac{T_l + T_a}{2}$ ([21], Eq. 7.2).

Integrated over the whole leaf, the average Reynolds number (N_{Re_L}) is given by Eq. 7. For an isothermal flat plate with a fully laminar boundary layer, the average Nusselt number is given as ([21], Eq. 6.23):

$$N_{Nu_L} = 0.664 N_{Re_L}^{1/2} N_{Pr}^{1/3}$$
(9)

where N_{Pr} is the dimensionless Prandtl number ($N_{Pr} \approx 0.71$ for air).

At a certain distance from the leading edge, N_{Re_x} can reach a critical number (N_{Re_c}) and flow transitions from laminar to turbulent flow. This critical Reynolds number depends on the surface roughness and the turbulence level of the free stream but is known to vary from about 10^5 to 3×10^6 ([21], P. 361). As opposed to purely laminar flow, where $N_{Re_L} < N_{Re_c}$, cases where a part of the boundary layer is turbulent ($N_{Re_L} > N_{Re_c}$) are referred to as mixed flow. Since turbulent convection is stronger than laminar convection, lower values of N_{Re_c}, implying earlier transition to turbulent flow, would lead to enhanced sensible heat flux. For mixed flow over an isothermal plate ($N_{Re_c} < N_{Re_L} < 10^8$), Incropera et al. gives the following empirical formulation ([21], Eq. 7.38):

$$N_{Nu_L} = (0.037 N_{Re_L}^{4/5} - C_1) N_{Pr}^{1/3}$$
(10)

With

$$C_1 = 0.037 N_{Re_c}^{4/5} - 0.664 N_{Re_c}^{1/2} \tag{11}$$

Incropera et al. ([21], P. 412) states that N_{Re_c} can be as low as 0 if the flow is "tripped" at the leading edge of the object using some mechanical turbulence promotor. However, we found that the equation does not give reasonable results for $N_{Re_c} < 3000$, as the resulting Nusselt number would be lower than that for fully laminar flow (Eq. 9). Eq. 10 is identical with Eq. 9 if $N_{Re_L} = N_{Re_c}$. Thus, to make it valid across the whole range, we modified N_{Re_c} such that it takes values of N_{Re_L} if $N_{Re_L} < N_{Re_c}$. This was achieved by substituting N_{Re_c} by the term $\dfrac{N_{Re_L} + N_{Re_c} - |N_{Re_c} - N_{Re_L}|}{2}$ in Eq. 11.

It is interesting to note that experiments with real leaves revealed an enhanced forced convection by a factor of up to 2.5 compared to flat plates of similar dimensions in laminar flow[22], [23]. This was largely attributed to the level of turbulence already present in canopy wind. However, this does not seem to be consistent with the variation of critical Reynolds numbers attributed to the level of turbulence by Incropera et al., which was estimated to be in the range of 10^5 to 3×10^6 ([21], P. 361). Within this range, a leaf of 5 cm width would only start experiencing turbulence at wind velocities of above 31 m s^{-1} (Eq. 8). Even the lowest critical Reynolds number of 3000, for which Eq. 10 is still applicable, would only lead to an onset of turbulence at wind velocities of above 1 m s^{-1}, which is still above the maximum wind velocity of 0.4 m s^{-1} used in the experiment by Parlange and Waggoner [24], so the observed enhancement in sensible heat flux cannot be simulated using the formulations given above. To get as close as possible to real leaves while using the established relationships for heated plates, we used a critical Reynolds number of 3000 rather than the 10^5 suggested by Incropera et al. [21].

Latent heat flux.

Evaporation from a wet leaf was formulated as a function of the concentration of water vapour inside the leaf (C_{wl}, mol m^{-3}) and in the free air (C_{wa}, mol m^{-3}) ([21], Eq. 6.8):

$$E_{l,mol} = g_{tv}(C_{wl} - C_{wa}) \tag{12}$$

where $E_{l,mol}$ (mol m^{-2} s^{-1}) stands for a flux of matter and g_{tv} (m s^{-1}) is the total conductance for water vapour.

For transpiration through stomata, g_{tv} is the combination of boundary layer and stomatal conductances (g_{bv} and g_{sv} respectively), derived from the assumption that stomatal and boundary layer resistances are in series and using the definition of conductances as the inverse of resistances:

$$g_{tv} = \frac{g_{bv} g_{sv}}{(g_{bv} + g_{sv})}$$ (13)

The concentration difference in Eq. 12 is a function of the temperature and the vapour pressure differences between the leaf and the free air. Assuming that water vapour behaves like an ideal gas, we can express its concentration as:

$$C_{va} = \frac{P_{va}}{R_{mol} T_a}$$ (14)

where P_{va} is the vapour pressure, R_{mol} is the universal gas constant and T_a is the temperature. In this study the vapour pressure inside the leaf is assumed to be the saturation vapour pressure at leaf temperature, which is computed using the Clausius-Clapeyron relation (Eq. B.3 in [25]):

$$P_{vl} = 611 \exp\left(\frac{\lambda_E M_w}{R_{mol}} \left(\frac{1}{273} - \frac{1}{T_l}\right)\right)$$ (15)

where λ_E is the latent heat of vaporisation and M_w is the molar mass of water. The conversion of the vapour flux in molar units to latent heat flux in energetic units was done by multiplying $E_{l,mol}$ by the molar mass of water and the latent heat of vaporisation:

$$E_l = E_{l,mol} M_w \lambda_E$$ (16)

Note that $E_{l,mol}$ is commonly expressed as a function of the vapour pressure difference between the free air (P_{va}) and the leaf (P_{vl}), in which the conductance ($g_{tv,mol}$) is expressed in molar units (mol m^{-2} s^{-1}):

$$E_{l,mol} = g_{tv,mol} \frac{P_{vl} - P_{va}}{P_a}$$ (17)

For $P_{vl} = P_{va}$, Eq. 12 can still give a flux, whereas Eq. 17 gives zero flux. This is because the concentrations of vapour in air (mol m^{-3}) can differ due to differences in temperature, even if the partial vapour pressures are the same (see Eq. 14). Therefore, the relation between g_{tv} and $g_{v,mol}$ has an asymptote at the equivalent temperature. It can be obtained by combining Eqs. 12 and 17 and solving for $g_{tv,mol}$:

$$g_{tv,mol} = g_{tv} \frac{P_a(P_{va} T_l - P_{vl} T_a)}{(P_{va} - P_{vl}) R_{mol} T_a T_l)}$$ (18)

For $T_l = T_a$, the relation simplifies to:

$$g_{tv,mol} = g_{tv} \frac{P_a}{R_{mol} T_a}$$

$$(19)$$

which, for typical values of P_a and T_a amounts to $g_{tv,mol} \approx 40$ mol m^{-3} g_{tv}. For all practical purposes, we found that Eqs. 12 and 17 with $g_{tv,mol} = g_{tv} \frac{P_a}{R_{mol} T_a}$ give similar results when plotted as functions of leaf temperature.

Boundary layer conductance to water vapour.

The boundary layer conductance in Eq. 13 is equivalent to the mass transfer coefficient for a wet surface ([21], Eq. 7.41):

$$g_{bv} = N_{Sh_L} D_{va} / L_l$$

$$(20)$$

where N_{Sh_L} is the dimensionless Sherwood number and D_{va} is the diffusivity of water vapour in air. If the convection coefficient for heat is known, the one for mass (g_{bv}) can readily be calculated from the relation ([21], Eq. 6.60):

$$g_{bv} = \frac{a_s h_c}{\rho_a c_{pa} N_{Le}^{1-n}}$$

$$(21)$$

where a_s is the fraction of one-sided transpiring surface area in relation to the surface area for sensible heat exchange, c_{pa} is the constant-pressure heat capacity of air, n is an empirical constant ($n = 1/3$ for general purposes) and N_{Le} is the dimensionless Lewis number, defined as ([21], Eq. 6.57):

$$N_{Le} = \alpha_a / D_{va}$$

$$(22)$$

where α_a is the thermal diffusivity of air. The value of a_s was set to 0.5 for leaves with stomata on one side only, and to 1.0 for stomata on both sides. Other values could be used for leaves only partly covered by stomata.

Model closure.

Progressively inserting Equations 14, 15, 13, 21, 6, 22, 10 (or 9) and 7 into Equation 12 gives an expression for the transpiration flux as a function of leaf temperature, where we still need to calculate P_a, D_{va}, α_a, k_a, and v_a, while L_l, Re_c and g_{sv} are prescribable leaf properties, and P_{va} and v_w (vapour pressure and wind speed) are part of the environmental forcing. D_{va}, α_a, k_a and v_a were parameterised as functions of boundary layer temperature only, by fitting linear curves to published data ([20], Table A.3):

$$D_{va} = (1.49 \times 10^{-7}) T_b - 1.96 \times 10^{-5}$$

$$(23)$$

$$\alpha_a = (1.32 \times 10^{-7})T_b - 1.73 \times 10^{-5} \tag{24}$$

$$k_a = (6.84 \times 10^{-5})T_b + 5.62 \times 10^{-3} \tag{25}$$

$$v_a = (9 \times 10^{-8})T_b - 1.13 \times 10^{-5} \tag{26}$$

Assuming that air and water vapour behave like an ideal gas, and that dry air is composed of 79% N2 and 21% O2, we calculated the density as a function of temperature, vapour pressure and the partial pressures of the other two components using the ideal gas law:

$$\rho_a = \frac{n_a M_a}{V_a} = M_a \frac{P_a}{R_{mol} T_a} \tag{27}$$

where n_a is the amount of matter (mol), M_a is the molar mass (kg mol^{-1}), P_a the pressure, T_a the temperature and R_{mol} the molar universal gas constant. This equation was used for each component, i.e. water vapour, N2 and O2 , where the partial pressures of N2 and O2 are calculated from atmospheric pressure minus vapour pressure, yielding:

$$\rho_a = \frac{M_w P_v + M_{N_2} P_{N_2} + M_{O_2} P_{O_2}}{R_{mol} T_a} \tag{28}$$

where M_{N_2} and M_{O_2} are the molar masses of nitrogen and oxygen respectively, while P_{N_2} and P_{O_2} are their partial pressures, calculated as:

$$P_{N_2} = 0.79(P_a - P_{va}) \tag{29}$$

And

$$P_{O_2} = 0.21(P_a - P_{va}) \tag{30}$$

Simulation of Observed Leaf Temperature Dynamics

To test whether the leaf energy balance model produces reasonable results and how leaf heat capacity could affect leaf temperature dynamics in a natural environment, we simulated the dynamics of leaf temperature of *Shorea leprosula* seedlings in response to observed fluctuations in solar irradiance in a rainforest understory and compared the results with observed leaf temperature fluctuations [19].

The forcing data set consisted of air temperature measured in two minute intervals and solar radiation measured in 10 second intervals. The observed leaf temperatures were also reported in 10 second intervals, all for a single day from 8:30am to 6pm (Fig. 1 in [19]). Andrew Leakey kindly provided the original data for the analysis. To convert photosynthetically active photon flux density (P_{PFD}, μmol m^{-2} s^{-1}) recorded by the quantum sensor SKP 215 (Skye Instruments) to shortwave irradiance (R_s, W m^{-2}), we used a conversion coefficient of 4.57×10^{-6} mol J^{-1} [26]. Then we expanded from the photosynthetically active range of 400–700 nm to the full shortwave range of 200–4000 nm by using a conversion coefficient of 0.45, which was derived from an online database [27]: $R_s = 1 \times 10^{-6} P_{PFD}/(4.57 \times 10^{-6} \times 0.45)$. Air temperature was linearly interpolated to obtain values at the same time steps as P_{PFD}. The resulting data set is shown in Fig. 2.

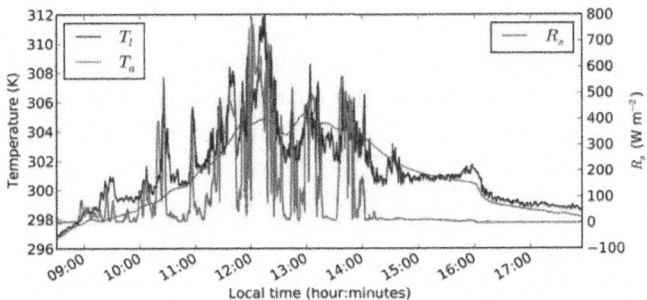

Figure 2. Observed irradiance (R_n), air temperature (T_a) and leaf temperature (T_l) in the understorey of a tropical rainforest. Data converted from [19]. doi:10.1371/journal.pone.0054231.g002

The leaves of the *Shorea leprosula* seedlings had dimensions of approximately 130×45 mm and a specific leaf area of 19 mm^2 mg^{-1} dry matter (pers. comm. Andrew Leakey). Leaf thickness of *Shorea leprosula* in the understorey was reported in the range of 83 ± 9 μm (P. 370 in [28]). Assuming a 1:1 partitioning between leaf dry matter and water content, we found that a water content of 0.05 kg m^{-2} would be reasonable, leaving 0.03 mm of the leaf thickness for dry matter and air. Any higher water content would have to result in greater leaf thickness. As a consequence we used 0.05 m as the characteristic length scale of the leaf and a heat capacity equivalent to a leaf water content (m_w) of 0.05 kg m^{-2} for the simulations.

Reference Threshold for Time to Heat Damage

Exposure of living plant tissue to excessive heat can cause immediate (direct) or delayed (indirect) damage. Heat damage not only depends on exposure temperature, but also on the duration of the exposure. Heat vulnerability can vary between species, and also over time, due to acclimation and so-called hardening in response to prior non-lethal exposures to high temperatures [29], [30]. In order to establish a realistic reference for heat damage as a result of dynamic exposure to high leaf temperatures, we used results obtained from experiments on black spruce (*Picea mariana*) twigs, performed by Colombo & Timmer [30]. It is not the purpose of this study to assess the heat vulnerability of a particular or representative species; we just use this one example as a reference for assessing potential heat damage risks related to rapid and short-lived leaf temperature rises due to sunflecks.

Colombo et al. [30] conducted extensive heat exposure experiments on black spruce needles and found that the critical exposure time and temperature are related exponentially. In the experiments, spruce twigs were submerged in water of varying temperatures for varying time periods and the percentage of damage was recorded. We have to consider that the submersion itself had a damaging effect in addition to the heat, as it is clear that a twig submerged for long enough would get damaged no matter what the temperature is. To separate these effects, we used the following formulation for the critical exposure temperature (T_e) as a function of submersion time (t):

$$T_e = T_{crit} + c_T/t - c_w t \qquad (31)$$

where T_e (K) is the exposure temperature, T_{crit} (K) is the critical temperature below which no damage occurs, c_T (K s) is a constant determining the effect of exposure time (t, s) and c_w (K s^{-1}) is a constant representing the effect of the submersion alone. A least-square fit of this model to the exposure temperature and duration data presented by Colombo et al. ([30], Tab. 2) revealed $T_{crit} = 322$ K, $c_T = 148$ K s, and $c_w = 0.000826$ K s^{-1} (Fig. 3). This suggests that the critical temperature for heat damage is around 49°C and the damaging time amounts to 148 seconds per Kelvin above that threshold, i.e. damage happens when $(T_l - T_{crit})t > 148$ K s. It further suggests that the submersion effect lowers the recorded damaging temperature by 0.0008 K per second of submersion time. Using these values as a reference, we computed the critical time (t_{crit} to heat damage as the time when the integral of $(T_l - T_{crit})t$ reaches 148 Ks, starting when $T_l = T_{crit}$.

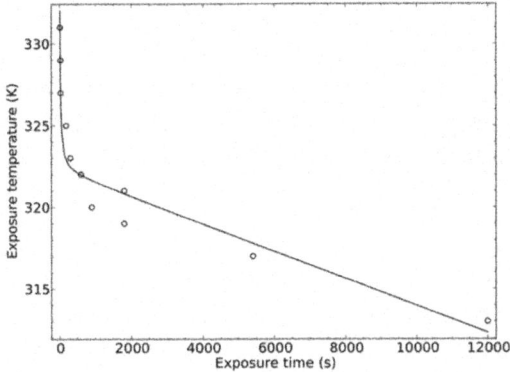

Figure 3. Fit of Eq. 31 **to data in** [30, Tab. 2]. $T_{crit} = 322$ K, $c_T = 148$ K s, and $c_w = 0.000826$ K s^{-1}, standard root mean square deviation: 0.07.

doi:10.1371/journal.pone.0054231.g003

Table 2. Natural and experimental light fluctuations vs. stomatal conductances.

Obs. R_s	Obs. t_{sun}	Exp. R_s	Min. g_{sv}	Max. g_{sv}	$t_{g90\%}$	Reference
50–750	300–1200	150–850	0.0047	0.01	300	[46][1]
20–750	180 ± 120	10–750	0.0019 ± 0.0013	0.0025 ± 0.0022	157 ± 133	[47][2]
50–750	1080 ± 840	10–750	0.003 ± 0.0008	0.004 ± 0.0006	65 ± 19	[47][3]
25–750	230 ± 275	2.5–850	0.0014 ± 0.0007	0.006 ± 0.002	900	[48][4]
25–750	2332 ± 1983	2.5–850	0.023 ± 0.017	0.025 ± 0.016	·	[48][5]
300–1050	300	150–900	0.0095	0.012	60	[49][6]
5–500	> 300	0–500	0.003	0.005	·	[50][7]
< 50–500	300–1200	25–500	0.0006	0.0029	720	[51][8]

Obs. R_s: typical irradiance in shade and sunfleck (W m^{-2}); Obs. t_{sun}: typical sunfleck duration (s); Exp. R_s: experimental range in irradiance (W m^{-2}); Min. g_{sv}: observed minimum stomatal conductance (m s^{-1}); Max. g_{sv}: observed maximum stomatal conductance (m s^{-1}); $t_{g90\%}$: time to 90% of max. g_{sv} (s). Literature values of g_{sv} reported in units of mol m^{-2} s^{-1} were converted to m s^{-1} using Equation 19.
[1]*Sorghum sp.*, lower leaves.
[2]*Nothofagus cunninghamii*, coppice leaves.
[3]*Nothofagus cunninghamii*, upper canopy leaves.
[4]*Psychotria micrantha*, canopy gaps.
[5]*Isertia haenkeana*, clearings.
[6]*Triticum sp.*, Fig. 3.
[7]*Pteridium aquilinum*.
[8]*Acer rubrum*.
doi:10.1371/journal.pone.0054231.t002

Note that the critical temperature of 49°C derived from the water submersion experiments is consistent with experimental results on the same species performed using heating in air [31]. It is remarkable that even for desert plants, extensive heat tissue damage commonly occurs close to the 50°C mark (up to 53°C, [32], [33]). This suggests that the function derived in this study from experimental data on black spruce may also be relevant for species in generally warmer habitats.

Reference Threshold for Time to Turgor Loss after Step Change in Irradiance

Assuming that the water supply rate from the xylem equals the steady-state leaf transpiration rate (E_{lm0}, kg m^{-2} s^{-1}) before the step change, we held this xylem supply rate constant and calculated the change in leaf water content (m_w, kg m^{-2}) as the time integral of the dynamic transpiration rate (E_{lm}, kg m^{-2} s^{-1}) minus the initial steady-state transpiration rate. The time to turgor loss was taken as the time (t, s) when the leaf water reservoir was depleted by 10%, i.e. when:

$$\int_0^t E_{lm}(t) - E_{lm0} dt = 0.1 m_w$$

(32)

The assumption that the xylem supply rate does not adjust within this time is likely to lead to an under-estimation of the critical time, whereas the assumption that the leaf is initially fully saturated and only loses turgor after 10% loss of its mass is likely to lead to an over-estimation of the critical time.

RESULTS

Leaf Temperature Dynamics in a Natural Setting

Using observations of diurnal variations in irradiance and air temperature in a tropical rainforest understorey [19], we simulated leaf temperature dynamics throughout the day considering a constant wind speed of 0.5 m s^{-1}, and a constant atmospheric vapour pressure corresponding to 90% saturation at 8:30am, while varying irradiance and air temperature every 10 seconds. Using observed leaf temperature at 8:30am as an initial condition, we simulated three scenarios, one with fully closed stomata throughout the day ($g_s = 0$ m s^{-1}), one with a constant stomatal conductance of $g_s = 0.01$ m s^{-1} and one with a non-limiting stomatal conductance ($g_s = 1.0$ m s^{-1}). The simulation using closed stomata tracked the observed leaf temperatures at the beginning and the end of the day, whereas the simulation with moderately open stomata tracked the observed leaf temperatures in the middle of the day (Fig. 4). The simulation with non-limiting stomatal conductance resulted in leaf temperatures well below observations throughout the day (data not shown). Note the large difference in simulated leaf temperatures in the middle of the day, depending on whether stomata are assumed open or closed.

Simulations are conducted for fully closed stomata (red) and a stomatal conductance of 0.01 m s^{-1} (blue). Observed leaf temperatures (yellow dots) and air temperatures (green dashed line) are taken from [19] and plotted against local time. doi:10.1371/journal.pone.0054231.g004

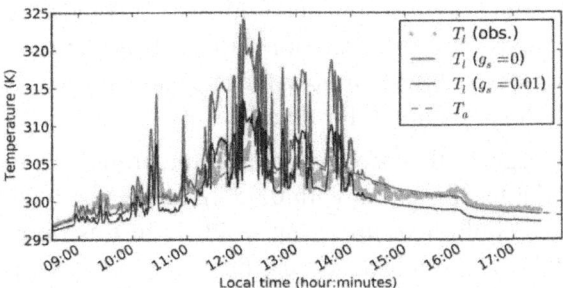

Figure 4. Observed and simulated leaf temperatures for an understorey plant in a tropical rain forest.

To assess how different leaf heat capacities could influence spikes in leaf temperature when stomata are closed, the same simulation were performed with different leaf water contents (0.025, 0.1 and 1.0 kg m^{-2}). Results reveal that halving or doubling the estimated leaf water content at this site (0.05 kg m^{-2}) did not have a large impact on simulated leaf temperature peaks (≤ 1 K), whereas a 20-fold increase in leaf water content to 1 kg m^{-2} could lead to a considerable reduction of simulated leaf temperature peaks by up to 5 K (Fig. 5). This suggests that the characteristic sunfleck durations are longer than the temperature time constants of the leaves at this site.

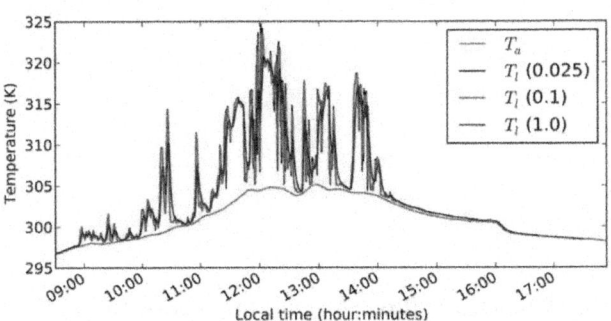

Figure 5. Simulated leaf temperatures in a rainforest understorey for closed stomata and different leaf water contents. Black: 0.025, red: 0.1 and blue: 1.0 kg m^{-2} leaf water content. The green line represents the observed air temperature [19], plotted against local time. doi:10.1371/journal.pone.0054231.g005

Temperature Dynamics for Closed Stomata

To understand the effect of a sudden increase in irradiance on a very hot day ($T_a = 313$ K or 40°C), we simulated the leaf temperature dynamics in response to a sudden increase in irradiance from 0 W m^{-2} (assuming that leaf temperature equals air temperature) to 400, 600 and 900 W m⁻². We also plotted the critical temperature and exposure time relationship for heat damage in black spruce twigs as a reference, to assess in how far leaf heat capacity could delay heat damage. See Methods section 0 for details.

The results suggest that steady-state temperatures are reached very fast (in less than a minute) for leaves with 0.05 kg m⁻² water content and that non-transpiring leaves could heat up by up to 20 K in this time. For irradiances greater than 400 W m⁻², sunflecks of less than two minutes duration could lead to heat damage (excursion into the shaded area in Fig. 6A). A 10-fold increase in leaf water content (from 0.1 kg/m2 to 1 kg/m2) could roughly quadruple the time to heat damage, from half a minute to two minutes for a sunfleck of 600 W m⁻² intensity (Fig. 6B).

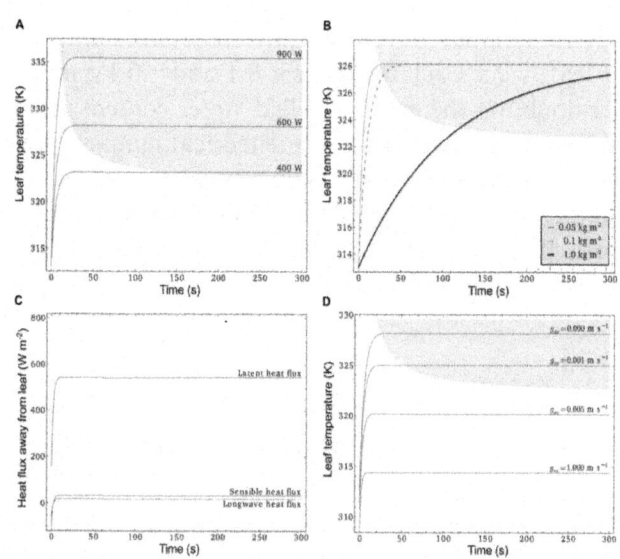

Figure 6. Leaf temperature and flux dynamics in response to sudden illumination. A: Temperature evolution of a non-transpiring leaf at different illumination intensities. B: Temperature evolutions of non-transpiring leaves with different water contents. C: Dynamics of latent, sensible and longwave heat flux from a leaf with non-limiting stomatal conductance ($g_{sv} \gg g_{bv}$). D: Temperature evolution of a transpiring leaf with different stomatal conductances (g_{sv}). Common environmental conditions for all simulations: $T_a = 313$ K, $v_w = 0.5$ m s⁻¹, 70% relative humidity, 0 W m⁻² irradiance prior to arrival of sunfleck. Unless otherwise indicated, simulations are performed assuming a 5 cm wide leaf with 0.05 kg m⁻² water content, exposed to $R_s = 600$ W m²sunfleck irradiance. The shaded area represents critical combinations of leaf temperatures and exposure times that are expected to cause considerable heat damage. It is computed using the equation $T_l = T_{crit} + c_T / t$, with $T_{crit} = 322$ K and $c_T = 148$ K s. This

equation was derived from experimental data for black spruce needles (see Methods). In Panel (c), the calculated boundary layer conductance is $g_{bv} = 0.012$ m s^{-1} and a stomatal conductance of 0.0029 m s^{-1}, resulting in latent heat flux of 63 W m^{-2} prior to illumination and 248 W m^{-2} at steady state during the sunfleck, would be sufficient to keep leaf temperatures below T_{crit}.

doi:10.1371/journal.pone.0054231.g006

Increasing wind speeds (or decreasing leaf sizes) would have an increasing effect on sensible heat flux and a reducing effect on the steady-state temperatures but no effect on the time constants (data not shown).

Temperature Dynamics at Constant Stomatal Conductance

When a leaf with open stomata is exposed to a sunfleck, the increase in leaf temperature may increase latent heat flux (Fig. 6C). However, evaporative cooling may concurrently suppress the rise in leaf temperature, leading to a lower steady-state leaf temperature than if stomata were closed. Fig. 6D illustrates the effect of evaporative cooling on the steady-state temperature of a leaf, for different stomatal conductances. Even low stomatal conductance (0.001 m s^{-1}) could substantially reduce steady-state leaf temperature, and thus delay heat damage. Intermediate stomatal conductance values (0.005 m s^{-1}) may reduce steady-state leaf temperature sufficiently to avoid risk of heat damage altogether. Note that the resulting latent heat flux at 600 W m^{-2} irradiance would be 126 and 319 W m^{-2} for the low and intermediate stomatal conductances respectively. This is equivalent to a transpiration of 2.2 and 5.6 mm respectively if integrated over 12 hours. In comparison, for fully open stomata, i.e. when stomatal conductance greatly exceeds boundary layer conductance ($g_{bv} = 0.012$ m s^{-1} in this case), the steady-state latent heat flux would be 547 W m^{-2}, amounting to 9.6 mm of transpiration integrated over 12 hours (Fig. 6C). At a stomatal conductance of 0.0029 m s^{-1}, steady-state leaf temperature would not exceed the critical temperature for heat damage of 322 K at 600 W m^{-2} illumination. The respective latent heat flux would be 248 W m^{-2}, compared with 63 W m^{-2} for the same stomatal conductance in darkness (data not shown).

Environmental Conditions Necessitating Evaporative Cooling

Next, we wanted to know under what environmental conditions evaporative cooling is necessary for avoiding heat damage during very long sunflecks. Taking a leaf temperature of 322 K (49°C) as a critical temperature for heat damage (see Methods), we estimate the necessary cooling rate for different environmental conditions to maintain leaf temperatures below this critical value, considering a planar leaf with a characteristic length scale of 5 cm. Fig. 7Asuggests that for low wind speeds (0.5 ms⁻¹) and sunfleck intensity of less than 600 Wm⁻², evaporative cooling would only be needed at air temperatures of more than 307 K (34°C). On the other hand, at air temperatures larger than 314 K (41°C), evaporative cooling is necessary for irradiance values as low as 300 W m⁻² (Fig. 7A). Either increasing air temperature or relative humidity would require increasing values of stomatal conductance to achieve the necessary evaporative cooling.

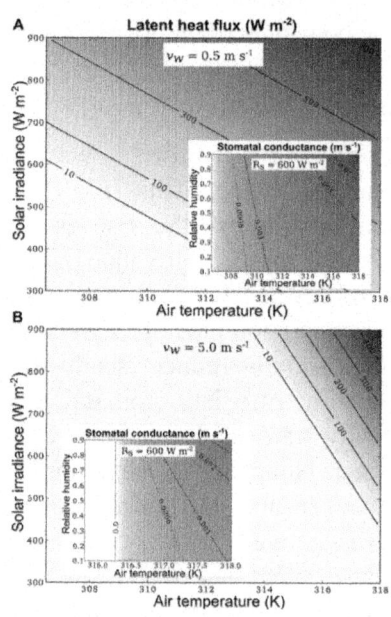

Figure 7. Rates of evaporative cooling and associated stomatal conductances to avoid heat damage.

Contour lines in main panels represent rates of latent heat flux (W m⁻²) necessary to keep leaf temperatures at or below 322 K (49°C), for different combinations of air temperatures and solar irradiances (R_s). Panel A: assumed wind speed $v_w = 0.5$ m s⁻¹; Panel B: $v_w = 5.0$ m s⁻¹. Insets: stomatal conductances that would achieve the latent heat fluxes computed for 600 Wm⁻² irradiance in main panels, for differrent relative humidities. Dashed contour lines mark the lowest stomatal conductance values observed in shaded leaves (Table 2). doi:10.1371/journal.pone.0054231.g007

For higher wind speeds (5.0 ms⁻¹) cooling by sensible heat flux may become more vigorous and greatly reduce the need for evaporative cooling. This is expressed in Fig. 7B, where the need for evaporative cooling is limited to combinations of very high air temperatures and irradiance intensities. This in combination with a greatly increased leaf boundary layer conductance under high wind speeds also results in largely reduced stomatal conductances necessary to keep leaf temperatures below the critical value (inset in Fig. 7B).

Critical Arrival Times to Heat Damage or Turgor Loss Conditions

Leaf water content affects the slope of leaf temperature fluctuations, while stomatal conductance affects the amplitude. Hence, both affect the time to heat damage due to sudden illumination. However, increasing stomatal conductance also results in increasing additional water loss during illumination and increasing risk of turgor loss. The risk of turgor loss, on the other hand, can again be reduced by increasing leaf water content. It follows that leaf water content has a beneficial effect for both time to heat damage and time to turgor loss in response to a sudden increase in illumination. Here we ask the question about the relative importance of leaf water content for delaying heat damage or turgor loss.

Assuming initial steady-state between water loss by transpiration and leaf water supply by the xylem at 10 W m^{-2} irradiance, we abruptly increased irradiance to 600 W m^{-2} and considered the resulting increase in transpiration rate (E_l) to be drawn from water stored in the leaf tissue. Note that the increase in latent heat flux (E_l) at constant conductance (g_{sv}) due to a step increase in radiation can be substantial, e.g. roughly 4-fold in 10 seconds for a leaf with 0.05 kg m^{-2} water content (Fig. 6C). Assuming that xylem water supply remains constant, the cumulative leaf water deficit was computed as the difference between the cumulative transpiration rate under the new radiation level and the transpiration rate at the initial level of 100 W m^{-2}. For different values of constant stomatal conductance (g_{sv}), the time ($t_{crit}(W_l)$) when water deficit reaches 10% of the intial leaf water content (W_l) is plotted as a function of W_l in Fig. 8. For the same values of g_{sv}, the critical time to heat damage ($t_{crit}(T_l)$) is also plotted as a function of initial leaf water content. Increasing g_{sv} from 0.0015 to 0.0025 m s^{-2} could increase the time to heat damage very effectively, and when $g_{sv} > 0.0029$ m s^{-1} heat damage would be avoided altogether in this case. At the same time, increasing g_{sv} decreases the time to critical water loss, however much less effectively. Leaf water content has a much larger effect on the critical time to turgor loss than on the critical time to heat damage (different slopes of the respective red and blue lines). At $g_{sv} = 0.0029$ m s^{-1}, which would be the necessary conductance for avoiding heat damage altogether, the resulting E_l would be 66 W m^{-2} prior to the sunfleck and 248 W m^{-2} at steady state during sunfleck illumination, roughly half of the maximum possible E_l of 547 W m^{-2} at non-limiting g_{sv}.

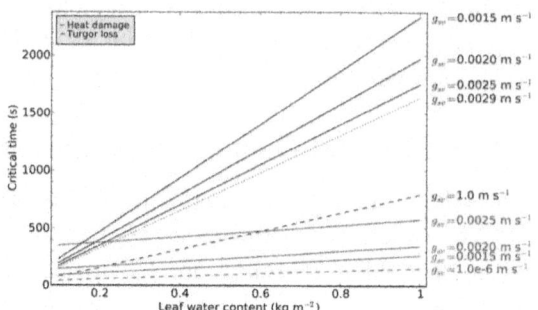

Figure 8. Critical exposure times to a sunfleck of 600 W m⁻² **light intensity for heat damage (red) or turgor loss (blue) as a function of initial leaf water content.** Environmental conditions: $T_a = 313$ K, $v_w = 0.5$ m s⁻¹, 70% relative humidity, 100 W m⁻² irradiance prior to arrival of light fleck. The steady-state transpiration rate at the pre-sunfleck light intensity was taken as a constant xylem water supply rate during the light fleck. Simulations were performed for different values of stomatal conductance, as indicated for each line on the right hand side. The dashed lines represent extreme cases of unlimited stomatal conductance (blue dashed) and negligible stomatal conductance (red dashed line). The blue dotted line represents the time to turgor loss if evaporative cooling is just sufficient to prevent heat damage altogether. In this case, latent heat flux rises from 90 W m⁻² before sunfleck arrival to 248 W m⁻² during the sunfleck.

doi:10.1371/journal.pone.0054231.g008

DISCUSSION

Simulated and Observed Leaf Temperatures

Rapid fluctuations in leaf-incident irradiance, e.g. due to moving sunflecks within a canopy, can result in large and rapid increase in leaf temperatures [19] to critical levels (see also Fig. 2). The leaf energy balance model presented here was capable of reproducing the observed diurnal variation in leaf temperature in the canopy of a tropical rainforest, when stomatal conductance was set to 0 early and late in the day, and a moderate value in the middle of the day (Fig. 4). Some deviations between simulated and observed leaf temperatures were expected, as stomatal conductance and wind velocities were not measured, so the simulations were run with a constant wind velocity ($v_w = 0.5$ m s⁻¹) and constant stomatal conductance. Correspondence between the red line ($g_{sv} = 0$) and observations in the morning suggests that stomata are closed in the morning and confirms the correct representation of sensible heat flux in the model. Correspondence between the blue line ($g_{sv} \gg g_{bv}$) and observations in the middle of the day suggests that stomatal conductance did not vary much between sunflecks. It is interesting to note that on some occasions, low leaf temperatures are best captured by the red line, while high leaf temperatures during sunflecks are better captured by the blue line (e.g. 10–11am in Fig. 4). This could suggest that stomata open during sunflecks and close in between. In the middle of the day (between 11:30am and 2:30pm) the red line stays well above the observed leaf temperatures, which could suggest that stomata stay open all the time, or that wind velocities are higher than the assumed 0.5 m s⁻¹. The effect of wind fluctuations could also be responsible for the leaf temperature jump around 2:30pm, when the observed leaf temperatures alternate between the red and the blue lines. The leaf temperature jump around 4pm cannot be explained based on the available data, as solar irradiance is near 0 and the jump exceeds air temperature.

Protection from Over-Heating

Simulated leaf temperatures for closed stomata have reached values of up to 325 K (52°C) in the middle of the day (Fig. 4). Given observed durations of the sunflecks [4], the leaf intrinsic heat capacity is incapable of significantly reducing leaf temperature peaks. Reducing the leaf water content by 50% would not significantly increase peak leaf temperatures, whereas a 20-fold increase in leaf water content could reduce peak leaf temperatures by ~ 5 K (Fig. 5).

Theoretical modelling results confirm that the increase in leaf temperature as a result of a step increase in irradiance on a hot summer day can be rapid enough to reach potentially damaging leaf temperatures even for relatively short-lived sunflecks (> 20 K increase in half a minute for non-transpiring leaves, Fig. 6A). The increase in heat capacity related to an increase in leaf water content from 0.05 to 0.1 kg m⁻² would only delay critical leaf temperatures by a few seconds (Fig. 6B). Only leaf water contents of 1 kg m⁻² and more would slow down the temperature rise sufficiently to delay heat damage by two minutes or more in our example (Fig. 6B). However, such thick leaves are not common in closed canopies where rapid variations in irradiance are most pronounced. This suggests that the leaf heat capacity is not commonly used by plants to mitigate increases in leaf temperature due to sun flecks.

Thick and watery leaves are commonly found in deserts, among succulent plants with crassulacean acid metabolism (CAM), which keep their stomata closed during the day to conserve water. Desert plants are usually fully exposed to sunlight and rely on cooling by mainly radiative and sensible heat flux. In a recent study, Leigh et al. [34] investigated the protective role of leaf heat capacity against thermal damage in four desert plant species during short periods with low wind speeds. They simulated leaf temperatures for 0.2 mm thick leaves in comparison to realistic leaves of 0.4–0.6 mm thickness and found that with thinner leaves, two out of the four species could become heat damaged. However, the authors did not consider exposure times to high leaf temperatures as a damaging factor while the differences in maximum leaf temperatures between thin and thick leaves were less than 0.5 K, consistent with our results.

Our analysis suggests that transpiration-induced cooling is much more effective in avoidance of high leaf temperatures than capacitive delays following exposure to sunflecks. However, for this mechanism to be effective, leaves must either keep their stomata open even in the shade, or be able to open stomata rapidly following sunfleck exposure. If stomata are already open, the sunfleck-induced increase in leaf temperature can result in substantial increase in leaf latent heat flux, which in return suppresses overall leaf temperature increase (Figs. 6C and 6D). It is primarily the reduction in steady-state leaf temperature and not the time to maximum temperature, that determines the effectiveness of transpirative cooling on the critical time to heat damage (Fig. 6D). In contrast, leaf water content does not affect the steady-state temperature during sunfleck exposure, but the rate of temperature rise and therefore arrival time and duration of exposure to damaging temperatures. Even if steady-state temperatures are reached immediately, e.g. in a leaf with negligible heat capacity, heat damage does not happen immediately, but is a function of the exposure time [30]. Therefore, the red lines in Fig. 8 intersect the ordinate at a level determined by the steady-state leaf temperature. Increasing leaf water content (i.e. heat capacity) increases the time until a critical leaf temperature (322 K in our examples) is reached in a roughly linear fashion (see Fig. 6B). This time should be added to critical exposure duration that is largely determined by steady-state temperature. For situations where steady-state temperature greatly exceeds the critical leaf temperature, the capacitive delay time to critical temperature becomes a significant factor in the onset of heat damage. For lower steady-state temperature, the thermal capacity becomes less significant, as the critical exposure duration is much longer than the time to critical leaf temperature.

Protection from Hydraulic Failure

As shown previously, a rapid increase in transpiration rate in response to a sunfleck is an effective protection mechanism against over-heating, but it also exposes plant leaves to the risks of turgor loss and/or cavitation [9]. Zimmermann [35] proposed that the hydraulic system of trees is segmented to prevent cavitation in trunks by imposing preferred cavitation in leaf petioles and roots before pressure drops propagate into the trunks. Zwieniecki et al. [36] found that the hydraulic conductance in petioles declines in response to a drop in leaf water potential, but recovers quickly when leaf water potential is restored, as long as the metabolism of living cells is not inhibited. This was regarded as an indication of active refilling of emobilised vessels, which requires the expenditure of energy by living tissues [36]. The proximity of living and photosynthesising tissues in leaf petioles may allow easier recovery from cavitation than in the trunk xylem, which could implicate petioles as "safety valves" [18] for accommodating cavitation before effects of rapid pressure drops can propagate into the trunk xylem. The hydraulic conductivity in the petiole determines the pressure jump between the leaf and the trunk xylem for a given flux rate. Thus, a decrease in petiole conductivity with decreasing leaf water potential could have a stabilising effect on the xylem pressure, as it would make the difference between leaf and xylem water potential increase (at a constant flux rate). Note that the role of a "safety valve" could equally be performed by the leaf tissue itself, if the shrinkage of parenchyma tissue at the end of xylem elements due to tissue water loss resulted in a reduction in hydraulic conductivity between xylem and parenchyma tissue. Based on rehydration experiments, Zwieniecki et al. [37] proposed different levels of leaf compartmentalisation that determine the connectivity of different leaf tissues with the xylem: (1) xylem is separated from leaf tissues by a low conductivity barrier, (2) xylem is linked to epidermis but mesophyll is separated by low conductivity barrier and (3) all leaf tissues are linked to the xylem. The low conductivity barriers are zones where the largest pressure drops occur during steady flow, so the three different scenarios determine which tissues are relatively depleted of water before turgor-induced stomatal closure. For case (1), the pressure drop would occur between the leaf xylem and all other tissues, i.e. stomata would close autonomously when the entire leaf tissue reaches a critical water depletion and potential, whereas the leaf xylem potential would remain relatively unchanged. In case (2), autonomous stomatal closure would be expected when water depletion in the epidermis becomes critical, while the mesophyll can maintain higher water potential. In this case, the leaf xylem potential would be expected to decline together with the water potential of the epidermis. In case (3), like in case (1), all leaf tissues would reach a critical water depletion and potential before autonomous stomatal closure, but in this case, the leaf xylem potential would also decline.

We investigated the role of hydraulic capacitance determined by leaf water content as an autonomous reservoir supplying the increased transpiration rate without affecting xylem status, i.e. considering unperturbed xylem water supply. Results show that even for fully open stomata, the increase in transpiration rate induced by a 600 W m^{-2} sunfleck could be accommodated for several minutes in leaves with water content > 0.1 kg m^{-2} (blue dashed line in Fig. 8). This is in contrast with Zwieniecki et al. [37], who assumed that the leaf mesophyll would only support transpiration for tens of seconds. The critical time to turgor loss at constant xylem water supply is a linear function of the leaf water content (m_w), with a slope that scales with the inverse of transpiration rate (E_l). Therefore, the lines in Fig. 8 become steeper with decreasing stomatal conductance (g_{sv}).

Our analysis relates to leaf compartmentalisation scenarios (1) or (3) in the above description, as we assumed that all turgid tissues in the leaf can contribute up to 10% of their water content to the transpiration stream. The analysis is not applicable to leaves of design (2), where stomatal closure is expected already when the leaf epidermis becomes water-depleted. Such leaves either have to be coupled to a very efficient water supply system that responds to pressure drops by increased supply rate, or avoid the combination of sunflecks and high air temperatures. Note that the water content of the leaves described in Figure 4 was only near 0.05 kg m^{-2}, and they were exposed to sunflecks of more than 5 minutes duration without signs of stomatal closure in the leaf temperature data (Fig. 4). This suggests that these plants do have an efficient xylem water supply that can adjust to fluctuating leaf water demand within minutes (the time scale of leaf water depletion according to our analysis).

Implications for Stomatal Adjustments

The numerical experiments revealed that keeping stomata partly open in shaded leaves provides effective protection from over-heating when a leaf is suddenly exposed to a sunfleck (Figs. 6D and 8). For a step exposure to 600 W m^{-2} irradiance intensity and conditions as inFig. 8, the stomatal conductance (g_{sv}) should be roughly one quarter the leaf boundary layer conductance (g_{bv}) to provide effective protection. With this stomatal conductance in the shade, water would be lost at a rate of 90 W m^{-2}, which is equivalent to 16% of the maximum possible transpiration rate at 600 W m^{-2} irradiance (547 W m^{-2}, Fig. 6C). Considering the low photosynthetic rates in the shade, such transpiration rate (16% of the maximum possible rate) represents considerable water loss, particularly if shaded leaves are exposed to only a few short sunflecks in a day. Thus, in a water-limited environment, it may be beneficial for plants to maintain closed stomata in the shade and only open during sunfleck exposure. For conditions as in Fig. 8, stomata must start opening within a minute and reach values of roughly $g_{sv} = 1/4 g_{bv}$ within 5 minutes to avoid heat damage. Considering typical stomatal response times of 2–60 minutes [2], our analysis implies that keeping stomata open is critical for avoidance of sunfleck-induced thermal damage on hot days with little wind. We found this confirmed in the observed leaf temperature data in Fig. 4, which was consistent with our simulations assuming open stomata even during low light periods in the middle of the day.

Using the red dashed line in Fig. 8 as a reference, we searched the literature for observations of g_{sv} in sun and shade in environments with sunflecks of > 600 W m^{-2} intensity and > 100 s duration. The results are summarised in Table 2. Note that the minimum stomatal conductances reported in Table 2 may under-estimate the stomatal conductance of a leaf just before it is hit by a strong sunfleck, as in many environments strong sunflecks are preceded by a series of weaker sunflecks in the morning, which already induce stomatal opening [38]. Keeping this in mind, it is remarkable that the minimum conductances observed in the shade are generally high enough to avoid critical leaf temperatures at air temperatures of more than 309 K (36°C), as implied by the dashed lines in Fig. 7 B. In the extensive data compilation by Vico et al. [2], initial values of stomatal conductance range between 0.00002 and 0.075 m s^{-2}, with a median value of 0.0035 m s^{-2}. Note that $g_{sv} = 0.0035$ m s^{-2} is close to the conductance necessary to completely avoid the danger of heat damage under the conditions simulated in Fig. 8.

Unfortunately, the growth conditions in the different studies were not documented in sufficient detail with respect to sunfleck intensities and durations as well as wind velocities, air temperatures and humidity to correlate observed shade conductances with those necessary to survive naturally occurring sunflecks. Furthermore, the measurements were usually performed under conditions that did not pose a risk of over-heating to the leaves, as air temperatures were not very high, while air flow in the leaf cuvettes was relatively high. To shed more light onto the links between avoidance of heat damage and stomatal adjustments, more experimental research is needed under potentially temperature-stressed conditions, i.e. high air temperatures and low wind speeds.

To date, open stomata in the shade have been generally regarded as a measure to alleviate stomatal limitations to CO_2 uptake in the early periods of sunflecks, given the restrictions on stomatal opening rates (see e.g. [2], [39]). Our modelling results suggest that on hot days with temperatures above 308 K (35°C), open stomata in the shade may have another, potentially much more vital role, namely protection from high leaf temperatures during sunflecks. The former function is expected to become relatively more important for short sunflecks (e.g. < 1min), where closed stomata would not result in overheating anyway, but in very low total sunfleck light use, whereas the latter function is expected to become relatively more important for sunflecks that are long enough to lead to critical leaf temperatures in leaves with closed stomata (e.g.

> 2 min). The examplary sunfleck durations mentioned here are deduced from the red dashed line in Fig. 8, but note that these critical times would vary with different levels of leaf heat tolerance, different wind velocities and air temperatures as well as different sunfleck light intensities. We do not imply that leaf temperature control is the major driver for stomatal adjustments, as the need to achieve sufficient CO_2 uptake during sunflecks may result in sufficiently high stomatal conductances to avoid the danger of heat damage anyway. However, our simulations suggest that water stress on hot summer days may not only have a negative impact on the leaf carbon balance and lead to starvation, but in fact is likely to have a much more immediate effect by leading to heat damage.

Heat Damage Under Water Stress

As discussed above, the potential for a single sunfleck of sufficient intensity and duration to damage non-transpiring leaves (Fig. 6A), combined with the relatively slow stomatal response suggest that open stomata in shaded canopies should be relatively common. On the other hand, under limited soil water supply, keeping stomata open throughout the day may not be feasible and thus limit the ability to simultaneously avoid heat damage and hydraulic failure.

A potential adaptation is increased heat tolerance in response to drought. In fact, drought preconditioning has been found to improve heat resistance in a range of plants (see [40] and references therein), suggesting that stronger limitation in evaporative cooling necessitates greater heat tolerance. In a review of mechanisms of drought damage to trees, Hartmann [41]quotes evidence that trees grown in higher temperatures had a higher mortality in response to drought than plants grown under normal temperatures. Conventional explanations attribute this to higher respiration rates under elevated temperatures and thus higher risk of carbon starvation for trees that must close stomata under drought. So far, these explanations have not yet been supported by evidence [42]. Hartmann [41] recommended analysing alternative hypotheses, such as symplastic failure or inhibition of the redistribution of assimilates, both due to low tissue water potentials.

In view of our study, we would propose to also look at heat damage as a result of reduced evaporative cooling under drought. This might explain increased mortality under elevated temperatures, whereas tissue water potential-related mechanisms alone cannot easily explain these observations. More evidence supporting our heat damage hypothesis was provided by Warren et al. [43], who found that a heat wave combined with drought led to increased leaf senescence under elevated CO_2 treatments compared to ambient CO_2 concentrations. If elevated CO_2 leads to lower stomatal conductances per leaf area or increased carbon gain (or both), then it should be expected to alleviate starvation issues and/or increase the heat damage risk. Warren et al. [43] documented a strong decrease in canopy conductance under elevated CO_2 and no increased carbon gain. The increased leaf senescence under elevated CO_2 during the drought was attributed to stomatal closure, increased leaf temperatures and reduced carbon gain [43]. Our study suggests that the increased leaf senescence and reduced carbon gain may also be explained by direct heat damage, particularly as it occurred during the "hottest time of the year, as T_{air} reached 38°C" [43].

Other protective measures from heat damage in times of inadequate water supply could include reduced absorption of sunlight due to wilting [9], vertical leaf inclination or high leaf reflectivity, and enhanced sensible heat flux by very small leaves.

Okajima et al. [44] documented a decreasing trend of leaf size with increasing mean annual temperatures within the same species, and argued that this correlation may be a result of optimising steady-state leaf temperatures for maximising photosynthesis. For species with an increasing lack of occurrences of large leaves at higher mean annual temperatures, but no lack of small leaves at low temperatures (at least half of the examples presented in [44]), we would argue that avoidance of heat damage may be a better explanation of the pattern. Only for species with a lack of small leaves at low temperatures is the photosynthesis-based explanation more plausible.

CONCLUSIONS

Our analysis suggests that leaf water content has a dual protective role in leaves exposed to short but intense sunflecks. On the one hand, it can delay the onset of heat damage due to its effect on the leaf heat capacity, and on the other hand it provides a buffer for fluctuations in evaporative losses and thereby delays turgor loss when a leaf with open stomata is exposed to a sudden increase in illumination. Our analysis further suggests that keeping stomata open before a sunfleck arrives is likely a vital strategy to avoid heat damage during the sunfleck on a hot day. This finding is consistent with a wide range of studies where initial stomatal conductances prior to the arrival of sunflecks were documented. This may have implications for daily water use efficiencies, but also suggests that drought conditions may result in heat damage to leaf tissues on hot days offering an alternative explanation for the damaging effect of simultaneous drought and heat waves on vegetation. In this context, clouds or aerosols in the atmosphere should not only allow higher photosynthesis rates in deeper canopies due to more diffuse light [45] but also reduce the intensity of sunflecks and hence allow an overall higher water use efficiency and lower the risk of heat damage due to sunflecks. In conclusion, we can answer the questions formulated in the introduction as follows:

- Do natural fluctuations in leaf irradiance necessitate stomatal regulation to avoid heat damage or hydraulic failure?

On hot summer days, a sunfleck could cause heat damage to a non-transpiring leaf within a minute, whereas moderate stomatal conductance can result in sufficient evaporative cooling to avoid heat damage under most realistic conditions. Since observed time scales of stomatal adjustments are generally longer than a minute, stomata need to be already partly open when a sunfleck arrives, in order to allow for autonomous evaporative cooling as the leaf heats up. Common variations in leaf water content are sufficient to supply > 3 minutes worth of transpiration without propagating a pressure drop into the xylem, even for large stomatal conductances (Fig. 8). Since the combination of leaf water capacity and hydraulic xylem efficiency has to be able to support sufficient evaporative cooling on hot days, it is unlikely that stomatal down-regulation of evaporation would become necessary during a sunfleck.

- What is the role of leaf heat and water capacities in negotiating the trade-off between cavitation and over-heating?

In typical canopy leaves, leaf heat capacity contributes only little to extending the time to heat damage during a sunfleck. For a variation in thermal capacitance by one order of magnitude, the simulated time to heat damage only increased by ~ 100 s (Figs. 6B and 8). In contrast, the same range of variation in leaf water capacity extends the time to critical dehydration during a sunfleck roughly 10-fold, e.g. from 200 to 2000 s (Fig. 8).

Acknowledgments

We would like to thank Giulia Vico and Andrew Leakey for making their data available to us and for inspiring discussions of topics related to this study. We would also like to thank Graham Farquhar for helpful discussions and two anonymous reviewers for comments that helped improve the clarity of the manuscript. Christian Lenz is thankfully acknowledged for his help in fine-tuning to the final document.

Author Contributions

Interpreted results and put into context with existing literature: SJS DO MZ.. Conceived and designed the experiments: SJS DO MZ. Performed the experiments: SJS. Analyzed the data: SJS. Wrote the paper: SJS DO.

REFERENCES

1. Chazdon RL (1988) Sunflecks and their importance to forest understorey plants. Advances in Ecological Research 18: 1–63.

2. Vico G, Manzoni S, Palmroth S, Katul G (2011) Effects of stomatal delays on the economics of leaf gas exchange under intermittent light regimes. New Phytologist 192: 640–652.

3. Pearcy RW, Krall JP, Sassenrath-Cole GF (2004) Photosynthesis in fluctuating light environments. In: Baker NR, editor, Photosynthesis and the Environment, Springer Netherlands, number 5 in Advances in Photosynthesis and Respiration. pp. 321–346.

4. Leakey ADB, Scholes JD, Press MC (2005) Physiological and ecological significance of sunflecks for dipterocarp seedlings. Journal of Experimental Botany 56: 469–482.

5. Way DA, Pearcy RW (2012) Sunflecks in trees and forests: from photosynthetic physiology to global change biology. Tree Physiology.

6. Beerling DJ, Osborne CP, Chaloner WG (2001) Evolution of leaf-form in land plants linked to atmospheric CO2 decline in the late palaeozoic era. Nature 410: 287–394.

7. Pearcy RW (1990) Sunflecks and photosynthesis in plant canopies. Annual Review of Plant Biology 41: 421–453.

8. Knapp AK, Smith WK (1990) Stomatal and photosynthetic responses to variable sunlight. Physiologia Plantarum 78: 160–165.

9. Schultz HR, Matthews MA (1997) High vapour pressure deficit exacerbates xylem cavitation and photoinhibition in shade-grown piper auritum H.B. & k. during prolonged sunflecks. Oecologia 110: 312–319.

10. Brodribb TJ, Holbrook NM (2003) Stomatal closure during leaf dehydration, correlation with other leaf physiological traits. Plant Physiology 132: 2166–2173.

11. Roderick ML, Canny MJ (2005) A mechanical interpretation of pressure chamber measurements - what does the strength of the squeeze tell us? Plant Physiology and Biochemistry 43: 323–336.

12. Myers BA, Duff GA, Eamus D, Fordyce IR, O'Grady A, et al. (1997) Seasonal variation in water relations of trees of differing leaf phenology in a wet-dry tropical savanna near darwin, northern australia. Australian Journal of Botany 45: 225–240.

13. Zweifel R, Item H, Ha¨sler R (2000) Stem radius changes and their relation to stored water in stems of young norway spruce trees. Trees - Structure and Function 15: 50–57.

14. Bauerle WL, Whitlow TH, Setter TL, Bauerle TL, Vermeylen FM (2003) Ecophysiology of acer rubrum seedlings from contrasting hydrologic habitats: growth, gas exchange, tissue water relations, abscisic acid and carbon isotope discrimination. Tree Physiology 23: 841–850.

15. Johnson DM, McCulloh KA, Meinzer FC, Woodruff DR, Eissenstat DM (2011) Hydraulic patterns and safety margins, from stem to stomata, in three eastern US tree species. Tree Physiology 31(6): 659–668.

16. Hao GY, Hoffmann W, Scholz F, Bucci S, Meinzer F, et al. (2008) Stem and leaf hydraulics of congeneric tree species from adjacent tropical savanna and forest ecosystems. Oecologia 155: 405–415.

17. Chen JW, Zhang Q, Li XS, Cao KF (2009) Independence of stem and leaf hydraulic traits in six euphorbiaceae tree species with contrasting leaf phenology. Planta 230: 459–468.

18. Chen JW, Zhang Q, Li XS, Cao KF (2010) Gas exchange and hydraulics in seedlings of hevea brasiliensis during water stress and recovery. Tree Physiology 30: 876–885.

19. Leakey ADB, Press MC, Scholes JD (2003) High-temperature inhibition of photosynthesis is greater under sunflecks than uniform irradiance in a tropical rain forest tree seedling. Plant Cell and Environment 26: 1681–1690.

20. Monteith J, Unsworth M (2007) Principles of Environmental Physics, Third Edition. Academic Press, 3 edition.

21. Incropera FP, DeWitt DP, Bergman TL, Lavine AS (2006) Fundamentals of Heat and Mass Transfer. John Wiley & Sons, 6th edition.

22. Parlange JY, Waggoner PE, Heichel GH (1971) Boundary layer resistance and temperature distribution on still and flapping leaves. Plant Physiology 48: 437–442.

23. Schuepp PH (1993) Tansley review no. 59. leaf boundary layers. New Phytologist 125: 477–507.

24. Parlange JY, Waggoner PE (1972) Boundary layer resistance and temperature distribution on still and flapping leaves. Plant Physiology 50: 60–63.

25. Hartmann DL (1994) Global physical climatology. Academic Press.

26. Thimijan RW, Heins RD (1983) Photometric, radiometric, and quantum light units of measure - a review of procedures for interconversion. Hortscience 18: 818–822.

27. Pinker RT, Laszlo I (1997). Photosynthetically active radiation (PAR) and conversion factors (CF). http://www.ngdc.noaa.gov/ecosys/cdroms/

ged_iib/ datasets/b05/pl.htm.

28. Edwards DS, Booth WE, Choy SC (1996) Tropical rainforest research–current issues: proceedings of the conference held in Bandar Seri Begawan, April 1993. Springer.

29. Kozlowski TT, Pallardy SG (2002) Acclimation and adaptive responses of woody plants to environmental stresses. Botanical Review 68: 270–334.

30. Colombo S, Timmer V (1992) Limits of tolerance to high temperatures causing direct and indirect damage to black spruce. Tree Physiology 11: 95–104.

31. Way DA, Sage RF (2008) Elevated growth temperatures reduce the carbon gain of black spruce [Picea mariana (Mill.) B.S.P.]. Global Change Biology 14: 624– 636.

32. Krause G, Winter K, Krause B, Jahns P, Garcı́a M, et al. (2010) High temperature tolerance of a tropical tree, ficus insipida: methodological reassessment and climate change considerations. Functional Plant Biology 37: 890–900.

33. Ortiz C, Cardemil L (2001) Heat-shock responses in two leguminous plants: a comparative study. Journal of Experimental Botany 52: 1711–1719.

34. Leigh A, Sevanto S, Ball MC, Close JD, Ellsworth DS, et al. (2012) Do thick leaves avoid thermal damage in critically low wind speeds? New Phytologist 194: 477–487.

35. Zimmermann MH (1983) Xylem structure and the ascent of sap. SpringerVerlag.

36. Zwieniecki MA, Hutyra L, Thompson MV, Holbrook NM (2000) Dynamic changes in petiole specific conductivity in red maple (Acer rubrum l.), tulip tree (Liriodendron tulipifera l.) and northern fox grape (Vitis labrusca l.). Plant Cell and Environment 23: 407–414.

37. Zwieniecki MA, Brodribb TJ, Holbrook NM (2007) Hydraulic design of leaves: insights from rehydration kinetics. Plant, Cell & Environment 30: 910–921.

38. Pearcy RW (1987) Photosynthetic gas exchange responses of australian tropical forest trees in canopy, gap and understory micro-environments. Functional Ecology 1: 169–178.

39. Hetherington AM, Woodward FI (2003) The role of stomata in sensing and driving environmental change. Nature 424: 901–908.

40. Peng Y, Xu C, Xu L, Huang B (2012) Improved heat tolerance through

drought preconditioning associated with changes in lipid composition, antioxidant enzymes, and protein expression in kentucky bluegrass. Crop Science 52: 807– 817.

41. Hartmann H (2011) Will a 385 million year-struggle for light become a struggle for water and for carbon? – how trees may cope with more frequent climate change-type drought events. Global Change Biology 17: 642–655.

42. Zeppel MJB, Adams HD, Anderegg WRL (2011) Mechanistic causes of tree drought mortality: recent results, unresolved questions and future research needs. New Phytologist 192: 800–803.

43. Warren JM, Norby RJ, Wullschleger SD (2011) Elevated CO2 enhances leaf senescence during extreme drought in a temperate forest. Tree Physiology 31: 117–130.

44. Okajima Y, Taneda H, Noguchi K, Terashima I (2011) Optimum leaf size predicted by a novel leaf energy balance model incorporating dependencies of photosynthesis on light and temperature. Ecological Research 27: 333–346.

45. Roderick ML, Farquhar GD, Berry SL, Noble IR (2001) On the direct effect of clouds and atmospheric particles on the productivity and structure of vegetation. Oecologia 129: 21–30.

46. Fay P, Knapp A (1995) Stomatal and photosynthetic responses to shade in sorghum, soybean and eastern gamagrass. Physiologia Plantarum 94: 613–620.

47. Tausz M, Warren CR, Adams MA (2005) Dynamic light use and protection from excess light in upper canopy and coppice leaves of nothofagus cunninghamii in an old growth, cool temperate rainforest in victoria, australia. New Phytologist 165: 143–156.

48. Valladares F, Allen MT, Pearcy RW (1997) Photosynthetic responses to dynamic light under field conditions in six tropical rainforest shrubs occuring along a light gradient. Oecologia 111: 505–514.

49. Fay PA, Knapp AK (1993) Photosynthetic and stomatal responses of avena sativa (Poaceae) to a variable light environment. American Journal of Botany 80: 1369–1373.

50. Hollinger DY (1987) Photosynthesis and stomatal conductance patterns of two fern species from different forest understoreys. Journal of Ecology 75: 925–935.

51. Naumburg E, Ellsworth DS (2000) Photosynthetic sunfleck utilization potential of understory saplings growing under elevated CO 2 in FACE. Oecologia 122: 163–174.

Chapter 5

BIOCHAR-INDUCED CHANGES IN SOIL HYDRAULIC CONDUCTIVITY AND DIS-SOLVED NUTRIENT FLUXES CONSTRAINED BY LABORATORY EXPERIMENTS

Rebecca T. Barnesa, Morgan E. Gallagher, Caroline A. Masiello, Zuolin Liu, Brandon Dugan

Department of Earth Science, Rice University, Houston, Texas, United States of America

ABSTRACT

The addition of charcoal (or biochar) to soil has significant carbon sequestration and agronomic potential, making it important to determine how this potentially large anthropogenic carbon influx will alter ecosystem functions. We used column experiments to quantify how hydrologic and nutrient-retention characteristics of three soil materials differed with biochar amendment. We compared three homogeneous soil materials (sand, organic-rich topsoil, and clay-rich Hapludert) to provide a basic understanding of biochar-soil-water interactions. On average, biochar amendment decreased saturated hydraulic conductivity (K) by 92% in sand and 67% in organic soil, but increased K by 328% in clay-rich soil. The change in K for sand was not predicted by the accompanying physical changes to the soil mixture; the sand-biochar mixture was less dense and more porous than sand without biochar. We propose two hydrologic pathways that are potential drivers for this behavior: one through the interstitial biochar-sand space and a second through pores within the biochar grains themselves. This second pathway adds to the porosity of the soil mixture; however, it likely does not add to the effective soil K due to its tortuosity and smaller pore size. Therefore, the addition of biochar can increase or decrease soil drainage, and suggests that any potential improvement of water delivery to plants is dependent on soil type, biochar amendment rate, and biochar properties. Changes in dissolved carbon (C) and nitrogen (N) fluxes also differed; with biochar increasing the C flux from organic-poor sand, decreasing it from organic-rich soils, and retaining small amounts of soil-derived N. The aromaticity of C lost from sand and clay increased, suggesting lost C was biochar-derived; though the loss accounts for only 0.05% of added biochar-C. Thus, the direction and magnitude of hydraulic, C, and N changes associated with biochar amendments are soil type (composition and particle size) dependent.

INTRODUCTION

Woolf et al. [1] estimate that 1.8 Pg CO_2-carbon equivalents can be sequestered each year through the sustainable production and application of 0.9 Pg of biochar to agricultural land which sequesters carbon (C), reduces CH_4 and N_2O emissions, and results in avoided CO_2 emissions. Biochar, charcoal intentionally produced by humans through pyrolysis for soil amendment, is a type of black carbon, like soot or charcoal [2], [3]. Adding 0.9 Pg of biochar to the agricultural landscape would correspond to a 4–20 fold increase in global black carbon production (0.04 to 0.194 Pg yr−1; [4]). Recent work illustrates the likely mobility of biochar added to soil: charcoal in soils can be released into rivers [5], [6], where, given its aromatic structure, it can be photo-oxidized [7] and/or transported downstream where it has the potential to alter ecosystem processes [6]. Given the potential magnitude of ecosystem perturbations from full-scale implementation of biochar-C sequestration, it is critical to determine the effects of biochar soil amendment on water and biogeochemical cycling.

Biochar soil amendment can increase crop productivity [3], [8], potentially by improving the hydrologic properties of the soils [9]. Biochar can also increase soil water-holding capacity, and available water content [8], [10]–[13], plant available water [14]–[16], alter soil hydrophobicity[17], and change soil hydraulic conductivity [18]–[22]. Biochar is predicted to cause sandy soils to drain more slowly [23] and clay-rich soils to drain more rapidly [24]. However, past results have not been consistent, likely due to confounding factors such as biochar characteristics (i.e. feedstock and pyrolysis temperatures), application rates, and soil characteristics. Given the importance of hydraulic conductivity in determining the partitioning of precipitation between infiltration and overland flow (i.e. infiltration rates; [25]), which impacts water storage in the subsurface and thus plant available water, it is necessary to understand the effects of biochar on the hydraulic properties of different soil types.

The C and nitrogen (N) content of biochar varies with feedstock and production conditions [26]. These conditions and the C:N ratio of biochar influence its stability [27], [28] as well as possible soil C and N losses [29]. While dependent on production conditions, biochar tends to have a high cation exchange capacity [30] and anion sorption ability [31], allowing for adsorption of dissolved organic matter (DOM) [32] and N [33], and can alter greenhouse gas emissions[34]–[36]. While biochar amendment adds C and N to soils (which may be available for leaching), it is also able to sequester additional C and nutrients in the soil due to its sorptive properties. Therefore the addition of biochar to soil could result in a net increase or decrease in dissolved C and N losses [37]. Worldwide analysis of dissolved black carbon (DBC), which includes derivatives of charcoal and biochar, exported from rivers indicates that, on average, DBC contributes 10% of the global total dissolved organic carbon (DOC) flux [6]. In addition, analysis of the bioavailability of biochar extracts in natural stream water suggests that some biochar-C molecules have turnover rates on the order of days to a month [38], indicating that at least a portion of biochar-C is not recalcitrant. In addition to the dissolution of biochar within soils, particles are also transported. Recent studies illustrate that the movement of biochar particles is related to particle size and surface chemistry, as well as pore water salt content and pH [39], [40]. Collectively this research points to the uncertainty in the fate of biochar and biochar-C and their down-gradient effects on aquatic ecosystems.

Using column experiments, we quantify the effects of a 10% (by mass) biochar amendment on the saturated hydraulic conductivity (K) of sand, clay-rich, and organic-rich soil materials as well as report the effects of this soil amendment on C and N leaching. Using simple, homogeneous soil materials with different grain sizes and surface chemistry allowed us to examine biochar-soil interactions and to compare our results to established soil hydrology models. These data begin to address an important knowledge gap by providing new quantitative constraints on how biochar amendments change K and the chemistry of soil leachate; this work points to the need for more mechanistic studies to examine biochar-soil-water interactions.

MATERIALS AND METHODS

Soil and Biochar

Three soil materials were used to gauge soil property responses to biochar amendment. Sand (Pavestone Natural Play Sand) and organic-rich soil (Micro-Gro Organic Rich Garden Soil, with no N fertilizer) were purchased from Home Depot. The organic-rich soil was texturally similar to a sandy loam. A clay-loam Hapludert, characterized by its poor drainage [41], was collected from Rice University's campus. All materials were oven dried at 60°C to remove any moisture prior to dry sieving. Dry materials were mixed, and then oven dried at 100°C for 24 hours to create homogenous mixtures with initial water content of zero; 100°C facilitates water loss but minimizes chemical impacts as it is significantly lower than our pyrolysis temperature [42]. We determined the grain size of the three soils and the biochar using seven sieves (38 μm to 500 μm).

We produced biochar from mesquite wood (*Prosopis* sp.), ground to smaller than 20 mesh (850 μm). Batches of mesquite (70–80 g) were pyrolyzed using the reactor design described in Kinney et al. [17], by heating in a muffle furnace at 6°C min^{-1} and holding at 400°C for 4 hours. On average these pyrolysis conditions provided a biochar mass yield of 40.4%. The produced biochar had a pH of 6.5±0.1, ash content of 3.33%±0.04%, and liming equivalent of 4%, determined using protocols outlined by the International Biochar Initiative [43]–[45].

Column Experiments

To test the response of K and dissolved nutrient fluxes to biochar amendment we conducted falling head experiments [46] across six materials: sand, sand+biochar, organic soil, organic soil+biochar, clay, and clay+biochar. We packed 150 g of each mixture into three replicate columns, 50 g per column, with 54 μm polyester mesh screen (Small Parts Inc.) at the bottom. Materials were packed with a consistent force into columns in four equal increments to achieve uniform bulk density [47] and the initial soil length was recorded. Bulk density for soil materials and soil+biochar mixtures was determined using the dry mass and column dimensions (height of soil materials, diameter of column) at the start of the experiment. Grain size distributions of soil+biochar mixtures were estimated using the proportional masses of each material (i.e. 10% biochar, 90% soil material) and appropriate grain size data (Table 1).

Table 1. Physical and elemental properties of soil materials and biochar.

Material	Grain Size			Bulk Density	%C	%N
	d_{10} (μm)	d_{50} (μm)	d_{90} (μm)	ρ_d (g cm^{-3})		
sand	70	160	380	1.68±0.18	0.4	0.01
organic-rich	95	400	480	0.43±0.002	37.9	0.54
clay-rich	45	115	460	1.72±0.04	0.9	0.03
biochar (mesquite)	75	320	470	0.36±0.03	71.6	0.84

doi:10.1371/journal.pone.0108340.t001

Biochar constituted 10% of the total mass in the columns that contained biochar. This represents a 133 tons ha^{-1} (95 Mg C ha^{-1}) application rate with a 10 cm tillage depth. We chose a high amendment rate to ensure that we altered the soil-water system in a way that would allow us to detect any effects across the three contrasting soil materials. Specifically, we were concerned about the potential of soil biochar amendments increasing C exports to surface waters, as well as any unforeseen consequences on soil hydrologic properties. This amendment rate is above what is likely to be added to an agricultural field, though within the range reported for positive or neutral productivity effects (up to 140 Mg C ha−1; [3]), similar to the application made by Chan et al. [48], and well below what was shown to be an upper limit to biochar-induced benefits to plants (200 tons C ha−1; [15]).

To saturate the soils, we capped the column bottoms, added 150 mL of 18 MΩ-cm MilliQ water, and allowed the columns to sit for 48 h before drainage. Six consecutive falling head experiments [i.e. flushing events; [46] were conducted on saturated soils using 150 mL of MilliQ water, with leachate collected at the end of each experiment for all columns. The same experimental set up was used for all six materials to allow for inter-comparison. The leachate was weighed, filtered through a pre-combusted glass fiber filter (Whatman GF/F) and kept at 4°C until analysis. This process was repeated without allowing columns to dry between flushing events. Evaporation was monitored daily (via the net change in the mass of water in a beaker) and water throughput (leachate volume) was corrected for evaporation by adding the product of the daily evaporative loss (mg hr^{-1}) and duration of flushing event (hrs).

Saturated hydraulic conductivity (K) from falling head data was calculated using equation 1[49]:

$$K = (L/\Delta t) * ln(h_2/h_1) \tag{1}$$

where L is soil sample length (m), t is time elapsed (s), and h_1 and h_2 are the initial and final water heights (m), respectively (data available in Table S1). Separate experiments on sand-only systems confirm consistent measures of K from top-saturated and bottom-saturated falling head and constant head experiments, suggesting full saturation and steady-state conditions were achieved using this falling-head technique.

Soil and Sample Analysis

We used a Shimadzu TOC-VCN to determine the DOC and total dissolved nitrogen (TDN) concentrations of the filtered leachate. Sample replicates indicate a 0.08 mg L^{-1} and 0.04 mg L^{-1} precision for DOC and TDN, respectively. Mass loss of C and N (dissolved flux) were determined by multiplying the DOC and TDN concentrations by the water volume of the sample, respectively. In contrast, water throughput was calculated using the evaporation-corrected water volumes. The aromaticity of the dissolved C was determined by calculating the specific UV absorbance at 254 nm (SUVA254; [50]). The UV absorption (m^{-1}) of filtered leachate was measured on a Cary UV-Vis spectrophotometer and divided by the DOC concentration (mg C L^{-1}) following the protocol developed by Weishaar et al. [50] to calculate $SUVA_{254}$ (L mg C^{-1} m^{-1}).

After six flushing events, soil mixtures were removed from the columns, weighed, dried at 100°C for at least 24 hours, and reweighed to determine water content at field capacity by mass. We measured C and N content on original soils, biochar, and dried, post-experiment soils using a Costech 4010 CHNS/O Elemental Analyzer. Replicate analysis shows a precision of 0.6% and 0.02% for C and N measurements, respectively.

Statistical Analyses

We used two-sample t-tests to determine statistical differences in the soil and leachate characteristics between treatments and controls. Paired t-tests were used to determine the statistical differences of K over the course of the experiment. Finally, a general linear model with flush number as a covariate, was used to determine if K changed significantly over the course of the experiment as a result of the biochar amendment. All statistical tests were done in the RStudio environment (v0.98.507, 2014 RStudio, Inc.) and results were considered significant when $p < 0.05$.

Results and Discussion

Soil Physical Characteristics

The addition of biochar to the soil materials changed a number of physical properties, e.g. grain size distribution and bulk density, which likely affected water movement. The grain size distribution of biochar differed from that of the three soils and was most similar to the organic soil (Table 1). When biochar was added to sand and clay soils, the d_{50} of the mixtures increased; however, when biochar was added to the organic soil the d_{50} decreased. Given the similarity in d_{10} of sand and biochar, the addition of biochar did not appreciably change the proportion of fines in the sand versus the sand+biochar mixture. The addition of 10% biochar changed soil bulk density (ρ_b) (Table 2), though these changes were not always significant; the addition of biochar decreased the ρ_b of sand and clay by 17% (p=0.056) and 20% (p=0.052), respectively. Biochar addition to clay lowered ρ_b enough to bring it within the range recommended by the National Soil Conservation Service to allow for adequate root growth (<1.47 g cm^{-3}, USDA, 2008). In contrast, when biochar was added to organic soil the mixture ρ_b increased 10% (p=0.018), despite the biochar having a lower

ρ_b (organic: 0.43±0.002 g cm^{-3}; biochar: 0.36±0.03 g cm^{-3}; Table 1); this is likely related to the smaller relative grain size of biochar and grain arrangement during packing of the columns (Table 1).

Table 2. The mean (and standard deviation) of physical, hydraulic, and nutrient properties of the three replicates of each soil and soil+biochar treatment.

Soil	Hydraulic conductivity (K) (m s⁻¹)		Bulk Density (ρb) (g cm⁻³)		Water Content at Field Capacity (fraction water)		Cumulative DOC loss (mg)		Cumulative TDN loss (mg)		SUVA₂₅₄ (L mg C⁻¹ m⁻¹)	
	Soil	+biochar	Soil	+biochar	Soil	+biochar	Soil	+biochar	Soil	+biochar	Soil	+biochar
sand	2.9×10^{-6} (6.3×10⁻⁷)	2.3×10^{-7} (5.9×10⁻⁸)	1.69 (0.18)	1.39 (0.06)	0.15 (0.02)	0.30 (0.03)	1.71 (0.08)	3.29 (0.35)	0.25 (0.04)	0.19 (0.01)	1.86 (0.83)	2.75 (0.41)
p-value	*<0.001*		*0.056*		*0.002*		*0.017*		*0.081*		*0.043*	
organic-Rich	2.1×10^{-6} (1.9×10⁻⁶)	7.8×10^{-7} (7.0×10⁻⁷)	0.43 (0.002)	0.47 (0.008)	0.65 (0.01)	0.66 (0.02)	105.45 (3.74)	95.6 (3.07)	5.52 (0.26)	5.99 (1.25)	3.48 (0.96)	3.38 (0.73)
p-value	*0.038*		*0.018*		*0.757*		*0.039*		*0.565*		*0.442*	
clay-rich	3.2×10^{-8} (1.9×10⁻⁸)	1.2×10^{-7} (1.2×10⁻⁸)	1.72 (0.04)	1.38 (0.14)	0.27 (0.01)	0.33 (0.004)	4.86 (2.22)	4.03 (0.13)	0.66 (0.30)	0.25 (0.01)	2.54 (0.70)	3.75 (0.74)
p-value	*<0.001*		*0.052*		*0.003*		*0.587*		*0.078*		*<0.001*	

Two-tailed t-tests were conducted to determine statistical differences between control and +biochar treatments; p-values are shown in italics below mean and standard deviation values for each treatment.
doi:10.1371/journal.pone.0108340.t002

Soil Hydraulic Characteristics

Saturated hydraulic conductivity (K) describes the ease of fluid flow through saturated porous media and it can be directly measured with flow through experiments [46], [51] or estimated using theoretical or empirical models. Theoretical models, such as the Kozeny equation [52],[53] require significant knowledge of porosity, tortuosity, pore shape, grain density, and specific surface area of solid grains. Because of the difficulty constraining all these parameters, others have developed empirical models that relate porosity and K [54] or grain size and K [55]. Here we compare our K results with these empirical relationships using our bulk density, grain size, and porosity data. Changes in K accompanying biochar amendment for both clay- and organic-rich soils follow the empirical relationships discussed above; specifically the change in K is inversely related to the change in bulk density (or positively related to the change in porosity) caused by the biochar amendment. The addition of biochar to clay-rich soil resulted in a ρ_b decrease of 20% (a porosity increase), an estimated d$_{50}$ increase of 18%, and an increase in K of over 300% (3.26×10^{-8} m s^{-1} to 1.16×10^{-7} m s^{-1}; Figure 1c, Table 2). In contrast, when biochar was added to organic-rich soil, the mixture was 10% denser (lower porosity) with a 2% smaller d$_{50}$, and K decreased by 67% (2.23×10^{-6} m s^{-1} to 7.79×10^{-7} m s^{-1}; Figure 1b, Table 2). Similarly, experiments using silt loams reported an increase in K with biochar amendment, though there were not always corresponding decreases in ρ_b (Table 3, [20]). Previous work in organic-rich soils has not documented changes in K in response to biochar amendment (0.5 to 2% by mass) (Table 3; [56]). Differences in results are attributable to different amendment rates, biochar grain size, soil properties, and/or threshold effects of amendment rate or grain size.

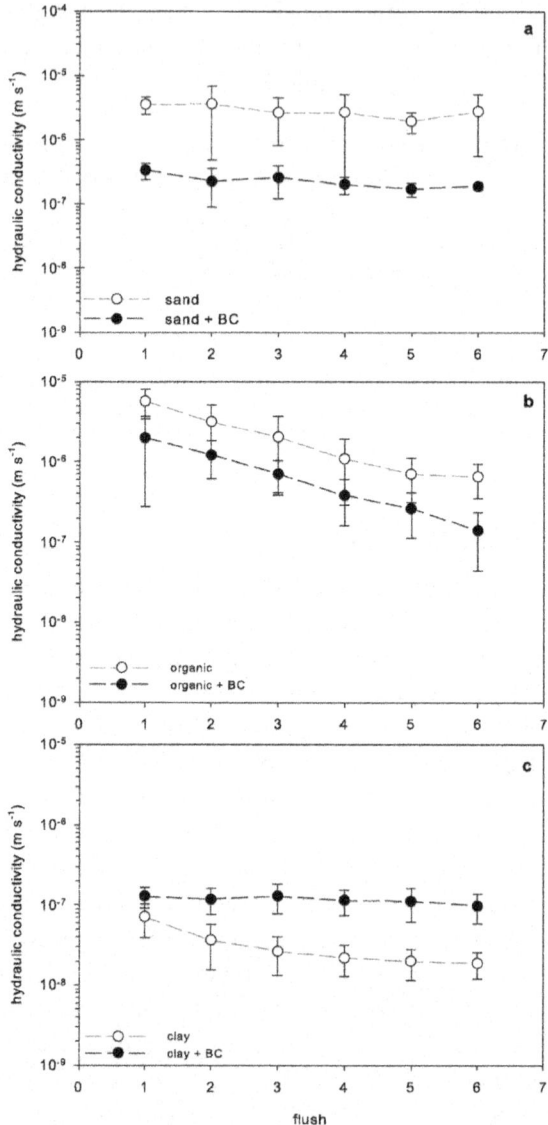

Figure 1. Impact of biochar amendment on saturated soil hydraulic conductivity. The saturated hydraulic conductivity (K), as measured using falling head experiments, for six soil treatments over subsequent flushing events: (a) sand and sand+biochar, (b) organic and organic+biochar, and (c) clay and clay+biochar. Note: the different soil treatment flushing events varied in duration with the clay (c) taking up to 10× longer to drain than the sand (a) or organic soil (b). Saturated hydraulic conductivity data and flushing duration for each flushing experiment available in Table S1. doi:10.1371/journal.pone.0108340.g001

Table 3. A comparison of studies that examined the impact of biochar amendments on soil saturated hydraulic conductivity (K) and bulk density (ρ_d). doi:10.1371/journal.pone.0108340.t003

Feedstock	Temperature (°C)	Application rate[a]	Application rate (tons biochar ha⁻¹)[b]	Experiment duration (d)	Soil type and/or % sand/silt/clay	Response of soil bulk density to biochar addition	Response of K to biochar addition	Reference
mixed hardwood lump charcoal (e.g. oak & hickory)	NR	5, 10, 20 g kg⁻¹	5.5, 11, 22 t ha⁻¹	500	Mollic Typic Haploxeroll	decreased	no effect detected	Laird et al 2010
mesquite	400	10% by dry weight mass	133 t ha⁻¹	1–20	organic rich	increased	decreased	this study
commercial biochar, wood residue (e.g. teak and rosewood)	NR	4, 8, 16 t ha⁻¹	4, 8, 16 t ha⁻¹	60	11/34/48	NR	increased	Asai et al 2009
					27/45/28	NR	no significant change	
corn stover	350 & 550	7.18 t C ha⁻¹	350°C 11.3 t ha⁻¹	205	Alfisol (silt loam)	decreased	increased with both biochars	Herath et al 2013
			550°C = 10.0 t ha⁻¹		Andisol (silt loam)	no change	increased with 350°C, no change for 550°C	
mesquite	400	10% by dry weight mass	133 t ha⁻¹	3–23	clay-loam Haplustert	decreased	increased	this study
dairy manure	300, 500, 700	5% by dry weight mass	61 t ha⁻¹	180	loamy, 40/35/25	decreased	increased, greater increases for biochars at higher temperatures	Lei & Zhang 2013
woodchip						decreased	increased, greater increases for biochars at higher temperatures	
charcoal produced in kilns	~400–500	ambient levels - beneath charcoal kilns	NA	NA	Haplic Acrisols	no significant change	increased	Oguntunde et al 2008
black locust	300, 400, 500	10, 20 Mg ha⁻¹	10, 20 t ha⁻¹	27	sandy	no change	decreased, greatest decrease seen for 300°C biochar & higher application rate	Uzoma et al 2011
powdered wood charcoal	NR	0.5%, 1.5%, 2.5%, 5% by dry weight mass	63, 188, 313, 62.5 t ha⁻¹	60	sandy loam	decreased	decreased with increasing biochar concentration	Devereaux et al 2012
acacia green waste	NR	47 Mg ha⁻¹	47 t ha⁻¹	900	Planosol, 72.8/16.6/10.4	decreased	increased near saturated K, no effect on unsaturated K	Hardie et al 2014
mesquite	400	10% by dry weight mass	133 t ha⁻¹	1–4	sand	decreased	decreased	this study

[a] Biochar application rate reported in paper.
[b] Biochar application rate converted using bulk density of soils or column materials provided in the paper and assuming tillage depth of 10 cm.
NR: not reported.
NA: not applicable.
Biochar amendment rates are provided two ways: the units provided by the study and in tons biochar ha⁻¹. The conversion assumed a tillage depth of 10 cm and the bulk density of the soil or column materials provided in the paper. Studies are organized by soil type, top to bottom: organic rich soils, clay and silt rich soils, and sandy soils. Results from this study are in italics.
doi:10.1371/journal.pone.0108340.t003

Several studies have documented threshold effects on porosity and permeability when fine-grained particles are added to a soil. Boadu [57] found that porosity decreased with increasing fines up to a threshold of ~5% (by mass) at which point fine grain additions increased the porosity. Similarly, Crawford et al. [58] noted that porosity is lowest when the fine particle volume equals the pore space of the coarse grains; however, if fine grains are removed or added, the porosity increases.

Models relating K to porosity and grain size suggest that amending sand with biochar should increase K given the greater porosity, overall grain size, and decreased ρ_b. A review of research reveals that this inverse relationship between ρ_b and K is preserved in most cases (Table 3). However, in our experiments biochar amendment to sand decreased the ρ_b and decreased K by 92% (2.88×10^{-6} m s⁻¹ to 2.28×10^{-7} m s⁻¹; Figure 1a, Table 2). Similar changes in K and ρ_b were reported by Deveraux et al. ([22]; Table 3), though this may be due to a decrease in porosity attributed to the relatively smaller d_{50} of the biochar. In our experiments, the median grain size of biochar was larger than the sand and therefore the decrease in ρ_b was accompanied by an increase in porosity. In addition to grain size, how the biochar is mixed into the soils is also important. The incorporation of charcoal into sandy, Haplic Acrisols underlying kilns, resulted in a non-significant change (9%) in ρ_b and an 88%

increase in K ([18]; Table 3). It is likely that the location of the charcoal within the soil column (only in the surface layer), the soil-charcoal layering, and the charcoal properties, significantly affected infiltration rates and other hydrologic properties. Furthermore, in cases where biochar amendment results in bulk density decreases and/or porosity increases, concomitant with a decrease in K (e.g. our results and [21]; (Table 3), changes in physical soil properties are not sufficient to explain the observed hydrologic changes.

The observed decreases in K despite the increased porosity and decreased ρ_b are likely due to the internal structure of biochar. The biochar had an average pore volume of 1.18 cm³ g⁻¹, porosity of 0.62, and surface area of 6×10^5 cm² g⁻¹ [59]. Thus the biochar has a greater porosity (typical sand has a porosity between 0.17 and 0.33 [60]) and surface area than our sand (based on d_{50}, sand=140 cm² g⁻¹). The highly porous structure of biochar [59], [61] creates two theoretical flow pathways, one in the interstitial space within the biochar-sand matrix and a second connecting the pores within the biochar. According to measurements made by Brewer et al. [59] the biochar we used is dominated (99%) by macropores (0.05 to 1000 µm), and therefore includes many pores larger than the diameter of a water molecule (0.28 nm). However, this second pathway likely has greater tortuosity and smaller median pore throat size due to the size of the smallest pores as well as their lack of complete connectivity[59]. While these pores contribute to the bulk density and total porosity of the mixture, they may not contribute to the effective porosity. In addition, the biochar grains likely create torturous interstitial space between the sand and biochar grains, further decreasing K. This mechanism assumes that the internal pores and surface of biochar are not hydrophobic. While we did not make hydrophobicity measurements on the mesquite biochar, multiple studies have reported that biochar amendments do not result in greater soil hydrophobicity [17], [20]. Furthermore, Briggs et al. [12] found that laboratory-produced charcoals leached with distilled water and naturally-produced charcoal collected beneath leaf litter were less water repellent than non-leached and surficial charcoal, respectively; suggesting that the hydrophobic surface compounds may be easily removed.

A second mechanism driving decreases in K could be related to the high field capacity of biochar [8]; i.e. water may have continued to sorb to biochar particles, contributing to the apparent decrease in K for some of the soil mixtures (Figure 1a). While partial saturation of column materials is possible, the measured K of sand-only systems from constant-head and falling-head experiments were similar, suggesting full saturation and equilibrium conditions for both experiments. Saturated hydraulic conductivity changed with repeated flushing events in four of the six soil treatments ($p < 0.01$)

(Figure 1). The change in K with flushing events was only dependent on the presence of biochar in the case of the organic-rich soil (F=7.366, p=0.01), i.e. while K changed over time in the unamended organic soil, the change was greater in organic+biochar columns (Figure 1b). These shifts in K could be related to physical mechanisms, such as swelling and grain segregation, leading to the clogging of pores, decrease in pore radii, and possibly a variation in bulk density and sample heterogeneity over the course of the experiment. Comparing soil column heights from before and after flushing events revealed that all materials swelled between 6% (sand) and 21% (organic+biochar) suggesting that bulk density changed over the course of the experiments; but biochar addition did not result in statistically different amounts of swelling. Visual observations indicate that biochar particles moved upwards within the columns during the experiment, consistent with observations made by Wang et al. [62]; however, these changes were not quantified and require further study to determine how this movement is related to flow in soil systems and how it impacts hydraulic properties.

The addition of biochar significantly (p<0.01) increased the water content at field capacity of the sand and clay. On average biochar amendment doubled the water content at field capacity in the sand and increased it by 20% in clay (Table 2). Several studies have reported increased water content in biochar-amended soils: e.g. organic-rich Mollisols [56], Amazonian *terra preta*[8], sandy loams [13], [16], [21], [22], and silt loams [20]. The increased water holding capacity is likely the result of the internal porosity of the biochar and grain-to-grain interactions; interactions that are highly dependent on soil type and biochar production conditions and are likely the reason that other studies point to mixed results [63]. The organic soil's water content at field capacity did not change with biochar addition, likely due to the already high water holding capacity of the organic soil (Table 2).

Soil Carbon and Nitrogen Dynamics

The addition of biochar significantly (p<0.05) increased the %C, by mass in all soil materials: sand increased from 0.40±0.35 to 6.98±0.5%C, clay-rich from 0.90±0.06 to 8.33±0.21%C, and organic-rich from 37.85±3.25 to 42.47±0.30%C. Similar patterns were seen for the N content of the soil materials, though not all changes were significant: %N, by mass, increased from 0.01±0.0 to 0.06±0.03%N for sand, 0.03±0.0 to 0.11±0.02%N for clay, and 0.51±0.01 to 0.54±0.06%N for organic-rich soil.

While biochar addition increased the amount of C in all soils, it did not have a universal effect on the C loss as soil leachate (i.e. DOC flux), suggesting that the release of DOC from biochar-amended soils varies with soil type. When added to the C-poor sand, biochar amendments significantly increased the aromaticity and cumulative loss of DOC (Table 2). The increase in aromaticity of the leached DOC from the biochar treatments suggests that biochar-derived C, and not native C, was lost. DOC fluxes decreased with each subsequent flushing event, suggesting that the easily leachable biochar-C was quickly depleted.

In contrast to sand, the addition of biochar to the organic- and clay-rich soils did not increase the cumulative DOC loss (Table 2). When added to C-rich organic soil, biochar significantly decreased the cumulative DOC loss and did not change the aromaticity (Table 2). This suggests that while the biochar added leachable C to the soil (as observed in the sand+biochar treatment); it is also capable of sorbing soil C. While there was no significant change in the magnitude of DOC lost when biochar was added to clay-rich soil, the aromaticity of the leachate was significantly greater when biochar was present, suggesting that while biochar-derived leachable DOC was lost, additional soil-derived C was retained within the soil-biochar matrix.

The partitioning of biochar-C between soil sequestration and leachate is partially dependent on soil type. Approximately 0.05% of the added biochar-C was lost from the sand+biochar mixture, an amount on the low end of the cumulative losses (0.1 to 1.7% of added C) reported by Mukherjee and Zimmerman [37] for similar short-term laboratory experiments. Furthermore, the $SUVA_{254}$ measurements indicate that while biochar-C is preferentially lost with water flow-through in sandy soils, there was also a net sorption of soil-DOC in both the clay and organic-rich soils. Thus the risk of a net increase of DOC export associated with biochar soil amendments may be minimal in soils with moderate amounts of clay, silt or organic material, a conclusion that mirrors field trials and modeling results [24], [64], [65].

Given the potential use of biochar to improve soil fertility in arid, drought-prone environments with C-poor sandy soils (e.g. [21] and references therein) it is important to quantify the partitioning of C between soil and leachate in sandy soils. In particular, while our results suggest that the majority of biochar-C remains in the sandy soil, other studies have shown that this partitioning is dependent on pore water and biochar characteristics [37], [66]. For example, Bruun and others [66] found that there is a tradeoff between increased water holding capacity and increased biochar-C leachate in sandy soils; the amount of C lost was dependent on production conditions, with greater losses seen when fast- versus slow-pyrolysis biochar was added to soils. Given the results from these experiments and recent analyses of DBC in rivers[5], [6] biochar amendments could change the chemical composition of DOC exported to downstream aquatic systems. The ecosystem impacts of this change in the chemical composition of the DOC pool remain unclear [6] and require further study.

Many studies have shown that biochar is able to retain inorganic N, reducing the nitrate flux from soils [29], [33], [67]–[69], and biochar-amended sand and clay retained more N in our experiments, resulting in a 24% (p=0.081) and 62% (p=0.078) reduction in N losses (Table 2); though these differences were not statistically significant. The well documented sorptive properties of biochar are attributed to surficial carboxylic and other acid functional groups that provide the cation and anion exchange capacity of biochar [32], [70], [71] and are dependent on production temperatures and feedstock [30]. The lack of statistical difference with biochar amendment in our study is likely due to the relatively low N content of the soil materials and biochar (<0.8%, Table 1), the only sources of N within our columns. Thus while the results of our experiments are in line with past research, it is important to note that they do not represent the full potential of biochar to mitigate N leaching from agricultural fields.

CONCLUSIONS

Biochar-associated changes in K and field capacity have implications for infiltration rates and plant water availability. As shown by our experiments, the addition of biochar to coarser soils decreases K, indicating the potential to decrease crop water stress and reduce nutrient loss below the rooting zone [29]. Conversely, biochar is able to increase porosity and permeability in fine-grained clay soils, making them more suitable for crop growth by increasing infiltration rates. Our results, combined with those of other studies, strongly support the argument that biochar addition increases the water holding capacity in coarse-grained soils, likely improving plant water availability. The saturated hydraulic conductivity (K) results of these laboratory and field experiments should be considered short-term, as reported changes in soil structure and hydrology are likely to evolve with time, impacting K. Brodowksi et al. [72] showed that biochar degrades into silt-sized particles over time likely changing the porosity and K of the amended soil. Other mechanisms may additional act to alter K on longer timescales, including increased microbial activity and increased bioturbation. Increased microbial activity and the addition of OM associated with biochar amendments has been shown to increase soil aggregation and macropore volume, thereby increasing K [73]. Increased bioturbation associated with biochar amendment can also increase K [74]. These examples provide further impetus for examining the effects of biochar particle size on soil hydraulic properties in the near- and long-term and in both laboratory/greenhouse and field settings. Our experiments also illustrate that biochar addition to soils can add and sorb leachable DOC, and potentially add aromatic

DOC to rivers [5]. Furthermore, biochar has the potential to reduce TDN in leachates, mitigating environmental impacts of agricultural N pollution. These changes in biogeochemical cycling, as well as alterations in greenhouse gas fluxes, plant productivity, and microbial activity are inherently linked to soil hydraulic properties. It is therefore crucial that future research addresses the complex interactions between biochar amendment and soil hydrology, C and N cycling, as well as how different soil types, biochars (varying feedstock, particle size, and production conditions) and amendment rates control these processes.

Supporting Information

Table S1: Average saturated hydraulic conductivity results for soil materials with and without biochar. The average and standard deviation saturated hydraulic conductivity (K) for each flushing experiment (n=3 for each soil material) in m/s. The duration of the experiment (in hours) is also provided.

doi:10.1371/journal.pone.0108340.s001

	1	2
1	Biochar-induced changes in soil hydraulic conductivity and dissolved nutrient fluxes constrained by laboratory experiments	
2	Rebecca T. Barnesa,b*, Morgan E. Gallaghera,c, Caroline A. Masielloa, Zuolin Liua, and Brandon Dugana	
3		
4	aDepartment of Earth Science, Rice University, Houston, TX, USA	
5		C.A. Masiello: masiello@rice.edu, (713) 348-5234
6		B. Dugan: dugan@rice.edu, (713) 348-5088
7		Z. Liu: zl17@rice.edu, (713) 348-5087
8	bPresent address: Environmental Program, Colorado College, Colorado Springs, CO, USA	

9		R.T. Barnes: becca.barnes@coloradocollege.edu, (203) 676-7285
10	cPresent address: Center on Global Change, Duke University, Durham, NC, USA	
11		M.E. Gallagher: morgan.gallagher@rice.edu, (919) 660-7289
12		
13		
14	* Corresponding author	

Acknowledgments

We thank F. Liljestrand, A. Woda, A. Hilton, and D. Leahy for help with experiments, soil processing and data collection; and V. Chuang for biochar production.

Author Contributions

Conceived and designed the experiments: RTB MEG CAM BD. Performed the experiments: RTB MEG ZL. Analyzed the data: RTB MEG. Contributed reagents/materials/analysis tools: CAM BD. Wrote the paper: RTB. Significant help revising the manuscript: MEG CAM BD.

REFERENCES

1. Woolf D, Amonette JE, Street-Perrott FA, Lehmann J, Joseph S (2010) Sustainable biochar to mitigate global climate change. Nature Communications 1: 1–9.

2. Masiello CA (2004) New directions in black carbon organic geochemistry. Marine Chemistry 92: 201–213.

3. Lehmann J, Gaunt J, Rondon M (2006) Bio-char sequestration in terrestrial ecosystems - A review. Mitigation and Adaptation Strategies for Global Change 11: 403–427.

4. Kuhlbusch TAJ, Crutzen PJ (1996) Black Carbon, the Global Carbon Cycle, and Atmospheric Carbon Dioxide. In: Levine JS, editor. Biomass Burning and Global Change: Remote Sensing, Modeling and Inventory Development, and Biomass Burning in Africa. Cambridge, MA: Massachusetts Institute of Technology. pp. 160–169.

5. Dittmar T, de Rezende CE, Manecki M, Niggemann J, Ovalle ARC, et al. (2012) Continuous flux of dissolved black carbon from a vanishing tropical forest biome. Nature Geoscience 5: 618–622.

6. Jaffe R, Ding Y, Niggemann J, Vahatalo AV, Stubbins A, et al. (2013) Global Charcoal Mobilization from Soils via Dissolution and Riverine Transport to the Oceans. Science 340: 345–347.

7. Stubbins A, Spencer RGM, Chen H, Hatcher PG, Mopper K, et al. (2010) Illuminated darkness: Molecular signatures of Congo River dissolved organic matter and its photochemical alteration as revealed by ultrahigh precision mass spactrometry. Limnology and Oceanography 55: 1467–1477.

8. Glaser B, Lehmann J, Zech W (2002) Ameliorating physical and chemical properties of highly weathered soils in the tropics with charcoal - a review. Biology and Fertility of Soils 35: 219–230.

9. Kammann C, Ratering S, Eckhard C, Mu¨ller C (2012) Biochar And Hydrochar Effects On Greenhouse Gas (carbon Dioxide, Nitrous Oxide, And Methane) Fluxes From Soils. J Environ Qual 41: 1052–1066.

10. Novak JM, Busscher WJ, Watts DW, Amonette JE, Ippolito JA, et al. (2012) Biochars Impact on Soil-Moisture Storage in an Ultisol and Two-Aridisols. Soil Science 177: 310–320.

11. Gaskin JW, Speir A, Morris LM, Ogden L, Harris K, et al. (2007) Potential for Pyrolysis Char to Affect Soil Moisture and Nutrient Status of a Loamy Sand Soil. March 27–29; University of Georgia.

12. Briggs C, Breiner JM, Graham RC (2012) Physical and chemical properties of Pinus ponderosa charcoal: implications for soil modification. Soil Science 177: 263–268.

13. Tryon EH (1948) Effects of Charcoal on Certain Physical, Chemical, and Biological Properties of Forest Soils. Ecological Monographs 18: 81–115.

14. Baronti S, Vaccari FP, Miglietta F, Calzolari C, Lugato E, et al. (2014) Impact of biochar application on plant water relations in Vitis vinifera (L.). European Journal of Agronomy 53: 38–44.

15. Kammann CI, Linsel S, Go¨ßling JW, Koyro H-W (2011) Influence of biochar on drought tolerance of Chenopodium quinoa Willd and on soil-plant relations. Plant and Soil.

16. Bruun EW, Petersen CT, Hansen E, Holm JK, Hauggaard-Nielsen H (2014) Biochar amendment to coarse sandy subsoil improves root growth and increases water retention. Soil Use and Management 30: 109–118.

17. Kinney TJ, Masiello CA, Dugan B, Hockaday WC, Dean MR, et al. (2012) Hydrologic properties of biochars produced at different temperatures. Biomass and Bioenergy 41: 34–43.

18. Oguntunde PG, Abiodun BJ, Ajayi AE, van de Giesen N (2008) Effects of charcoal production on soil physical properties in Ghana. Journal of Plant Nutrition and Soil Science 171: 591–596.

19. Asai H, Samson BK, Stephan HM, Songyikhangsuthor K, Homma K, et al. (2009) Biochar amendment techniques for upland rice production in Northern Laos 1. Soil physical properties, leaf SPAD and grain yield. Field Crops Research 111: 81–84.

20. Herath HMSK, Camps-Arbestain M, Hedley M (2013) Effect of biochar on soil physical properties in two contrasting soils: An Alfisol and an Andisol. Geoderma 209–210: 188–197.

21. Uzoma KC, Inoue M, Andry H, Zahoor A, Nishihara E (2011) Influence of biochar application on sandy soil hydraulic properties and nutrient retention. Journal of Food, Agriculture & Environment 9: 1137–1143.

22. Deveraux RC, Sturrock CJ, Mooney SJ (2012) The effects of biochar on soil physical properties and winter wheat growth. Earth and Environmental Science Transactions of the Royal Society of Edinburgh 103: 13–18.

23. Atkinson CJ, Fitzgerald JD, Hipps NA (2010) Potential mechanisms for achieving agricultural benefits from biochar application to temperate soils: a review. Plant and Soil 337: 1–18.

24. Major J, Lehmann J, Rondon M, Goodale CL (2010) Fate of soil-applied black carbon: downward migration, leaching and soil respiration. Global Change Biology 16: 1366–1379.

25. Mishera SK, Tyagi JV, Singh VP (2003) Comparison of infiltration models. Hydrological Processes 17: 2629–2652.

26. Krull ES, Baldock JA, Skjemstad JO, Smernik RJ (2009) Characteristics of biochar: organo-chemical properties. In: Lehmann J, editor. Biochar for environmental management science and technology. London: Earthscan. pp. 53–65.

27. Baldock JA, Smernik RJ (2002) Chemical composition and bioavailability of thermally altered Pinus resinosa (Red pine) wood. Organic Geochemistry 33: 1093–1109.

28. Schneider MPW, Hilf M, Vogt UF, Schmidt MWI (2010) The benzene polycarboxylic acid (BPCA) pattern of wood pyrolyzed between 200uC and 1000uC. Organic Geochemistry 41: 1082–1088.

29. Major J, Steiner C, Downie A, Lehmann J (2009) Biochar effects

on nutrient leaching. In: Lehmann J, Joseph S, editors. Biochar for environmental management, science, and technology. London: Earthscan. pp. 271–287.

30. Lehmann J (2007) Bio-energy in the black. Frontiers in Ecology and the Environment 5: 381–387.

31. Cheng C-H, Lehmann J, Engelhard MH (2008) Natural oxidation of black carbon in soils: Changes in molecular form and surface charge along a climosequence. Geochimica et Cosmochimica Acta 72: 1598–1610.

32. Liang B, Lehmann J, Solomon D, Kinyangi J, Grossman J, et al. (2006) Black carbon increases cation exchange capacity in soils. Soil Science Society of America Journal 70: 1719–1730.

33. Steiner C, Glaser B, Teixeira WG, Lehmann J, Zech W (2008) Nitrogen retention and plant uptake on a highly weathered Amazonian Ferralsol amended with compost and charcoal. Journal of Plant Nutrition and Soil Science 171: 893–899.

34. Yu L, Tang J, Zhang R, Wu Q, Gong M (2013) Effects of biochar application on soil methane emission at different soil moisture levels. Biology and Fertility of Soils 49: 119–128.

35. Mukome FND, Six J, Parikh SJ (2013) The effects of walnut shell and wood feedstock biochar amendments on greenhouse gas emissions from a fertile soil. Geoderma 200–201: 90–98.

36. Angst TE, Patterson CJ, Reay DS, Anderson P, Peshkur TA, et al. (2013) Biochar Diminishes Nitrous Oxide and Nitrate Leaching from Diverse Nutrient Sources. Journal of Environmental Quality 42: 672–682.

37. Mukherjee A, Zimmerman AR (2013) Organic carbon and nutrient release from a range of laboratory-produced biochars and biochar-soil mixtures. Geoderma 193–194: 122–130.

38. Norwood MJ, Louchouarn P, Kuo L-J, Harvey OR (2013) Characterization and biodegradation of water-soluble biomarkers and organic carbon extracted from low temperature chars. Organic Geochemistry 56: 111–119.

39. Zhang W, Niu J, Morales VL, Chen X, Hay AG, et al. (2010) Transport and retention of biochar particles in porous media: effect of pH, ionic strength, and particle size. Ecohydrology 3: 497–508.

40. Wang D, Zhang W, Hao X, Zhou D (2013) Transport of Biochar Particles in Saturate Granular Media: Effects of Pyrolysis Temperature and Particle Size. Environmental Science & Technology 47: 821–828.

41. Wheeler FF (1976) Soil Survey of Harris County, Texas. United States

Department of Agriculture, Soil Conservation Service. pp. 152.

42. ASTM International (2010) D2216-10, Standard Test Methods for Laboratory Determination of Water (Moisture) Content of Soil and Rock by Mass.

43. ASTM International (2007) D1762-84, Standard Test Method for Chemical Analysis of Wood Charcoal.

44. Rayment GE, Lyons DJ (2010) Soil Chemical Methods. Melbourne: CSIRO.

45. Rajkovich S, Enders A, Hanley K, Hyland C, Zimmerman A, et al. (2011) Corn growth and nitrogen nutrition after additions of biochars with varying properties to a temperate soil. Biology and Fertility of Soils.

46. ASTM International (2010) D5084-10 Standard Test Methods for Measurement of Hydraulic Conductivity of Saturated Porous Materials Using a Flexible Wall Perimeter.

47. Oliveira IB, Demond AH, Salehzadeh A (1996) Packing of sands for the production of homogeneous porous media. Soil Science Society of America Journal 60: 49–53.

48. Chan KY, Van Zwieten L, Meszaros I, Downie A, Joseph S (2007) Agronomic values of greenwaste biochar as a soil amendment. Australian Journal of Soil Research 45: 629–634.

49. Klute A (1986) Methods of soil analysis. Part 1. Physical and mineralogical methods. Madison, WI: American Society of Agronomy.

50. Weishaar JL, Aiken GR, Bergamaschi BA, Fram MS, Fujii R, et al. (2003) Evaluation of specific ultraviolet absorbance as an indicator of the chemical composition and reactivity of dissolved organic carbon. Environmental Science & Technology 37: 4702–4708.

51. ASTM International (2006) D2434, Standard Test Method for Permeability of Granular Soils (Constant Head) In: Materials ASfTa, editor. West Conshohocken, PA.

52. Kozeny J (1927) Über kapillare Leitung des Wassers im Boden –Aufstieg, Versickerung und Anwendung auf die Bewä sserung. Sitzungsberichte Akad Wiss Wien 136: 271–306.

53. Carmen PC (1937) Flow through a granular bed. Transactions of Institute of Chemical Engineers, London 15: 150–156.

54. Neuzil CE (1994) How permeable are clays and shales? Water Resources Research 30: 145–150.

55. Alyamani MS, Sen Z (1993) Determination of hydraulic conductivity from complete grain-size distribution curves. Groundwater 31: 551–555.

56. Laird DA, Fleming P, Davis DD, Horton R, Wang B, et al. (2010) Impact of biochar amendments on the quality of a typical Midwestern agricultural soil. Geoderma 158: 443–449.

57. Boadu FK (2000) Hydraulic conductivity of soils from grain-size distribution: New models. Journal of Geotechnical and Geoenvironmental Engineering 126: 739–746.

58. Crawford BR, Faulkner DR, Rutter EH (2008) Strength, porosity, and permeability development during hydrostatic and shear loading of synthetic quartz-clay fault gouge. Journal of Geophysical Research 113.

59. Brewer CE, Chuang VJ, Masiello CA, Gonnermann H, Gao X, et al. (2014) New approaches to measuring biochar density and porosity. Biomass and Bioenergy 66: 176–185.

60. Holtz RD, Kovacs WD (1981) An Introduction to Geotechnical Engineering. Upper Saddle River, NJ: Prentice Hall.

61. Sun H, Hockaday WC, Masiello CA, Zygourakis K (2012) Multiple Controls on the Chemical and Physical Structure of Biochars. Industrial and Engineering Chemistry Research 51: 3587–3597.

62. Wang C, Walter MT, Parlange J-Y (2013) Modeling simple experiments of biochar erosion from soil. Journal of Hydrology 499: 140–145.

63. Novak JM, Lima IM, Xing B, Gaskin JW, Steiner C, et al. (2009) Characterization of designer biochar produced at different tempeartures and their effects on a loamy sand. Annals of Environmental Science 3: 195–2006.

64. Jones DL, Rousk J, Edwards-Jones G, DeLuca TH, Murphy DV (2012) Biocharmediated changes in soil quality and plant growh in a three year field trial. Soil Biology and Biochemistry 45: 113–124.

65. Foereid B, Lehmann J, Major J (2011) Modeling black carbon degradation and movement in soil. Plant and Soil 345: 223–236.

66. Bruun EW, Petersen CT, Strobel BW, Hauggaard-Nielsen H (2012) Nitrogen and carbon leaching in repacked sandy soil with added fine particulate biochar. Soil Science Society of America Journal 76: 1142–1148.

67. Novak JM, Busscher WJ, Watts DW, Laird DA, Ahmedna MA, et al. (2010) Short-term CO2 mineralization after additions of biochar and switchgrass to a Typic Kandiudult. Geoderma 154: 281–288.

68. Bell MJ, Worrall F (2011) Charcoal addition to soils in NE England: A carbon sink with environmental co-benefits? Science of the Total Environment 409: 1704–1714.

69. Clough TJ, Condron LM (2010) Biochar and the Nitrogen Cycle: Introduction. Journal of Environmental Quality 39: 1218–1223.

70. Singh BP, Hatton BJ, Singh B, Cowie AL, Kathuria A (2010) Influence of biochars on nitrous oxide emission and nitrogen leaching from two contrasting soils. Journal of Environmental Quality 39: 1224–1235.

71. Asada T, Ishihara S, Yamane S, Toba T, Yamada A, et al. (2002) Science of bamboo charcoal: Study of carbonizing temperature of bamboo charcoal and removal capability of harmful gases. Journal of Health Sciences 48: 473–479.

72. Brodowski S, Amelung W, Haumaier L, Zech W (2007) Black carbon contribution to stable humus in German arable soils. Geoderma 139: 220–228.

73. Lei O, Zhang R (2013) Effects of biochars derived from different feedstocks and pyrolysis temperatures on soil physical and hydraulic properties. Journal of Soils and Sediments.

74. Hardie M, Clothier B, Bound S, Oliver G, Close D (2014) Does biochar influence soil physical properties and soil water availability? Plant and Soil 376: 347–361.

Chapter 6

STRESS/STRAIN-DEPENDENT PROPERTIES OF HYDRAULIC CONDUCTIVITY FOR FRACTURED ROCKS

Yifeng Chen and Chuangbing Zhou

State Key Laboratory of Water Resources and Hydropower Engineering Science, Key Laboratory of Rock Mechanics in Hydraulic Structural Engineering, Wuhan University, P. R. China

INTRODUCTION

In the last two decades there has seen an increasing interest in the coupling analysis between fluid flow and stress/deformation in fractured rocks, mainly due to the modeling requirements for design and performance assessment of underground radioactive waste repositories, natural gas/oil recovery, seepage flow through dam foundations, reservoir induced earthquakes, etc. Characterization of hydraulic conductivity for fractured rock masses, however, is one of the most challenging problems that are faced by geotechnical engineers. This difficulty largely comes from the fact that rock is a heterogeneous geological material that contains various natural fractures of different scales (Jing, 2003). When engineering works are constructed on or in a rock mass, deformation of both the fractures and intact rock will usually occur as a result of the stress changes. Due to the stiffer rock matrix, most deformation occurs in the fractures, in the form of normal and shear displacement. As a result, the existing fractures may close, open, grow and new fractures may be induced, which in turn changes the structure of the rock mass concerned and alters its fluid flow behaviors and properties. Therefore, the fractures often play a dominant role in understanding the flow-stress/deformation coupling behavior of a rock system, and their mechanical and hydraulic properties have to be properly established (Jing, 2003). Traditionally, fluid flow through rock fractures has been described by the cubic law, which follows the assumption that the fractures consist of two smooth parallel plates. Real rock fractures, however, have rough walls, variable aperture and asperity areas where the two opposing surfaces of the fracture walls are in contact with each other (Olsson & Barton, 2001). To simplify the problem, a single, average value (or together with its stochastic characteristics) is commonly used to describe the mechanical aperture of an individual fracture. A great amount of work (Lomize, 1951; Louis, 1971; Patir & Cheng, 1978; Barton et al., 1985; Zhou & Xiong, 1996) has been done to find an equivalent, smooth wall hydraulic aperture out of the real mechanical aperture such that when Darcy's law or its modified version is applied, the equivalent smooth fracture yields the same water conducting capacity with its original rough fracture. It is worth noting that clear distinction manifests between the geometrically measured mechanical aperture (denoted by b in the context) and the theoretical smooth wall hydraulic aperture (denoted by b^*), and the former is usually larger in magnitude than the latter due to the roughness of and filling materials in rock fractures (Olsson & Barton, 2001).

The ubiquity of fractures significantly complicates the flow behavior in a discontinuous rock mass. The primary problem here is how to model the flow system and how to determine its corresponding hydraulic properties for flow analysis. Theoretically, the representative elementary volume (REV) of a rock mass can serve as a criterion for selecting a reasonable hydromechanical model. This statement relates to the fact that REV is a fundamental concept that bridges the micro-macro, discrete-continuous and stochastic-determinate behaviors of the fractured rock mass and reflects the size effect of its hydraulic and mechanical properties. The REV size for the hydraulic or mechanical behavior is a macroscopic measurement for which the fractured medium can be seen as a continuum. It is defined as the size beyond which the rock mass includes a large enough population of fractures and the properties (such as hydraulic conductivity tensor and elastic compliance tensor) basically remain the same (Bear, 1972; Min & Jing, 2003; Zhou & Yu, 1999; Wang & Kulatilake, 2002). Owing to high heterogeneity of fractured rock masses, however, the REV can be very large or in some situations may not exist. If the REV does not exist, or is larger than the scale of the flow region of interest, it is no longer appropriate to use the equivalent continuum approach. Instead, the discrete fracture flow approach may be applied to investigate and capture the hydraulic behavior of the fractured rock masses. However, due to the limited available information on fracture geometry and their connectivity, it is not a trivial task to make a detailed flow path model. Thus, in practice, the equivalent continuum model is still the primary choice to approximate the hydraulic behavior of discontinuous rocks. The hydraulic conductivity tensor is a fundamental quantity to characterizing the hydromechanical behavior of a fractured rock. Various techniques have been proposed to quantify the hydraulic conductivity tensor, based on results from field tests, numerical simulations, and back analysis techniques, etc. Earlier investigations focused on using field measurements (e.g. aquifer pumping test or packer test (Hsieh & Neuman, 1985)) to estimate the three-dimensional hydraulic conductivity tensor. This approach, however, is generally time-consuming, expensive and needs well controlled experimental conditions. Numerical and analytical methods are also used to estimate the hydraulic properties of complex rock masses due to its flexibility in handling variations of fracture system geometry and ranges of material properties for sensitivity or uncertainty estimations. In the literature, both the equivalent continuum approach (Snow, 1969; Long et al., 1982; Oda, 1985; Oda, 1986; Liu et al., 1999; Chen et al., 2007; Zhou et al., 2008) and the discrete approach (Wang & Kulatilake, 2002; Min et al., 2004) are widely applied. In this chapter, however, only the equivalent continuum approach is focused for its capability of representing the overall behavior of fractured rock masses at large scales.

Among many others, Snow (1969) developed a mathematical expression for the permeability tensor of a single fracture of arbitrary orientation and aperture and considered that the permeability tensor for a network of such fractures can be formed by adding the respective components of the permeability tensors for each individual fracture. Oda (1985, 1986) formulated the permeability tensor of rock masses based on the geometrical statistics of related fractures. Liu et al. (1999) proposed an analytical solution that links changes in effective porosity and hydraulic conductivity to the redistribution of stresses and strains in disturbed rock masses. Zhou et al. (2008) suggested an analytical model to determine the permeability tensor for fractured rock masses based on the superposition principle of liquid dissipation energy. Although slight discrepancy exists between the permeability tensor and the hydraulic conductivity tensor (the former is an intrinsic property determined by fracture geometry of the rock mass, while the latter also considers the effects of fluid viscosity and gravity), when taking into account the flow-stress coupling effect, the above models presented, respectively, by Snow (1969), Oda (1985) and Zhou et al. (2008) were proved to be functionally equivalent for a certain fluid (Zhou et al., 2008). A common limitation with the above models lies in the fact that the hydraulic conductivity tensor of a fractured rock mass is all formulated to be either stress-dependent or elastic strain-dependent. Consequently, material nonlinearity and post-peak dilatancy are not considered in the formulation of the hydraulic conductivity tensor for disturbed rock masses. To address this problem, Chen et al. (2007) extended the above work and proposed a numerical model to establish the hydraulic conductivity for fractured rock masses under complex loading conditions. Based on the observation that natural fractures in a rock mass are most often clustered in certain critical orientations resulting from their geological modes and history of formation (Jing, 2003), characterizing the rock mass as an equivalent continuum containing one or multiple sets of planar and parallel fractures with various critical orientations, scales and densities turns out to be a desirable approximation. Starting from this point of view, the deformation patterns of the fracture network can be first characterized by establishing an equivalent elastic or elasto-plastic constitutive model for the homogenized medium. On this basis, a stress-dependent hydraulic conductivity tensor may be formulated for the former for describing the hydraulic behavior of the rock mass at low stress level and with overall elastic response; and a strain-dependent hydraulic conductivity tensor for the latter for demonstrating the influences of material non-linearity and shear dilatancy on the hydraulic properties after post-peak loading. This chapter mainly presents the research results on the stress/strain-dependent hydraulic properties of fractured rock masses under mechanical loading or engineering disturbance achieved by Chen et al. (2006), Zhou et al. (2006), Chen et al. (2007) and Zhou et al. (2008).

The stress-dependent hydraulic conductivity model (Zhou et al., 2008) was proposed for estimation of the hydraulic properties of fractured rock masses at relatively lower stress level based on the superposition principle of flow dissipation energy. It was shown that the model is equivalent to Snow's model (Snow, 1969) and Oda's model (Oda, 1986) not only in form but also in function when considering the effects of mechanical loading process on the evolution of hydraulic properties. This model relies on the geometrical characteristics of rock fractures and the corresponding fracture network, and demonstrates the coupling effect between fluid flow and deformation. In this model, the pre-peak dilation and contraction effect of the fractures under shear loading is also empirically considered. It was applied to estimate the hydraulic properties of the rock mass in the dam site of the Laxiwa Hydropower Project located in the upstream of the Yellow River, China, and the model predictions have a good agreement with the site observations from a large number of singlehole packer tests.

The strain-dependent hydraulic conductivity model (Chen et al., 2007), on the other hand, was established by an equivalent non-associative elastic-perfectly plastic constitutive model with mobilized dilatancy to characterize the nonlinear mechanical behavior of fractured rock masses under complex loading conditions and to separate the deformation of weaker fractures from the overall deformation response of the homogenized rock masses. The major advantages of the model lie in the facts that the proposed hydraulic conductivity tensor is related to strains rather than stresses, hence enabling hydro-mechanical coupling analysis to include the effect of material nonlinearity and post-peak dilatancy, and the proposed model is easy to be included in a FEM code, particularly suitable for numerical analysis of hydromechanical problems in rock engineering with large scales. Numerical simulations were performed to investigate the changes in hydraulic conductivities of a cube of fractured rock mass under triaxial compression and shear loading as well as an underground circular excavation in biaxial stress field at the Stripa mine (Kelsall et al., 1984; Pusch, 1989), and the simulation results are justified by in-situ experimental observations and compared with Liu's elastic strain-dependent analytical solution (Liu et al., 1999).

Unless otherwise noted, continuum mechanics convention is adopted in this chapter, i.e., tensile stresses are positive while compressive stresses are negative. The symbol (:) denotes an inner product of two second-order tensors (e.g., $a:b=a_{ij}b_{ij}$) or a double contraction of adjacent indices of tensors of rank two and higher (e.g., $c:d=c_{ijkl}d_{kl}$), and ($\otimes$) denotes a dyadic product of two vectors (e.g., $a\otimes b=a_i b_j$) or two second-order tensors (e.g., $c\otimes d=c_{ij}d_{kl}$).

STRESS-DEPENDENT HYDRAULIC CONDUCTIVITY OF ROCK FRACTURES

In this section, the elastic deformation behavior of rock fractures at the pre-peak loading region will be first presented, and then a stress-dependent hydraulic conductivity model will be formulated. The deformation model (or indirectly the hydraulic conductivity model) is validated by the laboratory shear-flow coupling test data obtained by Liu et al. (2002). The main purpose of this section is to provide a theory for developing a stress-dependent hydraulic conductivity tensor for fractured rock masses that will be presented later in Section 4.

Characterization of Rock Fractures

One of the major factors that govern the flow behavior through fractured rocks is the void geometry, which can be described by several geometrical parameters, such as aperture, orientation, location, size, frequency distribution, spatial correlation, connectivity, and contact area, etc. (Olsson & Barton, 2001; Zhou et al., 1997; Zhou & Xiong, 1997). Real fractures are neither so solid as intact rocks nor void only. They have complex surfaces and variable apertures, but to make the flow analysis tractable, the geometrical description is usually simplified. It is common to assume that individual fractures lie in a single plane and have a constant hydraulic aperture. When the fractures are subjected to normal and shear loadings, the fracture aperture, the contact area and the matching between the two opposing surfaces will be altered. As a result, the equivalent hydraulic aperture of the fractures varies with their normal and shear stresses/displacements, which demonstrates the apparent coupling mechanism between fluid flow and stress/deformation (Min et al., 2004).

The aperture of rock fractures tends to be closed under applied normal compressive stress. The asperities of the surfaces will be crushed when their localized compressive stresses exceed their compressive strength. As a large number of asperities are crushed under high compressive stress, the contact area between the fracture walls increases remarkably and the crushed rock particles partially or fully fill the nearby void, which decreases the effective flow area, reduces the hydraulic conductivity of the fracture, and even changes the flow paths through fracture plane. Fig. 1 depicts the increase in contact area of fractures under increasing compressive stresses modelled by boundary element method (Zimmerman et al., 1991).

The coupling process between fluid flow and shear deformation is more related to the roughness of fractures and the matching of the constituent walls. Fig. 2 shows the impact of the fracture structure on the shear stress-deformation coupling mechanism. In Fig. 2(a), the opposing walls of the fracture are well matched so that the fracture always dilates and the hydraulic conductivity increases under shear loading as long as the applied normal stress is not high enough for the asperities to be crushed. For the state shown in Fig. 2(c), shear loading will result in the closure of the fracture and the reduction in hydraulic conductivity. Fig. 2(b) illustrates a middle state between (a) and (c), and its shearing effect depends on the direction of shear stress. When the matching of a fracture changes from (a) to (b) then to (c) under shear loading, shear dilation occurs. On the other hand, shear contraction takes place from the movement of the matching from (c) to (b) then to (a). In a more complex scenario, shear dilation and shear contraction may happen alternately, resulting in the fluctuation of the hydraulic behavior of the fractures.

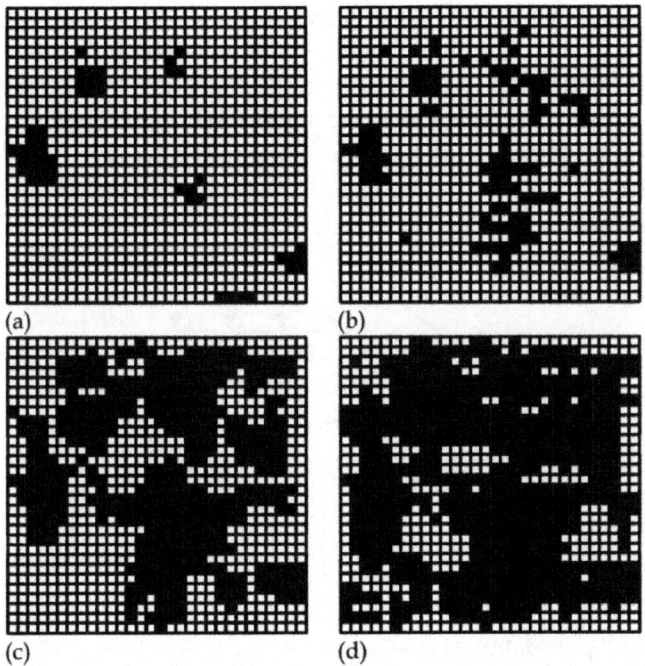

Figure 1. Variation of contact surface of fractures under increasing compressive stresses (after Zimmerman et al., (1991): (a) P=0 MPa; (b) P=20 MPa; (c) P=40 MPa and (d) P=60 MPa.

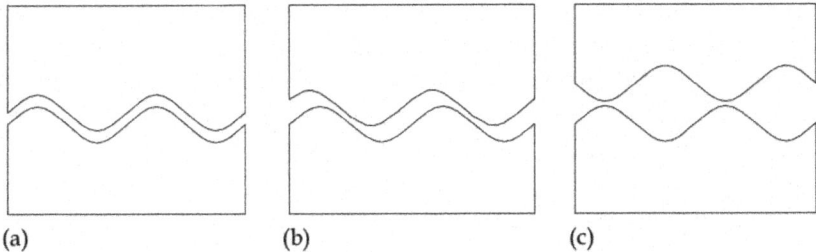

Figure 2. Shear dilation and shear contraction of fractures: (a) well-matched; (b) fair-matched; and (c) bad-matched.

An elastic constitutive model for rock fractures To formulate the stress-dependent hydraulic conductivity for rock fractures, we model the fractures by an interfacial layer, as shown in Fig. 3. The interfacial layer is a thin layer with complex constituents and textures (depending on the fillings, asperities and the contact area between its two opposing walls). Assumption is made here that the apparent mechanical response of the interfacial layer can be described by Lame's constant λ and shear modulus μ. Because the thickness of the interfacial layer (i.e., the initial mechanical aperture of the fracture) is generally rather small comparing to the size of rock matrix, it is reasonable to assume that $\varepsilon_x = \varepsilon_y = 0$ and $\gamma_{xy} = \gamma_{yx} = 0$ within the interfacial layer. Then according to the Hooke's law of elasticity, the elastic constitutive relation for the interfacial layer under normal stress σ_n and shear stress τ can be written in the following incremental form:

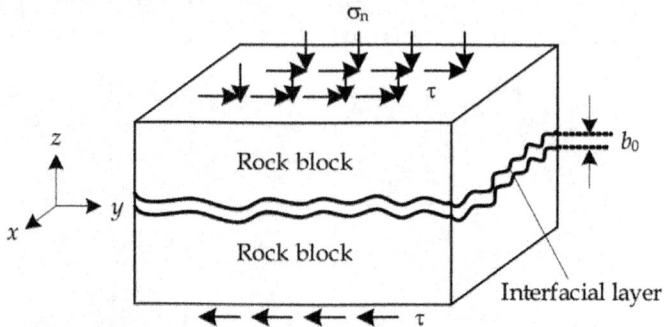

Figure 3. The interfacial layer model for rock fractures.

$$\left\{ \begin{array}{c} d\sigma_n' \\ d\tau \end{array} \right\} = \left[\begin{array}{cc} \lambda + 2\mu & 0 \\ 0 & \mu \end{array} \right] \left\{ \begin{array}{c} d\varepsilon_n \\ d\gamma \end{array} \right\} \tag{1}$$

For convenience, we use u_1 to denote the relative normal displacement of the interfacial layer caused by the effective normal stress σ'_n, δ to denote the relative tangential displacement caused by the shear stress τ, and u_2 to denote the relative normal displacement caused by shear dilation or contraction (positive for dilatant shear, negative for contractive shear). Hence, the total normal relative displacement u is represented as

$$u = u_1 + u_2 \tag{2}$$

The increments of strains, $d\varepsilon_n$ and $d\gamma$, can be expressed in terms of the increments of relative displacements, du_1 and $d\delta$, as follows:

$$\left\{ \begin{array}{l} d\varepsilon_n = du_1 / (b_0 + u) \\ d\gamma = d\delta / (b_0 + u) \end{array} \right. \tag{3}$$

where b_0 is the thickness of the interfacial layer or the initial mechanical aperture of the fracture. Substituting Eq. (3) in Eq. (1) yields:

$$\begin{Bmatrix} \mathrm{d}\sigma'_n \\ \mathrm{d}\tau \end{Bmatrix} = \begin{bmatrix} k_n & 0 \\ 0 & k_s \end{bmatrix} \begin{Bmatrix} \mathrm{d}u_1 \\ \mathrm{d}\delta \end{Bmatrix} \tag{4}$$

where k_n and k_s denote the tangential normal stiffness and tangential shear stiffness of the interfacial layer, respectively.

$$k_n = (\lambda + 2\mu)/(b_0 + u), \quad k_s = \mu/(b_0 + u) \tag{5}$$

Interestingly, k_n and k_s show a hyperbolic relation with normal deformation and characterize the deformation response of the interfacial layer under the idealized conditions that each fracture is replaced by two smooth parallel planar plates connected by two springs with stiffness values k_n and k_s. As can be seen from Eq. (5), as long as the initial normal stiffness and shear stiffness with zero normal displacement, k_{n0} and k_{s0}, are known, they can be used as substitutes for λ and μ.

Substituting Eq. (2) in Eq. (4) results in:

$$\mathrm{d}\sigma'_n = \frac{(\lambda + 2\mu)\mathrm{d}u_1}{b_0 + u_1 + u_2} \tag{6}$$

$$\mathrm{d}\tau = \frac{\mu\mathrm{d}\delta}{b_0 + u_1 + u_2} \tag{7}$$

Suppose normal stress σ_n is firstly applied before the loading of shear stress, u_1 can be obtained by directly integrating Eq. (6):

$$u_1 = (b_0 + u_2)\left[\exp\left(\frac{\sigma'_n}{\lambda + 2\mu} \right) - 1 \right] \tag{8}$$

Here, it is to be noted that the elastic constitutive model for the rock fracture leads to an exponential relationship between the fracture closure and the applied normal stress, which has been widely revealed in the literature, e.g., in Min et al. (2004).

On the other hand, the shear expansion caused by $\mathrm{d}\delta$ can be estimated from shear dilation angle d_m:

$$\mathrm{d}u_2 = \tan d_m \, \mathrm{d}\delta \tag{9}$$

By introducing two parameters, s and φ, pertinent to normal stress σ_n, we represent the dilation angle d_m under normal stress σ_n in the form of Barton's strength criterion for joints (Barton, 1976) ($\tau = \sigma_n \tan(2d_m + \varphi_b)$, where φ_b is the basic frictional angle of joints):

$$tan\, d_m = \frac{1}{2}\left[arctan\left(\frac{\tau}{s}\right) - \varphi \right] \tag{10}$$

Obviously, s is a normal stress-like parameter, and φ is a frictional angle-like parameter. But to make the above formulation still valid into pre-dilation state (i.e., shear contraction state), s and φ differ from their initial implications. Later, we will show how they can be back calculated from shear experimental data.

Substituting Eqs. (9) and (10) into (7) yields:

$$\frac{du_2}{b_0 + u_1 + u_2} = \frac{1}{2\mu}\left[arctan\left(\frac{\tau}{s}\right) - \varphi \right] d\tau \tag{11}$$

By integrating Eq. (11), we have:

$$u_2 = (b_0 + u_1)\left\{ exp\left[\frac{|\tau|}{2\mu}\left(arctan\frac{|\tau|}{s} - \varphi \right) - \frac{s}{4\mu}\ln\left(1 + \frac{\tau^2}{s^2} \right) \right] - 1 \right\} \tag{12}$$

By solving the simultaneous equations, Eqs. (8) and (12), we have:

$$\begin{cases} u_1 = \dfrac{A(1+B)}{1-AB}b_0 \\[2mm] u_2 = \dfrac{B(1+A)}{1-AB}b_0 \end{cases} \tag{13}$$

where

$$A = exp\left(\frac{\sigma'_n}{\lambda + 2\mu} \right) - 1 \tag{14}$$

$$B = exp\left[\frac{|\tau|}{2\mu}\left(arctan\frac{|\tau|}{s} - \varphi \right) - \frac{s}{4\mu}\ln\left(1 + \frac{\tau^2}{s^2} \right) \right] - 1 \tag{15}$$

Thus, the total normal deformation under normal and shear loading can be obtained,

$$u = u_1 + u_2 = \frac{A + B + 2AB}{1 - AB}b_0 \tag{16}$$

The actual aperture of the fracture, $b = b_0 + u$, is given by:

$$b = b_0 + u = (1 + \chi)b_0 \tag{17}$$

where

$$\chi = \frac{A + B + 2AB}{1 - AB} \tag{18}$$

Stress-Dependent Hydraulic Conductivity for Rock Fractures

Since natural fractures have rough walls and asperity areas, it is not appropriate to directly use the aperture derived by Eq. (17) for describing the hydraulic conductivity of the fractures. Instead, an equivalent hydraulic aperture is usually taken to represent the percolation property of the fractures, as demonstrated in Section 1. Based on experimental data, the relationship between the equivalent hydraulic aperture and the mechanical aperture has been widely examined in the literature, and the empirical relations proposed by Lomize (1951), Louis (1971), Patir & Cheng (1978), Barton el al. (1985) and Olsson & Barton (2001) are listed in Table 1. For example, if Patir and Cheng's model is used to estimate the equivalent hydraulic aperture that accounts for the flow-deformation coupling effect in pre-peak shearing stage, then there is

$$b^* = (1 + \chi)b_0 \left[1 - 0.9\exp(-0.56 / C_v)\right]^{1/3} \tag{19}$$

where C_v is the variation coefficient of the mechanical aperture of the discontinuities, which is mathematically defined as the ratio of the root mean squared deviation to the arithmetic mean of the aperture. For convenience, Eq. (19) is rewritten as:

$$b^* = b_0 f(\beta) \tag{20}$$

Obviously, f(β) is a function of the normal and shear loadings, the mechanical characteristics and the aperture statistics of the fractures.

Thus, the hydraulic conductivity of the fractures subjected to normal and shear loadings is approximated by the hydraulic conductivity of the laminar flow through a pair of smooth parallel plates with infinite dimensions:

$$k = \frac{g b^{*2}}{12v} \tag{21}$$

where k is the hydraulic conductivity, g is the gravitational acceleration, and v is the kinematic viscosity of the fluid.

An alternative approach to account for the deviation of the real fractures from the ideal conditions assumed in the parallel smooth plate theory is to adopt a dimensionless constant, ς, to replace the constant multiplier, $1/12$, in Eq. (21), where $0<\varsigma<1/12$ (Oda, 1986). In this manner, the hydraulic conductivity of the fractures is estimated by

$$k = \varsigma \frac{gb^2}{v} \tag{22}$$

Clearly, the constant, ς, approaches $1/12$ with increasing scale and decreasing roughness of the fractures.

Eqs. (21) and (22) show that the hydraulic conductivity of a rock fracture varies quadratically with its mechanical aperture. The latter depends, by Eq. (18), on the normal and shear stresses applied on the fracture. Hence, we call the established model, Eq. (21) or (22), the stress-dependent hydraulic conductivity model, and it is suitable to describe the hydraulic behavior of the fractures subjected to mechanical loading in the pre-peak stage.

Table 1. Empirical relations between equivalent hydraulic aperture and mechanical aperture

Authors	Expressions	Descriptions
Lomize (1951)	$b^* = b\left[1.0 + 6.0(e/b)^{1.5}\right]^{-1/3}$	b^* is the equivalent hydraulic aperture of fractures, b the mechanical aperture, e the absolute asperity height, e_m the average asperity height, D_H the hydraulic radius, C_v the variation coefficient of the mechanical aperture, JRC the joint roughness coefficient, JRC_0 the initial value of JRC, JRC_{mob} the mobilized JRC, δ the shear displacement and δ_p the peak shear displacement.
Louis (1971)	$b^* = b\left[1.0 + 8.8(e_m/D_H)^{1.5}\right]^{-1/3}$	
Patir & Cheng (1978)	$b^* = b\left[1 - 0.9\exp(-0.56/C_v)\right]^{1/3}$	
Barton, et al. (1985)	$b^* = b^2 JRC^{-2.5}$	
Olsson & Barton (2001)	$\begin{cases} b^* = b^2 JRC_0^{-2.5} & \delta \le 0.75\delta_p \\ b^* = b^{1/2} JRC_{mob} & \delta \ge \delta_p \end{cases}$	

Validation of the Elastic Constitutive Model

The key point of the stress-dependent hydraulic conductivity model is whether the established elastic constitutive model can properly describe the variation of mechanical aperture under normal and shear loadings at low stress level. Here, we use the results of the laboratory test performed by Liu et al. (2002) to validate the mechanical model. The test was conducted to study shear-flow coupling properties for a marble fracture with fillings of sand under low normal stresses and small shear displacements. The marble specimen for shear-flow coupling test is illustrated in Fig. 4, which was collected from the

Daye Iron Mine in China. The uniaxial compressive strength and density of the rock sample are 52.4 MPa and 2.66×10^3 kg/m³, respectively. The specimen was cut into round shape and the fracture surfaces were polished, with its size of 290 mm in diameter and 200 mm in height. The opposite walls of the fracture were cemented with a layer of filtered sands with their diameters ranged from 0.5 to 0.69 mm, and the fracture was further filled with the same sands. The initial aperture of the fracture, b_0, is about 1.31 mm.

The coupled shear-flow test were conducted by first applying a prescribed normal stress ranging between 0.1 and 0.5 MPa and then applying shear displacement in steps until a maximum displacement of about 0.4 mm was reached. During tests, steady-state fluid flow rate and normal displacement were continuously recorded.

With such low normal stresses and small shear displacements, it is reasonable to consider that the fracture behaves elastic during the coupled shear-flow test. According to the experimental results, the elastic parameters, λ and μ, of the fracture with fillings are estimated as λ=1.81 MPa and μ=3.62 MPa. In order to enable Eq. (16) to predict the mechanical aperture of the facture under normal and shear loads, the normal stress-like parameter, s, and the frictional angle-like parameter, φ, should be further determined. Fortunately, both of them can be derived by fitting the experimental curve between normal displacement and shear displacement, as plotted in Fig. 5, using Eq. (16) such that the least square error is minimized. With this approach, we obtain that for σ_n=0.1 MPa, s=0.062, φ=1.324, and for σ_n=0.4 MPa, s=0.046, φ=1.310.

Fig. 5 plots the experimental results as well as the model predictions of the relation between mechanical aperture and shear displacement of the fracture under constant normal stresses. Generally, the proposed elastic constitutive model manifests the behavior of the fracture with fillings during the shear-flow coupling test with low normal and shear loads. Shear contraction is observed in the initial 0.06-0.08 mm of shear displacement, which is followed by shear dilation in the remaining of the shear displacement. This property, which is actually ensured by the empirical relation assumed in Eq. (9), suggests that the resultant model is suitable for phenomenologically describing the pre-peak shear-flow coupling effect of fractures.

Fig. 6 further depicts the sensitivity of s and φ on the behavior of the fracture. In Fig. 6(a), φ is fixed to 1.324, while s varies from 0.02 to 0.08. As s increases, shear contraction more apparently manifests, and the mechanical aperture versus shear displacement curves become lower as a whole. On the other hand, the effect of varying φ from 0.524 to 1.222 but fixing s to 0.062 is plotted in Fig. 6(b). For small value of φ, shear contraction is trivial and the curve extends with a larger slope. As φ increases, however, shear contraction becomes relatively remarkable and the curve turns relatively flat. Thus, by adjusting s and φ, the mechanical and hydraulic behaviors of the fracture can be appropriately established.

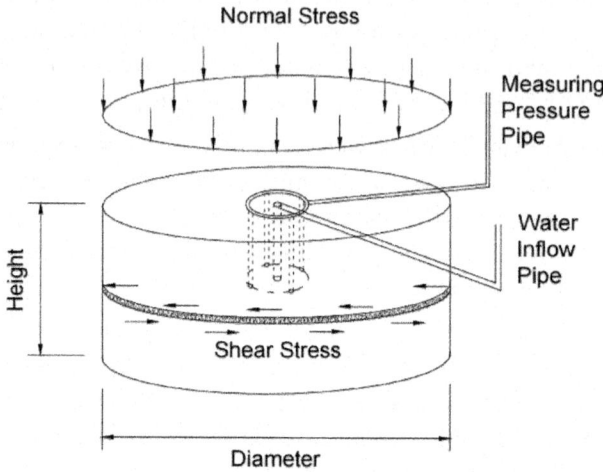

Figure 4. Sketch of the marble specimen for shear-flow coupling test.

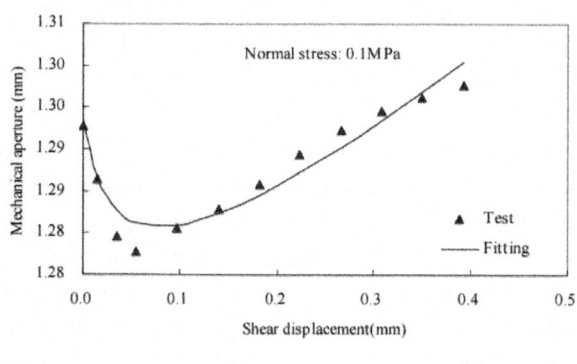

(a)

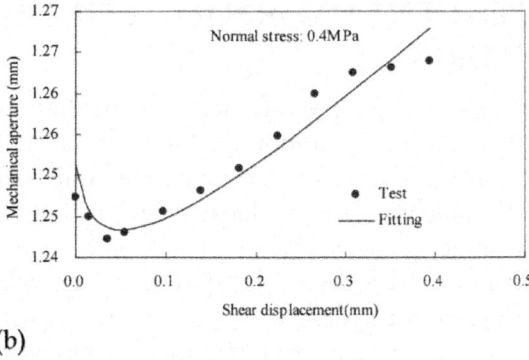

(b)

Figure 5. Mechanical aperture versus shear displacement curve under constant normal stress: (a) Normal stress: 0.1 MPa and (b) Normal stress: 0.4 MPa.

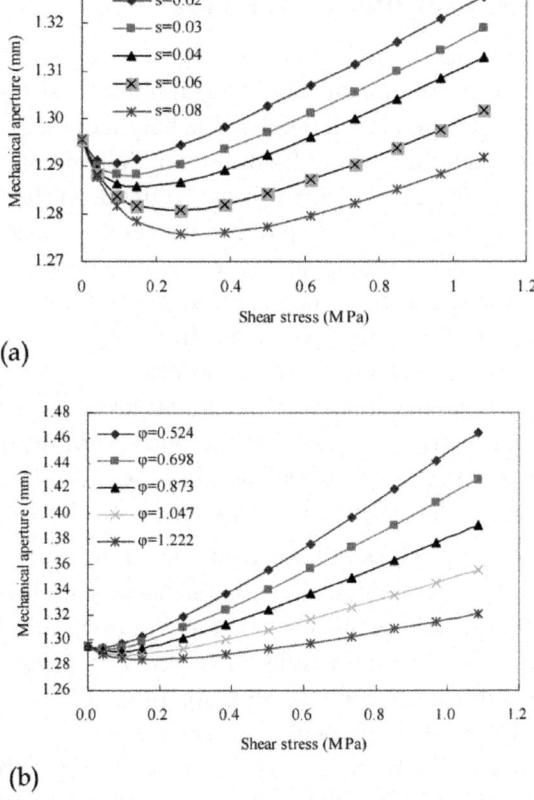

(a)

(b)

Figure 6. The sensitivity of s and φ on the behavior of the fracture: (a) φ=1.324 and (b) s=0.062.

STRAIN-DEPENDENT HYDRAULIC CONDUCTIVITY OF ROCK FRACTURES

In this section, we develop an elasto-plastic constitutive model for single hard rock fractures with consideration of nonlinear normal deformation and post-peak shear dilatancy, and then formulate the strain-dependent hydraulic conductivity for the fractures under dilated shear loading. Compared with the stress-dependent model presented in Section 2, one major difference is that the strain-dependent model is capable of describing the influence of postpeak mechanical response on the hydraulic properties of the fractures. This work is of paramount importance in the sense that the theoretical results are directly comparable with the experimental data of coupled shear-flow test, e.g. in Esaki et al. (1999). The straindependent hydraulic conductivity tensor can then be developed on this basis, which will be presented later in Section 5.

An Elasto-Plastic Constitutive Model for Rock Fractures

The underlying physical model considered is the same with the model plotted in Fig. 3, in which a fracture of hard rock is located at the mid-height of a specimen between two intact rock blocks. The height of the specimen is denoted by s, and the initial aperture of the fracture is b_0. When constant normal stress σ_n and increasing shear displacement δ are applied on the specimen, typical and idealized curves of shear displacement versus shear stress and shear displacement versus normal displacement (i.e. $\delta \sim \tau$ curve and $\delta \sim u$ curve) are plotted in Fig. 7. The shear stress increases linearly with the shear displacement (linked by the initial shear stiffness of the fracture, k_{s0}) until the shear stress approaches the peak, τ_p, which is then followed by a shear softening process in which the shear stress decreases to a residual level at a decreasing gradient with increasing shear displacement. For the purpose of deriving the hydraulic property of the fracture in post-peak loading section, however, an elastic-perfectly plastic $\delta \sim \tau$ relationship can be assumed, as shown in Fig. 7(a)

The deformation response of a rock fracture subjected to normal and shear loadings includes two components: one is the nonlinear closure of the fracture due to normal compression, and the other is the opening of the fracture due to shear dilation. Experimental results in Esaki et al. (1999) show that in the shearing process under constant normal loading, dilatancy will start when the shear stress approaches the peak and it increases at a decreasing gradient with increasing shear displacement, as illustrated in Fig. 7(b). As a result, the aperture of the fracture and then the hydraulic conductivity vary with increasing shear displacement.

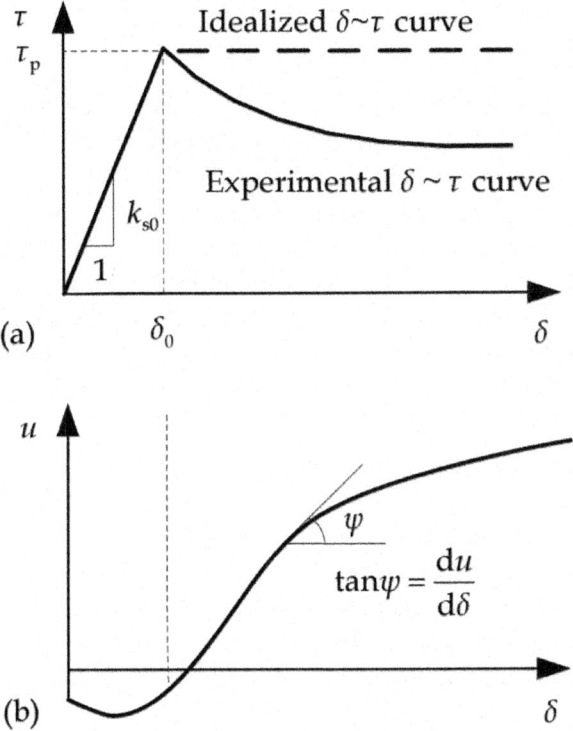

Figure 7. Typical and idealized curves of shear displacement versus shear stress and shear displacement versus normal displacement of a fracture subjected to normal and shear loads.

Therefore, we may consider that shear dilatancy as well as the change in hydraulic conductivity accompanies normal and plastic shear deformations of the fracture. To deduce the hydraulic conductivity of the fracture with an averaging method, which will be further used later for deriving the hydraulic conductivity tensor for fractured rocks, we view the specimen with fracture as an equivalent continuous medium, i.e. the hydromechanical properties of the fracture are averaged into the whole specimen. As can be seen later, such a treatment does not affect our final solution to a single fracture, but it renders valid the small strain assumption on the fractures in the presence of post-sliding plasticity.

For a one-dimensional problem with a single rock fracture, the elasto-plastic constitutive model can be represented in the following forms:

$$\gamma_p = \gamma - \gamma_e = \frac{\delta}{s} - \frac{\delta_0}{s} = \frac{\delta}{s} - \frac{\tau_p}{sk_{s0}} \tag{23}$$

$$\varepsilon_n = \frac{\sigma'_n}{sk_n} + \int \tan\psi\, d\gamma_p \tag{24}$$

where γ, γ_e and γ_p are the total shear strain, the elastic shear strain and the plastic shear strain of the fracture, respectively; ε_n is the normal strain of the fracture; τ_p is the peak shear stress of the fracture under effective normal stress σ'_n; k_n and k_{s0} are, respectively, the normal stiffness and the initial shear stiffness of the fracture; δ_0 is the maximum elastic shear displacement upon shear yielding, with $\delta_0 = \tau_p/k_{s0}$, as shown in Fig. 7(a); and ψ is the mobilized dilatancy angle of the fracture. Note that in Eq. (24), the first term on the right hand side denotes the nonlinear closure of the fracture subjected to effective normal stress σ'_n, while the second term denotes the opening of the fracture due to shear dilatancy.

Existing studies have indicated that shear dilatancy is highly dependent on the plasticity already experienced by the fractures and normal stress, and non-negligibly dependent on scale (Barton & Bandis, 1982; Yuan & Harrison, 2004; Alejano & Alonso, 2005). The decaying process of the dilatancy angle in line with plasticity can be described by the following negative exponential expression through the plastic shear strain, γ_p, or indirectly through the plastic shear displacement, δ, on the basis of Eq. (23):

$$\psi = \psi_{peak}\exp\left[-r(\delta - \delta_0)\right] \tag{25}$$

where r is a parameter for modelling the rate of decay that ψ undergoes as the plastic shear strain evolves. If r=0, then a constant dilatancy angle is recovered. As r→∞, the dilatancy angle quickly decays to zero. ψ_{peak} is the peak dilatancy angle of the fracture in the form of (Barton & Bandis, 1982)

$$\psi_{peak} = JRC \cdot \log_{10}\frac{JCS}{-\sigma'_n} \tag{26}$$

where JRC and JCS are the roughness coefficient and the wall compressive strength of fractures, respectively, and the actual values of them should be scale-corrected (Barton & Bandis, 1982). Thus, the dependencies of fracture dilatancy on plasticity, normal stress and scale are established through Eqs. (25) and (26).

Note that Eq. (25) shares the same shape with the asperity angle proposed for the description of shear dilatancy and surface degradation (Plesha, 1987), but the latter is represented as a function of the plastic tangential work. With the assumption of elasticperfectly plasticity, they are fully equivalent for monotonic loading (Jing et al., 1993). Cyclic loading is not a concern in this simple model, but when cyclic loading is involved, another independent function can be associated to the reverse loading that starts from the original point, just as the suggestion given in Plesha (1987) for asperity angles in two opposite directions, in order to satisfy the thermodynamic restriction condition presented in Jing et al. (1993).

Using the Mohr-Coulomb criteria, the peak shear stress τ_p of the fracture under effective normal stress σ'_n satisfies

$$\tau_p = -\sigma'_n \tan\varphi + c \tag{27}$$

where φ and c are the frictional angle and the cohesion of the fracture.

Differentiating Eq. (23) yields

$$d\gamma_p = d\gamma = \frac{1}{s}d\delta \tag{28}$$

Combining Eqs. (24) and (28) results in

$$\Delta b \approx s\varepsilon_n = \frac{\sigma'_n}{k_n} + \int_{\delta_0}^{\delta} \tan\psi(\delta)d\delta \tag{29}$$

An interesting phenomenon in Eq. (29) is, as described before, the change in the aperture of the fracture, Δb, is irrelevant to the height of the specimen, s. To conveniently use this formulation, two remedies can be further made:

First, suppose that the hyperbolic variation of k_n with the increase of aperture can be considered in the following (Huang et al., 2002):

$$k_n = \frac{-\sigma'_n + b_0 k_{n0}}{b_0} \tag{30}$$

where k_{n0} is the initial normal stiffness of the fracture.

Second, by employing the Taylor series expansion (truncated at the third order term), $\tan\psi$ can be adequately approximated by $\psi + \psi^3/3$ in radians for a rather large ψ_{peak}, e.g. 30°.

From Eq. (29) and the above two remedies, we have

$$\Delta b = \chi b_0 \tag{31}$$

$$b = b_0 + \Delta b = (1 + \chi)b_0 \tag{32}$$

with the parameter, χ, in the following form

$$\chi = \frac{\sigma'_n}{-\sigma'_n + b_0 k_{n0}} + \frac{1}{b_0} \left\{ \frac{\psi_{peak}}{r} \left[1 - e^{-r(\delta - \delta_0)} \right] + \frac{\psi_{peak}^3}{9r} \left[1 - e^{-3r(\delta - \delta_0)} \right] \right\}$$

(33)

Strain-Dependent Hydraulic Conductivity for Rock Fractures

Rewrite from Eq. (22) the initial hydraulic conductivity of the fracture, k_0, in the following form:

$$k_0 = \varsigma \frac{g b_0^2}{\nu}$$

(34)

Then, the hydraulic conductivity of the fracture under effective normal stress σ'_n and shear displacement δ can be described by

$$k = \varsigma \frac{g b^2}{\nu} = k_0 (1 + \chi)^2$$

(35)

Hence, a theoretical model of the hydraulic conductivity for a single rock fracture is finally formulated, which is totally determined by the effective normal stress σ'_n and the shear displacement δ, as well as a set of parameters characterizing the behavior of the fracture (i.e. b_0, ς, k_{n0}, k_{s0}, φ, c, JRC, JCS and r, which all can be deduced or back-calculated from experimental data).

Note that by Eqs. (35) and (33), the proposed hydraulic conductivity model for rock fractures subjected to normal and shear loadings with mobilized dilatancy behavior depends in form on the plastic shear displacement, but from Eq. (23), one observes that the model depends indirectly on the plastic shear strain. Thus, we classify the established model into the stain-dependent hydraulic conductivity model.

Validation of the Proposed Model

Esaki et al. (1999) systematically investigated the coupled effect of shear deformation and dilatancy on hydraulic conductivity of rock fractures by developing a new laboratory technique for coupled shear-flow tests of rock fractures. In this section, we validate the theory proposed in Section 3.2 using the experimental data reported in Esaki et al. (1999). For this purpose, we first briefly introduce the experiments, and then predict our analytical results through Eqs. (31) and (35) by directly comparing with the experimental data.

The Coupled Shear-Flow Tests

The coupled shear-flow tests were conducted with an artificially created granite fracture sample under various constant normal loads and up to a residual shear displacement of 20 mm (Esaki et al., 1999). The underlying specimen for coupled shear-flow tests is sketched in Fig. 3, with its size of 120 mm in length, 100 mm in width and 80 mm in height. The initial aperture of the created fracture, b_0, is about 0.15 mm. The value of JRC is 9, and the value of JCS is 162 MPa, respectively.

The coupled shear-flow tests were conducted by first applying a prescribed normal stress ranging between 1 MPa and 20 MPa and then applying shear displacement in steps at a rate of 0.1 mm/s until a maximum shear displacement of 20 mm was reached. During tests, steady-state fluid flow rate, shear loading and dilatancy were all continuously recorded. The hydraulic aperture and conductivity were back-calculated by applying the cubic law, with the flow equations solved by using a finite difference method.

Determination of the Parameters for the Proposed Model

Some of the experimental values of the mechanical parameters of the fracture specimen during the coupled shear-flow tests are listed in Table 2 (taken from Table 1 in Esaki et al. (1999)). Using the data as listed in Table 2, we plot the peak shear stress versus normal stress curve in Fig. 8, which can be fitted by a linear equation $\tau_p = 1.058\sigma_n + 0.993$ with a high correlation coefficient of 0.9999. Therefore, the shear strength of the specimen can be derived as $\varphi = 46.6°$ and $c = 0.99$ MPa, respectively.

Table 2. Mechanical parameters of the artificial fracture (After Esaki et al. (1999))

σ_n (MPa)	τ_p (MPa)	k_{s0} (MPa/mm)
1	2.06	3.37
5	6.16	10.65
10	11.74	11.97
20	22.10	17.97

The initial normal stiffness of the fracture of the specimen, k_{n0}, has to be estimated from the recorded initial normal displacement with zero shear displacement under different normal stresses. From the data plotted in Fig. 9 (which is taken from Fig. 7b in Esaki et al. (1999)), k_{n0} can be estimated as $k_{n0} = 100$ MPa/mm by considering the possible deformation of the intact rock under high normal stresses. It is to be noted that in the remainder of this section, the hard intact rock deformation of the small specimen is neglected, meaning that the normal displacement of the specimen mainly occurs in the fracture of the specimen and it is approximately equal to the increment of the mechanical aperture of the fracture. Theoretically, the decay coefficient of the fracture dilatancy angle, r, can be directly measured from the normal displacement versus shear displacement curves as plotted in Fig. 9. A better alternative, however, is to fit the experimental curves using Eq. (31) such that the least square error is minimized. By this approach, we obtain that r=0.13 with a correlation coefficient of 0.9538.

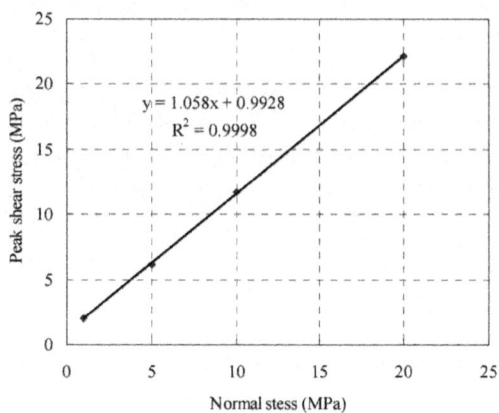

Figure 8. Peak shear stress versus normal stress curve of the fracture.

To obtain the dimensionless constant, ς, in Eq. (35) that relates the mechanical aperture to the hydraulic conductivity of the fracture under testing, further efforts are needed. A simple approach is to back-calculate ς directly using Eq. (34) with initial hydraulic conductivity, k0. But similarly, the better alternative is to fit the hydraulic conductivity versus shear displacement curves, as plotted in Fig. 11 (which is taken from Fig. 7c-f in Esaki et al. (1999)), using Eq. (35) such that the least square error is minimized. With such a method, we obtain that ς=0.00875. This means that the mechanical aperture, b, and the hydraulic aperture, b*, are linked with b*=0.324b, which is very close to the experimental result shown in Fig. 8 in Esaki et al. (1999).

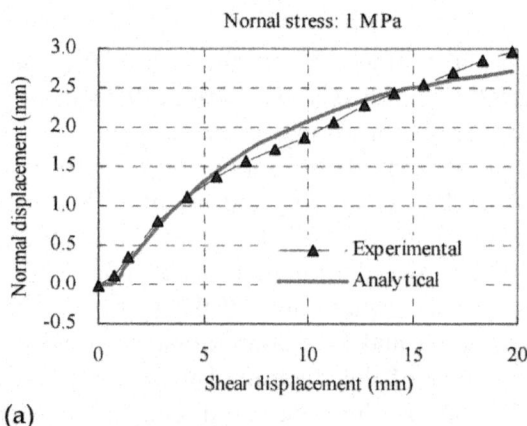

(a)

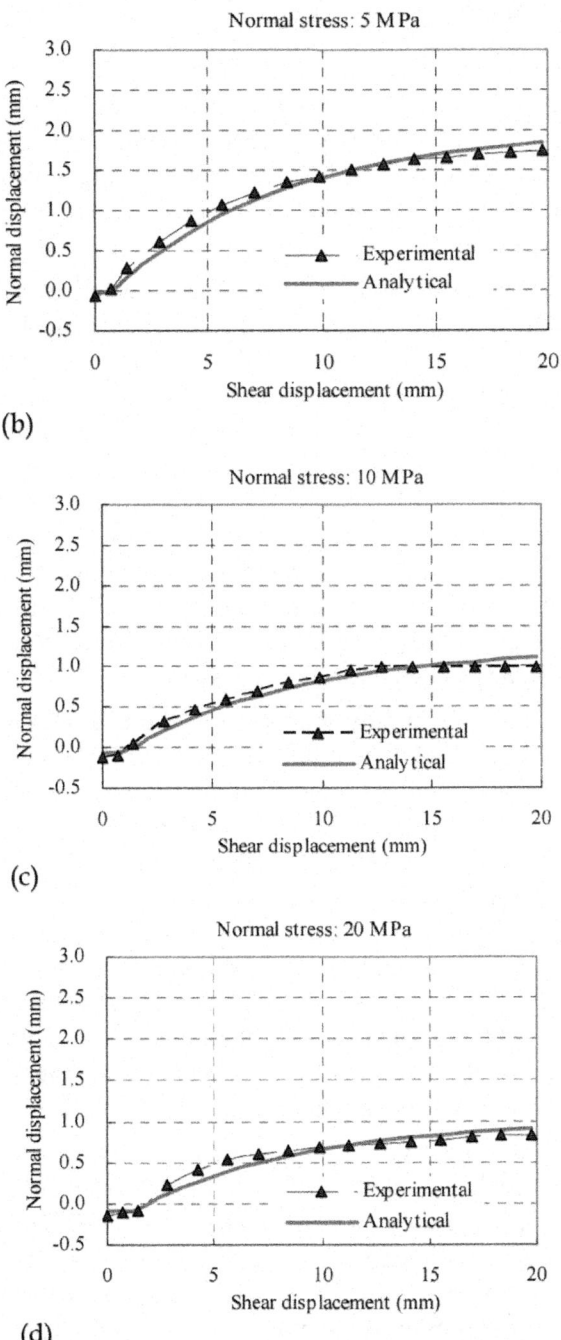

(b)

(c)

(d)

Figure 9. Comparison of the fracture aperture analytically predicted by Eq. (31) with that measured in coupled shear-flow tests.

Validation of the Proposed Theory

With the necessary parameters obtained in Section 3.3.2, we are now ready to compare the proposed model in Eqs. (31) and (35) with the experimental data presented in Esaki et al. (1999). Note that although the experimental data are available for one cycle of forward and reverse shearing, only the results for the forward shearing part are considered. The reverse shearing process, however, can be similarly modelled.

Fig. 9 depicts the relations between the mechanical aperture and shear displacement that were measured from the coupled shear-flow tests presented in Esaki et al. (1999) and predicted by using the proposed model given in Eq. (31) under different normal stresses applied during the testing. It can be observed from Fig. 9 that our proposed analytical model is able to describe the shear dilatancy behavior of a real fracture under wide range of normal stresses between 1 MPa and 20 MPa by feeding appropriate parameters. Even the fracture aperture increases by one order of magnitude due to shear dilation, the analytical model still fitted the experimental results well. For practical uses, the slight discrepancies between the analytical results and the experimental data are negligible and the proposed model is accurate enough to characterize the significant dilatancy behavior of a real fracture.

This performance is largely attributed to the dilatancy model introduced through Eqs. (25) and (26). The dilatancy angles of the fracture evolving with the plastic shear displacement under different normal stresses are illustrated in Fig. 10. The high dependencies of the dilatancy angle of the fracture on normal stress and plasticity are clearly demonstrated in the curves. The peak dilatancy angle, which can be rather accurately modelled by Barton's peak dilatancy relation (Barton & Bandis, 1982), decreases logarithmically with the increase of the applied normal stress. For normal stresses of 1 MPa, 5 MPa, 10 MPa and 20 MPa, the peak dilatancy angles are 19.9°, 13.6°, 10.9° and 8.2°, respectively. On the other hand, the dilatancy angle undergoes negative exponential decay with increasing plastic shear displacement, a process related to surface degradation of rough fractures.

Fig. 11 shows the hydraulic conductivity versus shear displacement relations that were back-calculated from fluid flow results using the finite difference method from the coupled shear-flow tests presented in Esaki et al. (1999) and that are predicted by the proposed model given in Eq. (35) under different normal stresses during testing. As shown in the semi-logarithmic graphs in Fig. 11, the proposed analytical model can well predict the evolution of hydraulic conductivity of the tested rock fracture, with the change in the magnitude of 2 orders, during coupled shear-flow tests under different normal stresses. The ratios of the predicted hydraulic conductivities to the corresponding experimental results all fall in between 0.3 and 3.0, indicating that they are rather close in orders of magnitude and the predicted results are suitable for practical use.

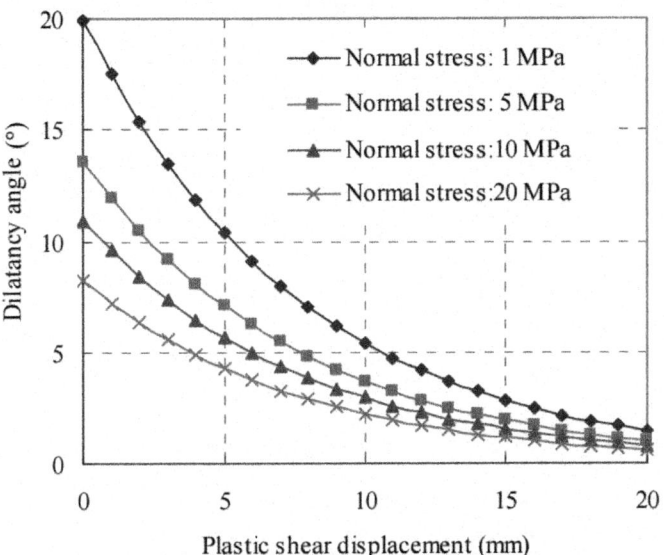

Figure 10. Dilatancy angles of the fracture evolving with the plastic shear displacement under different normal stresses.

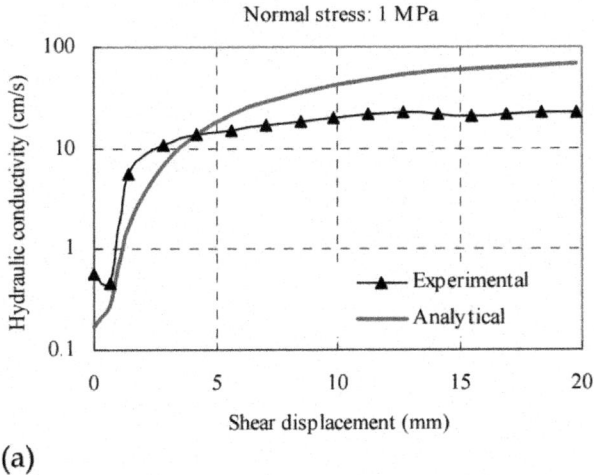

(a)

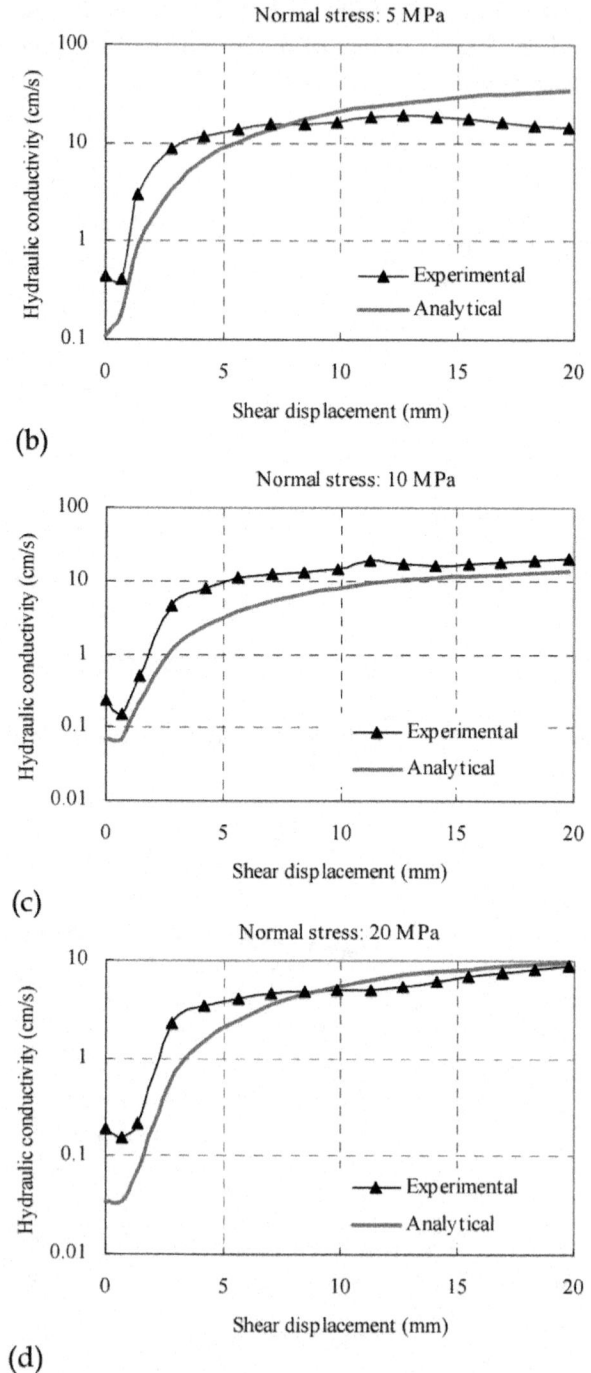

(b)

(c)

(d)

Figure 11. Comparison of the hydraulic conductivity analytically predicted by

Eq. (35) with that calculated from coupled shear-flow tests with finite difference method.

STRESS-DEPENDENT HYDRAULIC CONDUCTIVITY TENSOR OF FRACTURED ROCKS

When the response of each fracture under normal and shear loading is understood (see Section 2), the remaining problem is how to formulate the hydraulic conductivity for fractured rock mass based on the geometry of the underlying fracture network. Fig. 12 depicts a two-dimensional fracture network (taken after Min et al. (2004)) in a biaxial stress field. As shown in Fig. 12, each fracture plays a role in the hydraulic conductivity of the rock mass, and its contribution primarily depends on its stress state, its occurrence, as well as its connectivity with other fractures. Also shown in Fig. 12 is the scale effect of the rock mass on hydraulic properties. When the size of the rock mass is small, only a few number of fractures are included and heterogeneity of the hydraulic conductivity of the rock mass may dominate. As the population of factures grows with the increasing size, an upscaling scheme may be available to derive a representative hydraulic conductivity tensor for the rock mass at the macroscopic scale. Based on the above observations, in this section, we formulate an equivalent hydraulic conductivity tensor for fractured rock mass based on the superposition principle of liquid dissipation energy, in which the concept of REV is integrated and the applicability of an equivalent continuum approach is able to be validated.

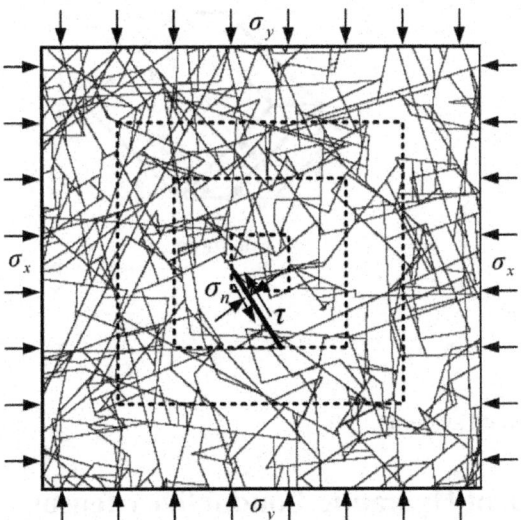

Figure 12. A fracture network (taken after Min et al. (2004)) in biaxial stress field and the scale effect of the rock mass.

Computational Model

Without loss of generality, the global coordinate system $X_1X_2X_3$ is established in such a way that its X_1-axis points towards the East, X_2-axis toward the North and X_3-axis vertically upward. A local coordinate system $x_1^f x_2^f x_3^f$ is associated with the *fth* set of fractures such that the x_1^f-axis is along the main dip direction, the x_2^f-axis is in the strike, and the x_3^f-axis is normal to the fractures, as shown in Fig. 13.

In order to formulate the stress-dependent hydraulic conductivity tensor for fractured rock masses using the aforementioned elastic constitutive model for rock fractures, the following assumptions, similar to Oda (1986), are made in this section:

- A cube of volume, V_p, is considered as the flow region of interest, which is cut by n sets of fractures. The orientation of each set of fractures is indicated by a mean azimuth angle β and a mean dip angleα. Other geometrical statistics of the fractures are assumed to be available through field measurements or empirical estimations.

- Even though the geometry of real fractures is complex, generally it can be simplified as a thin interfacial layer with radius r and aperture b*.

- The rock mass is regarded as an equivalent continuum medium, which means the representative elementary volume (REV) exists in the rock mass and its size is smaller than or equal to V_p.

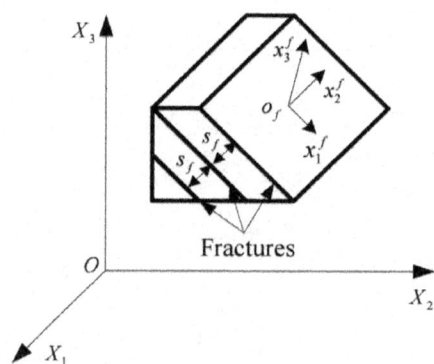

Figure 13. Coordinate systems.

Stress-Dependent Hydraulic Conductivity Tensor

Fluid flow through the equivalent continuum media can be described by the generalized 3- D Darcy's law as follows:

$$v = \mathbf{KJ} \tag{36}$$

where v denotes the vector of flow velocities, J denotes the vector of hydraulic gradients, and K is the hydraulic conductivity tensor for the rock mass.

For steady-state seepage flow, the dissipation energy density, $e(X_1, X_2, X_3)$, of fluid flow through the media can be represented as (Indelman & Dagan, 1993):

$$e = \frac{1}{2}\mathbf{J}^\mathrm{T}\mathbf{KJ} \tag{37}$$

Hence, the total flow dissipation energy, E, in the rock mass V_p can be calculated by performing an integration throughout the whole flow domain:

$$E = \int_{V_p} e\,d\Omega = \frac{1}{2}\int_{V_p} \mathbf{J}^\mathrm{T}\mathbf{KJ}\,d\Omega \tag{38}$$

If REV does exist in the rock mass and its size is smaller than or equal to V_p, by defining $\overline{\mathbf{J}}$ to be the vector of the average hydraulic gradient within V_p and $\overline{\mathbf{K}}$ to be the average hydraulic conductivity tensor, Eq. (38) can be reduced to:

$$E = \frac{1}{2}\overline{\mathbf{J}}^\mathrm{T}\overline{\mathbf{K}}\overline{\mathbf{J}}V_p \tag{39}$$

Suppose that the volume density of the ith set of fractures is J_{vi}. The number of this set of fractures can be estimated by $m_i = J_{vi} V_p$.

For permeable rock matrix, the flow dissipation energy shown in Eq. (39) consists of two components, i.e., the flow dissipation energy through rock matrix, E_r, and the flow dissipation energy through crack network, E_c:

$$E = E_\mathbf{r} + E_\mathbf{c} \tag{40}$$

E_r can be represented as:

$$E_\mathbf{r} = \frac{1}{2}\overline{\mathbf{J}}^\mathrm{T}\overline{\mathbf{K}}_\mathbf{r}\overline{\mathbf{J}}V_\mathbf{p} \tag{41}$$

where $\overline{\mathbf{K}}_\mathbf{r}$ denotes the hydraulic conductivity tensor for rock matrix. If rock matrix is impermeable, all elements in $\overline{\mathbf{K}}_\mathbf{r}$ vanish. To estimate E_c, we introduce a weight coefficient W_{ij} to describe the effect of the connectivity of the fracture network on fluid flow:

$$W_{ij} = \xi_{ij} / \overline{\xi}_i \tag{42}$$

where ξ_{ij} is a stochastic variable denoting the number of fractures intersected by the jth fracture belonging to the ith set; and $\bar{\xi}_i$ denotes the maximum number of fractures cut by the ith set of fractures. Obviously, $0 \le W_{ij} \le 1$ and when $\xi_{ij} = 0$, $W_{ij} = 0$. This implies that an entirely isolated fracture which does not intersect any other fracture effectively contributes nothing to the hydraulic conductivity of the total rock mass.

For the jth fracture belonging to the ith set, a void volume equal to $\pi r_{ij}^2 b_{ij}^*$ is associated with it. Then, the flow dissipation energy through it is described as:

$$E_{cij} = W_{ij} e_{ij} \pi r_{ij}^2 b_{ij}^*$$

(43)

where e_{ij} is shown as follows:

$$e_{ij} = \frac{1}{2} k_{ij} \bar{J}_{ci}^T \bar{J}_{ci}$$

(44)

where k_{ij} denotes the hydraulic conductivity of the jth fracture of the ith set, which can be calculated by the stress-dependent hydraulic conductivity model, Eq. (21).

$\bar{J}_{ci}$ denotes the hydraulic gradient within the ith set of fractures:

$$\bar{J}_{ci} = (\delta - \mathbf{n}_i \otimes \mathbf{n}_i) \bar{J}$$

(45)

where δ is the Kronecker delta tensor, and $\mathbf{n}_i$ denotes the unit vector normal to the ith set of fractures, with its components $n_1 = \sin\alpha\sin\beta$, $n_2 = \sin\alpha\cos\beta$, and $n_3 = \cos\alpha$.

Thus, E_c can be represented as

$$E_c = \frac{8\pi}{12\nu} \sum_{i=1}^{n} \sum_{j=1}^{m_i} W_{ij} r_{ij}^2 b_{ij}^{*3} \bar{J}^T (\delta - \mathbf{n}_i \otimes \mathbf{n}_i) \bar{J}$$

(46)

From Eqs. (39)-(41), (46) and (20), it can be referred that

$$\bar{K} = \bar{K}_r + \frac{8\pi}{12\nu V_p} \sum_{i=1}^{n} \sum_{j=1}^{m_i} W_{ij} f^3 (\beta_{ij}) r_{ij}^2 b_{0ij}^3 (\delta - \mathbf{n}_i \otimes \mathbf{n}_i)$$

(47)

In Eq. (47), n is determined by the orientation of the fractures, which reflects the effect of the orientation of the fractures on the fluid flow. r and b_0 represent the size or the scale of the fractures; they retrain the fluid flow through the fractures from their developing magnitude. W is a parameter introduced to show the impact of the connectivity of the fracture network on fluid flow. Finally, $f(\beta)$ is a function used to demonstrate the coupling effect between fluid flow and stress state.

The hydraulic tensor for fractured rock masses given in Eq. (47) is related to the volume of the flow region, V_p, which exactly shows the size effect of the hydraulic properties. Intuitively, the smaller the V_p size is, the less number of fractures is contained within the volume, and thus the poorer the representative of the computed hydraulic conductivity tensor. On the other hand, when V_p is increased up to a certain value, the fractures involved in the cubic volume are dense enough and the hydraulic conductivity tensor for the rock mass does not vary with the size of the volume. This V_p size is exactly the representative elementary volume, REV, of the flow region. The V_p size of the flow region is required to be larger than REV for estimating the hydraulic conductivity tensor for the fractured rock mass. Otherwise, treating the fractured rock mass as an equivalent continuum medium is not appropriate, and the discrete fracture flow approach is preferable.

Comparison with Snow's and Oda's Models

Now we make a comparison between the formulation of the hydraulic conductivity tensor presented in Eq. (47) and the formulation given by Snow (1969) as well as the formulation given by Oda (1986). The Snow's formulation is as follows:

$$\mathbf{K} = \frac{g}{12v} \sum_{i=1}^{n} \frac{b_i^3}{s_i} (\boldsymbol{\delta} - \mathbf{n}_i \otimes \mathbf{n}_i)$$

(48)

where s_i is the average spacing of the ith set of fractures. If we neglect the hydraulic conductivity of the rock matrix and the connectivity of the factures, and define

$$b_i = \frac{1}{m_i} \sum_{j=1}^{m_i} f(\beta_{ij}) b_{0ij} \quad \text{and} \quad s_i^{-1} = \frac{\pi}{V_p} \sum_{j=1}^{m_i} r_{ij}^2$$

(49)

Then, the formulation presented in Eq. (47) is totally equivalent to Snow's formulation, Eq. (48).

On the other hand, the Oda's formulation is described by

$$\mathbf{K} = \varsigma (P_{kk} \boldsymbol{\delta} - \mathbf{P})$$

(50)

where P is the fracture geometry tensor, with $P_{kk} = P_{11} + P_{22} + P_{33}$.

$$\mathbf{P} = \pi\rho\int_0^\infty \int_0^\infty \int_\Omega r^2 b^3 \mathbf{n} \otimes \mathbf{n} E(n,r,b) d\Omega dr db$$

(51)

where $E(n, r, b)$ is a probability density function of the geometry of the fractures, ρ is the number of fracture centers per unit of volume, with $\rho = m_v/V_p$, $m_v = \sum m_i$, and ς is the dimensionless scalar adopted to penalize the permeability of real fractures with roughness and asperities. Assuming that a statistically valid REV exists and being aware that the fracture orientation is a discrete event, the fracture geometry tensor may be empirically constructed by the following direct summation

$$\mathbf{P} = \frac{\pi}{V_p}\sum_{i=1}^{m_v} r_i^2 b_i^3 \mathbf{n}_i \otimes \mathbf{n}_i$$

(52)

Following a similar deduction, it can be inferred that all these three formulations are equivalent not only in form but also in function, though they are derived from different approaches and different assumptions. The formulation presented in Eq. (47) can be directly obtained from Snow's formulation by considering the connectivity and roughness of the fractures and integrating the aperture changes under engineering disturbance. The discretized form of the Oda's formulation is much closer to the current formulation, and the latter can also be directly achieved from the former by considering the connectivity of the fracture network. However, the proposed method for formulating an equivalent hydraulic conductivity tensor for complex rock mass based on the superposition principle of liquid dissipation energy is a widely applicable approach not only to equivalent continuum but also to discrete medium.

A Numerical Example: Hydraulic Conductivity of the Rock Mass in the Laxiwa Hydropower Project

In order to validate the theoretical model presented in Section 4.2, we investigated the hydraulic conductivity of a fractured rock mass at the construction site of the Laxiwa Hydropower Project, the second largest hydropower project on the upstream of the Yellow River. The selected construction site for a double curvature arch dam is a V-shaped valley formed by granite rocks, as shown in Fig. 14. The dam height is 250 m, the top elevation of the dam is 2460 m, the reservoir storage capacity is 1.06 billion m^3 and the total installed capacity is 4200 MW. A typical section of the Laxiwa dam site is illustrated in Fig. 15. Besides faults, four sets of critically oriented fractures are developed in the rock mass at the construction site. The geological characteristics of the fractures are described by spacing, trace length, aperture, azimuth, dip angle, the joint roughness coefficient, JRC, of the fractures as well as the connectivity of the fracture network (i.e., the number of fractures intersected by one fracture). According to site investigation, the statistics (i.e., the averages and the mean squared deviations, as well as the distribution of the characteristics) of the fractured rock mass on the right bank of the valley are listed in Table 3.

Figure 14. Site photograph of the Laxiwa valley.

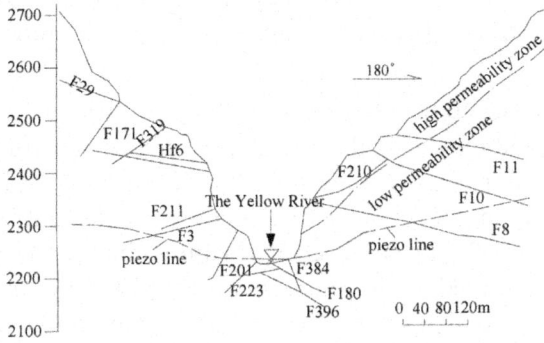

Figure 15. A typical section of the Laxiwa dam site.

Table 3. Characteristic variables of the fractured rock mass*

Set	Spacing (m)	Length (m)		Aperture (mm)		Azimuth (°)		Dip (°)		Connectivity	
		avg.	dev.	avg.	dev.	avg.	dev.	avg.	dev.	avg.	dev.
1	1.45	5	1.5	0.096	0.02	85.3	10	54.5	10	5	3
2	2.62	3	1.0	0.096	0.02	355.1	20	29.8	5	3	2
3	10.96	3	1.0	0.096	0.02	287.4	20	61.4	10	3	2
4	10.96	3	1.0	0.096	0.02	320.2	20	11.9	5	3	2
Distribution	logarithmic normal	negative exponential		Gama		normal		normal		normal	

*'avg.' denotes arithmetic mean of a variable,
'dev.' represents root mean squared deviation

At the construction site of the Laxiwa dam, a total number of 1450 single-hole packer tests were conducted to measure the hydraulic properties of the rock mass, with 113 packer tests for the shallow rock mass on the right bank in 0−80 m horizontal depth and 278 packer tests for the deeper rock mass. The measurements of the hydraulic conductivity range from 10^{-5} cm/s to 10^{-6} cm/s for the shallow rock mass and from 10^{-6} cm/s to 10^{-7} cm/s for the deeper rock mass, with in average 4.94×10^{-5} cm/s for the former and 3.80×10^{-6} cm/s for the latter, respectively (Liu, 1996). On the other hand, in-situ stress tests showed that the geostress in the base of the valley and in deep rock mass has a magnitude of 20−60 MPa, with the direction of the major principal stress pointing towards NNE. As a result of stress release, the release fractures are frequently developed and a high permeability zone of 0−80 m horizontal depth is formed in the bank slope, as shown in Fig. 15. The stress release fractures, however, become infrequent in deeper rock mass, and the measured hydraulic conductivity is generally 1−2 orders of magnitude smaller than the hydraulic conductivity of the rock mass in shallow depth away from the bank slope. Therefore, the hydraulic conductivity of the rock mass at the construction site of the Laxiwa arch dam is mainly controlled by the fracture network and the stress state.

Based on these statistics given in Table 3, fracture networks can be generated and calibrated for the rock mass at the construction site of the Laxiwa Hydropower Project using the MonteCarlo method by assuming that each fracture is a smooth, planar disc, with its center uniformly distributed in the simulated area. For each set of fractures, the geometrical parameters of any one are sampled by Monte-Carlo method until enough fractures are included in the simulated area. Then, a calibration procedure is invoked to check whether the generated model satisfies the distribution mode of the real fracture network. If doesn't, the fracture network will be regenerated until one matches the distribution mode. With the generated fracture network, the actual connectivity can be computed by spatial operation on the fractures. But for calibrated fracture network, a more convenient approximate approach to determine the connectivity of the fracture network, as it is adopted here, is to directly produce ξ_{ij} in Eq. (42) with the Monte-Carlo method and the characteristics presented in Table 3, then W_{ij} is derived from Eq. (42) with $\bar{\xi}_i$, the maximum number of fractures cut by the ith set of fractures. Field measurements are used to estimate $\bar{\xi}_i$, with $\bar{\xi}_1 = 11$, $\bar{\xi}_2 = 8$ and $\bar{\xi}_3 = \bar{\xi}_4 = 6$ for the four sets of fractures, respectively. Fig. 16 illustrates a simulated fracture network with size of $20 \times 20 \times 20$ m.

On the basis of the fracture network generated above, we compute the hydraulic conductivity tensor for the simulated cubic volume of rock mass with size of $20 \times 20 \times 20$ m using the method given by Snow (1969) and the method presented in Section 4.2, respectively. To show the coupling effect of stress/deformation on hydraulic properties, we consider two scenarios for examination. In the first scenario, we consider the fracture network located in the shallow depth away from the bank slope, where the impact of the in-situ stress is negligible. While in the second scenario, the fracture network is situated in larger depth, and a typical stress state with $\sigma_x = \sigma_z = 10$ MPa and $\sigma_y = 20$ MPa is associated with it. Based on laboratory test results, the shear modulus of the fractures is estimated as $\mu = 2$ MPa, and then by taking the Poisson's ratio as $\nu = 0.25$, the Lame's constant is derived with $\lambda = 2$ MPa. The kinematic viscosity of underground water is set to be $v_w = 1.14 \times 10^{-6}$ m2/s and the frictional angle-like parameter and the normal stress-like parameter are taken as $\varphi = 0.4363$ and $s = \sigma_n / 20$.

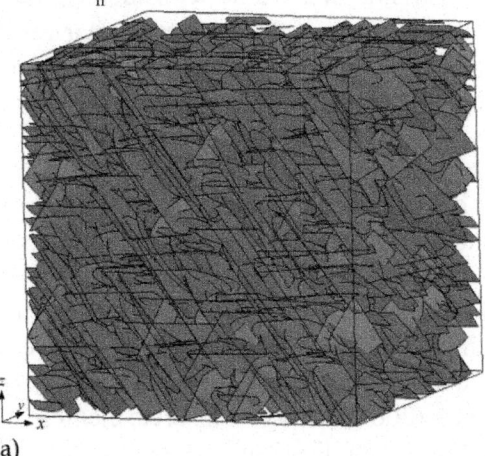

(a)

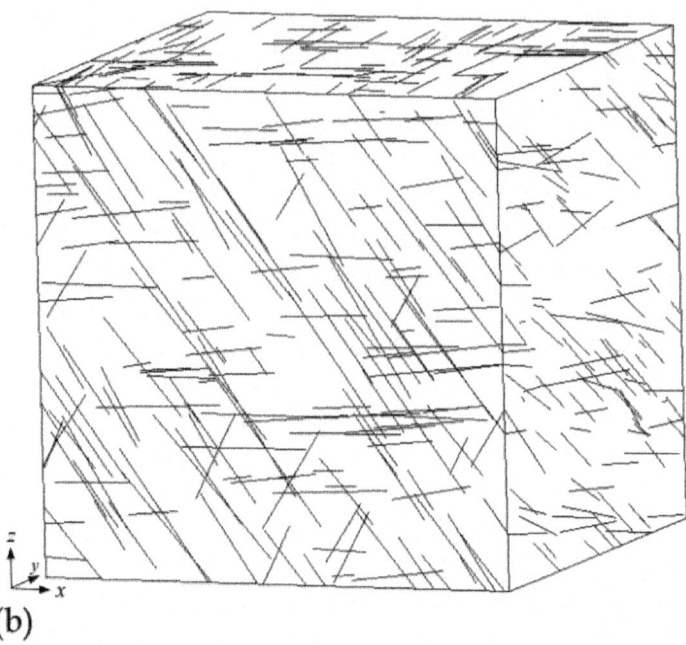

(b)

Figure 16. A three dimensional fracture network with size of 20×20×20 m generated by using the Monte-Carlo method for the rock mass in the Laxiwa Hydropower Project: (a) fracture network and (b) traces of the fractures on the surfaces of the simulated area.

The predicted hydraulic conductivity tensor for the examined rock mass is listed in Table 4. From Table 4, one observes that for shallow rock mass (where the effect of in-situ stress is not considered), the Snow's method and the method presented in Section 4.2 predict similar results and the predicted hydraulic conductivity is in the magnitude of 10^{-5} cm/s and close to in-situ hydraulic observations, but the anisotropy in hydraulic conductivity manifests due to non-uniform distribution of fractures. Compared with the hydraulic conductivity of the shallow rock mass, the predicted hydraulic conductivity for the rock mass in larger depth with the same fracture network decreases in 2 orders of magnitude due to the closure of the fractures applied by the in-situ stresses, but the anisotropic property of the hydraulic conductivity remains, which suggests that the occurrence of the fractures has a significant impact on permeability. Taking into consideration the applied stress level, the reduction of hydraulic conductivity in orders of magnitude is very close to the results achieved in Min et al. (2004) through a discrete element method, and generally agrees with the in-situ hydraulic observations.

Table 4. Predicted hydraulic conductivity tensor of the rock mass at the construction site of the Laxiwa dam (cm/s)

Snow's model (for shallow rock mass)		
4.78E–05	–4.76E–07	–1.71E–05
–4.76E–07	7.49E–05	–1.41E–05
–1.71E–05	–1.41E–05	4.08E–05
The proposed model (for shallow rock mass)		
1.93E–05	–1.75E–07	–6.39E–06
–1.75E–07	2.99E–05	–5.81E–06
–6.39E–06	–5.81E–06	1.64E–05
The proposed model (for deep rock mass)		
9.06E–08	–4.81E–09	–6.10E–08
–4.81E–09	1.85E–07	–1.92E–08
–6.10E–08	–1.92E–08	1.10E–07

Now, we take for example the rock mass in shallow depth to estimate the REV size of the rock mass. For this purpose, the scale of the rock mass is increased gradually from $3\times3\times3$ m to $40\times40\times40$ m with an increment of 1 m in each dimension. In each step, a fracture network with prescribed size is generated by using the Monte-Carlo method described above, and it is worth noting that this method is somewhat different from the method used by Min & Jing (2003) and Long et al. (1982). For each fracture network, the hydraulic conductivity tensor is calculated from Eq. (47) and then the principal hydraulic conductivities are further obtained from the hydraulic conductivity tensor. The relationship between the computed principal hydraulic conductivities and the sizes of the rock mass is illustrated in Fig. 17. As we can see from Fig. 17, when the block size of the rock mass is smaller than $18\times18\times18$ m, the population of fractures is not dense enough and the principal hydraulic conductivities fluctuate dramatically. On the other hand, as the size scales up to about $20\times20\times20$ m, the examined rock mass has included enough fractures and the computed principal hydraulic conductivities approach a rather steady state, with k_1, k_2, k_3 estimated to be 2.41×10^{-5} cm/s, 3.59×10^{-5} cm/s, 1.08×10^{-5} cm/s, respectively. This suggests that the REV does exist in the rock mass and the rock mass can be regarded as an equivalent continuum medium as long as its size is no less than, e.g., $20\times20\times20$ m or 8000 m^3.

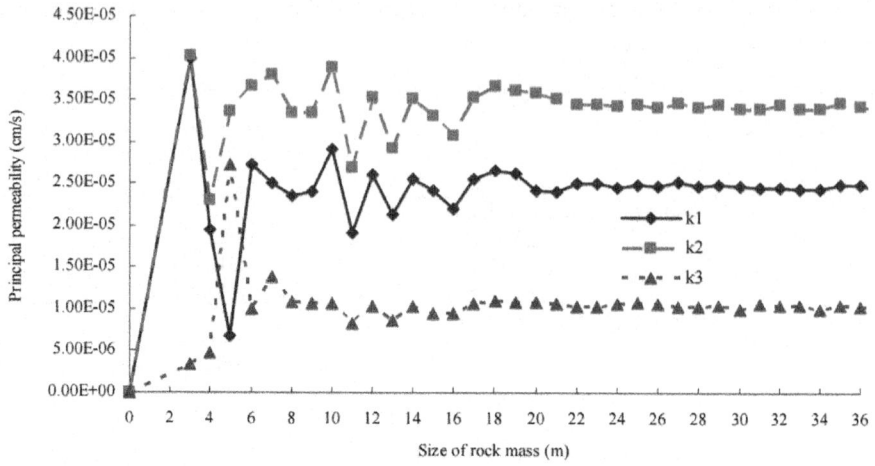

Figure 17. Hydraulic conductivity versus the volume size of the fractured rock mass.

STRAIN-DEPENDENT HYDRAULIC CONDUCTIVITY TENSOR OF FRACTURED ROCKS

On the basis of the strain-dependent model presented in Section 3 for rock fractures, this section formulates the strain-dependent hydraulic conductivity tensor for fractured rock masses cut by one or multiple sets of parallel fractures. The major difference between the model in this section and the stress-dependent model presented in Section 4 is that the former is capable of describing influence of the post-peak mechanical behaviors on the hydraulic properties of the rock masses, and is suited for modelling the coupled processes in rock masses at high stress level and in drastic engineering disturbance condition.

An Equivalent Elasto-Plastic Constitutive Model for Fractured Rocks

Consider a fractured rock mass cut by n sets of planar and parallel fractures of constant apertures with various orientations, scales and densities. The global response of the fractured rock mass under loading comes both from weak fractures and from stronger rock matrix. Based on this observation, an equivalent elasto-plastic constitutive model can be formulated by imposing assumptions on the interaction between fractures and rock matrix. The coordinate systems are defined in the same way with those defined in Section 4.1 (see Fig. 13). Denote the unit vector along X_i-axis of the global frame as e_i (i=1, 2, 3) and the unit vector along x_i^f-axis of the fth local frame as e_i^f (i=1, 2,

3). Then, a second order tensor, $l\,f$, can be defined for transforming physical quantities between the frames, with the components in the form of

$$l_{ij}^{f} = \mathbf{e}_{i}^{f} \cdot \mathbf{e}_{j}$$

(53)

Regarding the fractured rock mass as a continuous medium at the macroscopic scale, it is rational to assume that the global strain increment of the fractured rock mass is composed of the strain increments of rock matrix and fractures (Pande & Xiong, 1982; Chen & Egger, 1999), i.e.

$$d\varepsilon = d\varepsilon^{R} + \sum_{F} d\varepsilon^{F}$$

(54)

where $d\varepsilon$, $d\varepsilon^{R}$ and $d\varepsilon^{F}$ are the total incremental strain tensor, the incremental strain tensor of rock matrix and the incremental strain tensor of fth set of fractures measured in the global coordinate system, respectively. Note that a variable with a superscript in upper case (i.e. R or F) means that it is measured in the $X_1 X_2 X_3$ system, while a variable with a superscript in lower case (i.e. f) is measured in $x_1^f x_2^f x_3^f$ system, respectively. Unless otherwise specified, the superscripts F and f are not summing indices.

On the other hand, traction continuity has to be ensured across the fracture interfaces. In the global coordinate system, this condition can be strictly represented by (Pande & Xiong, 1982; Chen & Egger, 1999)

$$d\boldsymbol{\sigma} = d\boldsymbol{\sigma}'^{R} = d\boldsymbol{\sigma}'^{F}$$

(55)

where $d\sigma'$, $d\sigma'^{R}$ and $d\sigma'^{F}$ are the effective incremental stress tensor of the fractured rock mass, the effective incremental stress tensor of rock matrix and the effective incremental stress tensor of fth set of fractures, respectively. The effective stress tensor σ' is defined as

$$\boldsymbol{\sigma}' = \boldsymbol{\sigma} + \alpha p \boldsymbol{\delta}$$

(56)

where σ is the total stress tensor (positive for tension), p is the pore water pressure (positive for compressive pressure), and α ($\alpha \leq 1$) is an effective stress parameter.

Combining the plastic potential flow theory and the consistency conditions of rock matrix and fractures, an equivalent elasto-plastic constitutive model can be derived from Eqs. (54) and (55):

$$d\varepsilon = \mathbf{S}^{ep} : d\boldsymbol{\sigma}'$$

(57)

with

$$\mathbf{S}^{ep} = (\mathbf{C}^{R,ep})^{-1} + \sum_F (\mathbf{C}^{F,ep})^{-1}$$

(58)

where $\mathbf{S}^{ep}$ is the equivalent elasto-plastic compliance tensor of the fractured rock mass.

$\mathbf{C}^{R,ep}$ in Eq. (58) is the elasto-plastic modulus tensor of rock matrix. Neglecting the degradation of rock strength in the volume close to fracture intersections, $\mathbf{C}^{R,ep}$ can be written as

$$\mathbf{C}^{R,ep} = \mathbf{C}^R - \frac{\mathbf{C}^R : \dfrac{\partial Q_R}{\partial \boldsymbol{\sigma}'} \otimes \dfrac{\partial F_R}{\partial \boldsymbol{\sigma}'} : \mathbf{C}^R}{\dfrac{\partial F_R}{\partial \boldsymbol{\sigma}'} : \mathbf{C}^R : \dfrac{\partial Q_R}{\partial \boldsymbol{\sigma}'} + H_R}$$

(59)

where $\mathbf{C}^R$ is the fourth-order elastic modulus tensor of rock matrix, which can be represented in terms of the Lame's constants λ and μ:

$$C^R_{ijkl} = \lambda \delta_{ij} \delta_{kl} + \mu (\delta_{ik} \delta_{jl} + \delta_{il} \delta_{jk})$$

(60)

F_R, Q_R and H_R in Eq. (59) are the yield function, the plastic potential function and the hardening modulus of rock matrix, respectively. A non-associative flow rule with elasticperfectly plasticity (i.e. $H_R = 0$) is adopted for better modeling dilatant behavior of rock matrix by virtue of, for example, the Druker-Prager criterion with its cone fully inscribed by the Mohr-Coulomb hexagon, defined by functions

$$F_R = aI'_1 + \sqrt{J_2} - \kappa = 0$$

(61)

$$Q_R = \beta I'_1 + \sqrt{J_2}$$

(62)

with

$$\alpha = \sin \varphi_R / \sqrt{3(3 + \sin^2 \varphi_R)}$$

(63)

$$\kappa = 3c_R \cos \varphi_R / \sqrt{3(3 + \sin^2 \varphi_R)}$$

(64)

$$\beta = \sin\psi_R / \sqrt{3(3 + \sin^2\psi_R)} \qquad (65)$$

where c_R and φ_R are the cohesion and the friction angle of rock matrix, respectively. 1_1 and J_2 are the first invariant of the effective stress and the second invariant of the deviatoric stress of rock matrix, respectively. ψ_R is the mobilized dilatancy angle of rock matrix.

It should be noted here that in the literature, Drucker-Prager criterion has been used by many authors to model the elasto-plastic behavior of intact rock matrix, see Pande & Xiong (1982) and Chen & Egger (1999) for example. Although a modified Drucker-Prager yield function may be more suitable for this formulation in order to model plastic deformation properties of intact rock such as pressure dependency, strain hardening, transition from compressibility to dilatancy and stress path dependency (Chiarelli et al., 2003), the criterion given above may keep the formulation compact and does not lose generality. Other yield functions, such as the modified Drucker-Prager criterion (Chiarelli et al., 2003) or the modified Hoek-Brown criterion (Hoek et al., 1992), can also be integrated into the formulation without major mathematical difficulties.

With the researches conducted by Yuan & Harrison (2004) and Alejano & Alonso (2005), the decaying process of the rock dilatancy angle in line with plasticity can be described by the following negative exponential expression through the equivalent plastic strain of rock matrix, $\bar\varepsilon_R^p$ (Lai, 2002):

$$\psi_R = \psi_R^{peak}\exp(-r_R\bar\varepsilon_R^p) \qquad (66)$$

where $r_R \geq 0$ is a parameter for modelling the decaying process of the dilatancy angle, and ψ_R^{peak} is the peak dilatancy angle of rock matrix and the following expression has been proposed by recovering the shape of the peak dilatancy angle of fractures given by Barton & Bandis (1982) and by assuming $\psi_R^{peak} = \varphi_R$ for null confinement pressures (Alejano & Alonso, 2005):

$$\psi_R^{peak} = \frac{\varphi_R}{1 + \log_{10}\sigma_c}\log_{10}\frac{\sigma_c}{-\sigma_3' + 0.1} \qquad (67)$$

where σ_c is the unconfined compressive strength for intact rock. By Eqs. (66) and (67), the dependencies of rock dilatancy on plasticity, confining stress and scale are produced.

The equivalent plastic strain $\bar\varepsilon^p$ is computed by the following:

$$\bar{\varepsilon}^P = \int d\bar{\varepsilon}^P = \int \sqrt{\frac{2}{3} d\varepsilon^P : d\varepsilon^P}$$

(68)

Similarly, $C^{F,ep}$ in Eq. (58) is the elasto-plastic modulus tensor of fth set of fractures measured in the $X_1 X_2 X_3$ system, which can be calculated from its corresponding elastoplastic modulus tensor measured in the $x_1^f x_2^f x_3^f$ system, $C^{f,ep}$, with the assumption of small strain and by imposing the following tensor transformation:

$$C_{ijkl}^{F,ep} = l_{mi}^f l_{nj}^f l_{ok}^f l_{pl}^f C_{mnop}^{f,ep}$$

(69)

with

$$C^{f,ep} = C^f - \frac{C^f : \dfrac{\partial Q_f}{\partial \sigma'} \otimes \dfrac{\partial F_f}{\partial \sigma'} : C^f}{\dfrac{\partial F_f}{\partial \sigma'} : C^f : \dfrac{\partial Q_f}{\partial \sigma'} + H_f}$$

(70)

where Cf is the fourth-order tangential elastic modulus tensor of the fth set of fractures, with $C_{3333}^f = s_f k_{nf}$, $C_{2323}^f = C_{3131}^f = s_f k_{sf}$, and with all other elements equal to zero. The symbols k_{nf}, k_{sf} and s_f are the normal stiffness, the tangential stiffness and the spacing of the fth set of fractures, respectively. The expressions for the elements in Cf mean that the strain of fractures is evaluated over the fracture spacing, not over the fracture aperture, thus enabling the proposed model to consider the post-sliding plasticity of fractures and nonlinear variations of k_{nf} and k_{sf} with dilatancy caused by shear loading, without violating the small strain assumption.

F_f, Q_f and H_f in Eq. (70) are the yield function, the plastic potential function and the hardening modulus of the fth set of fractures, respectively. The elasto-plastic behavior of the fractures is treated in a similar fashion as that for the rock matrix, with a non-associative Mohr-Coulomb criterion:

$$F_f = \sqrt{\tau_{zxf}^2 + \tau_{zyf}^2} + \sigma_{zf}' \tan\varphi_f - c_f = 0$$

(71)

$$Q_f = \sqrt{\tau_{zxf}^2 + \tau_{zyf}^2} + \sigma_{zf}' \tan\psi_f$$

(72)

where σ_{zf}', τ_{zxf} and τ_{zyf} are the effective normal stress and the shear stresses on the fracture surfaces, respectively. c_f, φ_f and ψ_f are the cohesion, the friction angle and the mobilized dilatancy angle of the fth set of fractures, respectively. Similar to Eq. (66), ψ_f is also a shrinking function of the equivalent plastic strain of fractures $\overline{\varepsilon}_f^{\,p}$, and depends on normal stress and scale as well, in the following form:

$$\psi_f = \psi_f^{\text{peak}} \exp(-r_f \overline{\varepsilon}_f^{\,p})$$

(73)

where r_f is the decaying parameter and ψ_f^{peak} is the peak dilatancy angle of the fth set of fractures, respectively, with the latter calculated by Eq. (26).

Thus at any loading step, as long as the stress increment of the equivalent rock mass, $d\sigma'$, is obtained, the local strain pertinent to fth set of fractures can be derived as follows:

$$d\varepsilon^F = (C^{F,\text{ep}})^{-1} : d\sigma'$$

(74)

and

$$d\varepsilon_{ij}^f = l_{im}^f l_{jn}^f d\varepsilon_{mn}^F$$

(75)

The separation of the incremental strain of fractures from that of the rock mass through the proposed equivalent constitutive model plays a significant role in the present study. It enables the formulation of strain-dependent hydraulic conductivity that accounts for the mobilized dilatancy behavior, which will be demonstrated in the following section.

Strain-Dependent Hydraulic Conductivity Tensor for Fractured Rocks

Consider a domain of flow that has been discretized into several sub-domains according to rock quality classification. Suppose that each sub-domain contains n sets of fractures, with average initial aperture b_{f0} and spacing s_f for the fth set of fractures. Starting from Eq. (22) and using the averaging concept for the hydraulic conductivity over the whole sub-domain, the equivalent initial hydraulic conductivity of the fth set of fractures, k_{f0}, in the examined sub-domain can be represented as (Castillo, 1972; Liu et al., 1999)

$$k_{f0} = \varsigma \frac{g b_{f0}^3}{v s_f}$$

(76)

where ς, as pointed out before, is a dimensionless constant introduced to penalize the real water conducting capacity of natural fractures with rough walls, finite scales, asperity areas and filling materials. The validity of using a constant value of ς has been examined by Zhou et al. (2006).

Assuming that the change in spacing s_f during modeling is negligible, under normal and shear stress loadings we have

$$k_f = \varsigma \frac{g b_f^3}{v s_f} = \varsigma \frac{g(b_{f0} + \Delta b_f)^3}{v s_f}$$

(77)

where Δb_f and k_f are the increment of the aperture and the equivalent hydraulic conductivity of the fth set of fractures under loading, respectively. Suppose that strain localization (Lai, 2002; Vajdova, 2003) is not dominantly exhibited in the concerned fractures, it is approximately valid that

$$\Delta b_f = s_f \Delta \varepsilon_{zf}$$

(78)

where $\Delta \varepsilon_{zf}$ is the increment of the normal strain of the fth set of fractures, which can be directly obtained from Eq. (75).

Substituting Eq. (78) into Eq. (77) then yields

$$k_f = k_{f0} \left(1 + \frac{s_f}{b_{f0}} \Delta \varepsilon_{zf} \right)^3$$

(79)

Following the theory proposed by Snow (1969), a strain-dependent equivalent hydraulic conductivity tensor for fractured rock masses with n sets of fractures is represented by

$$\mathbf{K} = \sum_f k_f (\boldsymbol{\delta} - \mathbf{n}_f \otimes \mathbf{n}_f) = \sum_f k_{f0} \left(1 + \frac{s_f}{b_{f0}} \Delta \varepsilon_{zf} \right)^3 (\boldsymbol{\delta} - \mathbf{n}_f \otimes \mathbf{n}_f)$$

(80)

where K is the equivalent hydraulic conductivity tensor of the examined rock mass, and n_f is the unit vector normal to the fth set of fractures. The following significant implications can be observed from the formulation of K in Eq. (80):

- K is a cubic function of $\Delta\varepsilon_{zf}$, and any variation in ε_{zf} under loading will trigger the change in K, even in orders of magnitude. This exactly accounts for the coupling effect of mechanical loading (strain/stress) on hydraulic properties.

- K depends on incremental strains, rather than on stresses, which makes it possible to integrate various material nonlinearities in hydro-mechanical coupling analysis.

- In addition to cubic relation, the influence of $\Delta\varepsilon_{zf}$ on K is amplified by s_f / b_{f0}, indicating that K can be rather sensitive to b_{f0} and s_f. Therefore, techniques for estimating b_{f0} and s_f need to be carefully developed, on the basis of laboratory or in-situ hydraulic test data.

- The orientations of fractures possibly render K highly anisotropic, even if K is initially assumed isotropic, as has been systematically examined, e.g. by Liu et al. (1999).

- When implemented in a FEM code, a different K can be associated to each geological sub-domain or even to each element, as long as k_{f0}, b_{f0} and s_f for the sub-domains or elements can be estimated in advance.

- As a nature of the homogenized equivalent continuum approach, the size effect of fractures, especially the size-dependency of aperture, is not fully considered in the formulation of K for simplicity, even though it can be reflected to some degree through ς and scaled JRC and JCS values. The connectivity and the intersection effect of fractures, on the other hand, may have a more significant influence on K, but similarly, they cannot be properly considered in the equivalent continua without explicit representation of fractures. A rough remedy is to process the fracture system in such a way that only the connected fracture populations are included for conducting analyses.

To determine K of a fractured rock under any loading paths, a coupled hydro-mechanical process has to be invoked. With the assumption of incompressible rock matrix and fluid (e.g. groundwater), the governing equations for the coupled process of saturated fluid flow and deformation are given below as balance equation, geometric equation and fluid flow equation, respectively:

$$\sigma'_{ij,j} - \alpha p_{,i} + f_i = 0$$

$$(81)$$

$$\varepsilon_{ij} = \frac{1}{2}\left(u_{i,j} + u_{j,i}\right)$$
(82)

$$\frac{\partial}{\partial x_i}\left(k_{ij}\frac{\partial h}{\partial x_j}\right) = \frac{\partial \varepsilon_v}{\partial t}$$
(83)

where f_i and u_i are the components of the body force and displacement in the ith direction, $h = p/\gamma_w + z$ the water head, z the vertical coordinate, γ_w the unit weight of water, and ε_v the volume strain of the rock mass.

In the coupled process given above, mechanical loading or disturbance to the rock mass results in change in flow properties and flow behavior through Eqs. (80) and (83), while the change in flow behavior leads to change in mechanical response of the rock mass through Eq. (81). When the coupled process reaches a stable state, the solution to K is also available.

Now we briefly discuss how to determine k_{f0}, b_{f0} and s_f in Eq. (80) based on laboratory or insitu hydraulic test or site investigation data. Obviously, the initial hydraulic conductivity, k_{f0}, can be determined by in-situ hydraulic tests. Suppose the initial hydraulic conductivity tensor, K_0, is known through in-situ hydraulic test, as suggested by Hsieh & Neuman (1985), then K_0 can be rewritten, from Eq. (80), in the following form:

$$\mathbf{K}_0 = \sum_f k_{f0}(\boldsymbol{\delta} - \mathbf{n}_f \otimes \mathbf{n}_f)$$
(84)

By optimizing Eq. (84), k_{f0} (f=1, ..., n) can be estimated if the number of the sets of critically oriented fractures, n, is less than or equal to 6 (i.e. the number of the independent components of K_0), regardless K_0 is assumed to be isotropic or anisotropic.

The average spacing of the fth set of fractures, s_f, can be roughly estimated from the statistics of drill holes or scanlines. An alterative, however, is to use RQD (Rock Quality Designation) for determining s_f, as suggested by Liu et al. (1999), when the value of RQD for a specific rock mass is known a priori.

After the initial hydraulic conductivity, k_{f0}, and the average spacing, s_f, of the fractures are determined, the mean initial aperture of the fractures, b_{f0}, is ready to be back-calculated from Eq. (76).

Validation of the Proposed Model

Hydraulic Conductivity of the Surrounding Rock of a Circular Tunnel in the Stripa Mine

Here we compare the proposed method with results from a previous study as presented by Liu el al. (1999) by applying the method to an excavated circular tunnel with a biaxial stress field, σ_x and σ_z. The physical model is illustrated in Fig. 18, which is actually a manifestation of the reality of the Stripa mine in Sweden (Kelsall et al., 1984; Pusch, 1989). The following description about the tunnel is directly taken from Liu et al. (1999):

A Buffer Mass Test was conducted in Stripa Mine over the period 1981-1985 (Kelsall et al., 1984; Pusch, 1989) to measure the permeability of a large volume of low permeability fractured rock mass by monitoring water flow into a 33 m long section of the tunnel, as a large scale in-situ experiment for the research and development programs of underground geological disposal of nuclear wastes of the participating countries of the Stripa Project. The radius of the tunnel is about 2.5 m with two major sets of fractures striking obliquely to the tunnel axis, as shown in Fig. 18.

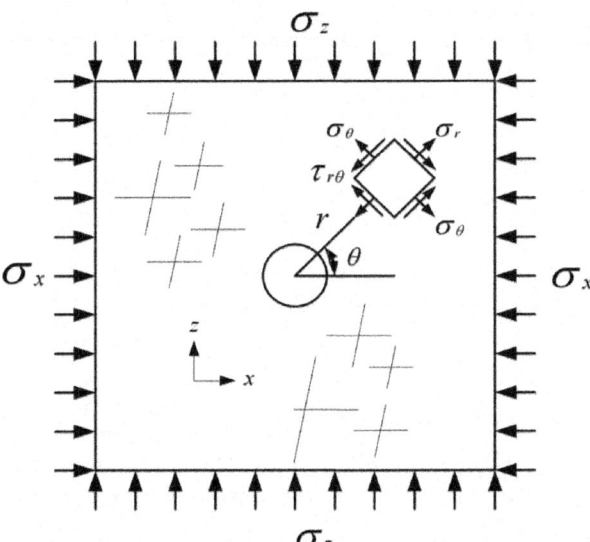

Figure 18. Sketch of a circular excavation in a biaxial stressed rock mass.

Fracture frequency measured in holes drilled from the tunnel was on average 4.5 fractures/m in inclined holes and 2.9 fractures/m in vertical holes. The initial stress field is anisotropic with high horizontal stress component and the conductivity of the virgin rock is about 10^{-10} m/s. The excavation of the test drift produced a dramatic increase in axial hydraulic conductivity in a narrow zone adjacent to the periphery of the drift. The conductivity increase is estimated to be 3 orders of magnitude.

The following assumptions are made in the calculations, with some of them similar to those in Liu et al. (1999):

- Statically uniform aperture and spacing distributions exist before excavation;
- Fracture spacing and continuity are not altered by the excavation;
- The high obliquity of the two major sets of fractures can be well approximated by two orthogonal sets of fractures;
- Excavation-induced strain redistribution may be adequately captured by the proposed equivalent elasto-plastic constitutive model.

Some of the parameters are directly taken from Liu et al. (1999), while other unavailable parameters are assumed, as listed in Table 5, in which the initial mechanical aperture of the fractures is back-calculated from Eq. (76) by taking $k_0 = 10^{-10}$ m/s. Consistent with Liu et al. (1999), the far-field stress components are taken as $\sigma_x = 20$ MPa and $\sigma_z = 10$ MPa, respectively.

Table 5. Geometrical and mechanical parameters for a circular tunnel

Category	Parameter	Setting
Intact rock matrix	Elastic modulus, E	37.5 GPa
	Poisson's ratio, ν	0.25
	Cohesion, c_R	5 MPa
	Friction angle, φ_R	46°
Fractures	Initial mechanical aperture, b_0	0.0075 mm
	Spacing, s	0.27 m
	Normal stiffness, k_n	200 GPa/m
	Shear stiffness, k_s	100 GPa/m
	Dimensionless constant, ς	0.0067
	Cohesion, c_f	0.4 MPa
	Friction angle, φ_f	40°

To avoid the difficulty in determining the initial dilatancy angles and the corresponding decay parameters of fractures and intact rock matrix, associative flow rule is used in this simulation. Again for simplicity, both the normal stiffness and the shear stiffness of the fractures are assumed constant during excavation. The finite element mesh of the model is shown in Fig. 19, and the FEM program was run to simulate the excavation effect of the tunnel. Fig. 20 shows the deformation zone and plastic zone of the rock mass after the tunnel excavation. Fig. 21 plots the excavation-induced changes in hydraulic conductivities around the circular tunnel, which are directly compared with the results presented in Liu et al. (1999).

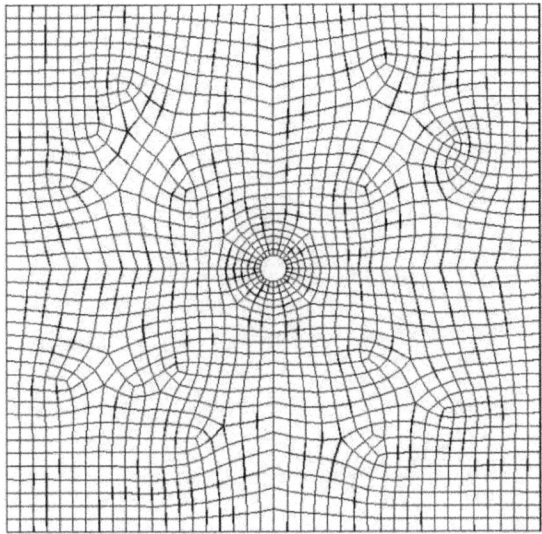

Figure 19. Finite element mesh for simulation of a tunnel excavation.

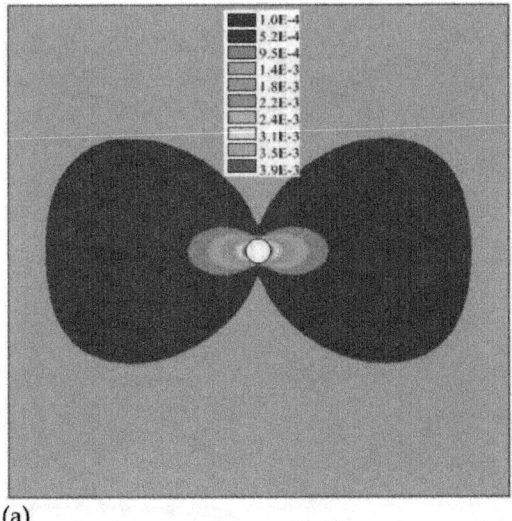

(a)

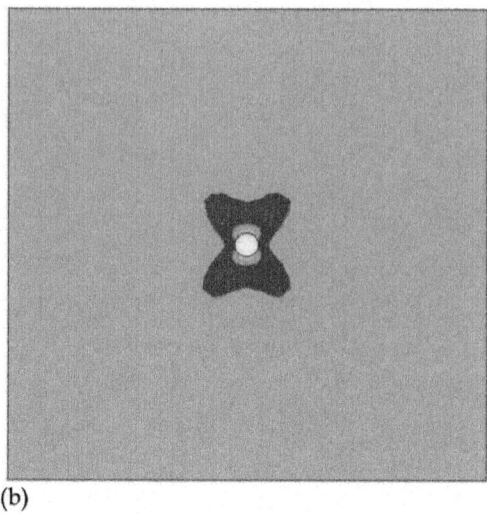

(b)

Figure 20. Deformation zone and plastic zone induced by the tunnel excavation: (a) deformation zone and (b) plastic zone.

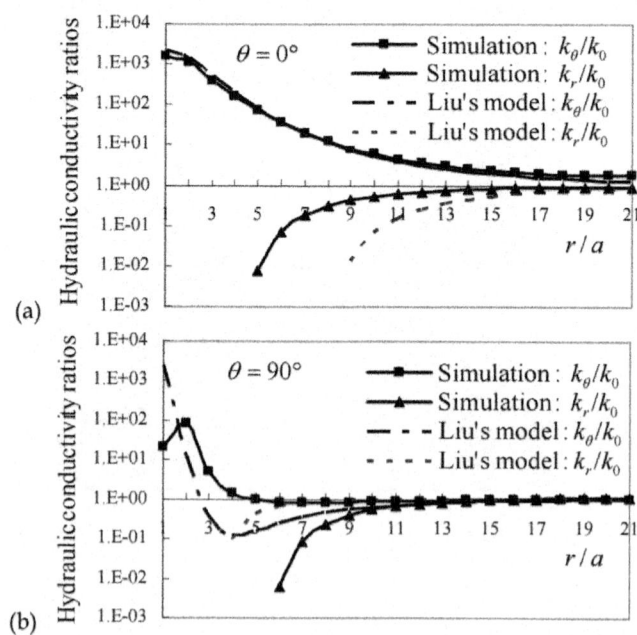

Figure 21. Excavation-induced hydraulic conductivity ratios around a circular tunnel in a biaxial stressed rock mass, where a is the radius of the tunnel and r is the distance away from the tunnel center. $\theta=0°$ denotes the horizontal direction while $\theta=90°$ the vertical direction.

It can be observed from Fig. 21 that generally tangential conductivities are found to increase greatly due to the formation of the excavation disturbed zone around the tunnel, while radial conductivities diminish greatly as a result of closure on related fractures. In the horizontal direction (i.e. $\theta=0°$), the excavation-induced tangential hydraulic conductivity ratios, k_θ/k_0, predicted by our model are very close to the results presented in Liu et al. (1999). For radial hydraulic conductivity ratios, k_r/k_0, however, deviation occurs in the vicinity of the excavation. Such a deviation is also found both for k_θ/k_0 and for k_r/k_0 in the vertical direction (i.e. $\theta=90°$).

Clearly, these deviations are largely resulted from the facts that (1) Different strain distribution patterns are assumed in the elastic model in Liu et al. (1999) and in our elastoplastic model; (2) Different methods are used to compute the strain increments of fractures. In Liu et al. (1999), normal strains of fractures were separated from rock matrix through a modulus reduction ratio empirically defined as a function of RMR, while in this simulation fracture strains were calculated by strain decomposition through an equivalent elasto-plastic constitutive model; (3) Radial and tangential fractures were assumed in Liu et al. (1999), leading to different background fracture networks; and (4) As mentioned above, some of the parameters, such as the shear strength of fractures and rock matrix, the shear stiffness and normal stiffness of the fractures, are unavailable in the literature (Kelsall et al., 1984; Pusch, 1989; Liu et al., 1999) and hence are empirically assumed in the calculations. If these parameters are determined based on in-situ or laboratory experiments, more convincing results may be achieved.

Despite the deviations, the trends of variation of the hydraulic conductivity ratios around the tunnel due to excavation are consistent between the two studies, and basically accord with the in-situ experimental observations, demonstrating the applicability of the present model in this section.

From Fig. 20, one observes that the excavation-induce deformation zone and plastic zone are asymmetric, due to the anisotropic initial stress field. As a result, the predicted hydraulic conductivities are highly anisotropic due to strain redistribution, as shown in Fig. 21. In the horizontal direction (i.e. $\theta=0°$), the deformation zone extends as far as more than 16 times of the tunnel radius and the plastic zone extends 2 times of the tunnel radius, while in the vertical direction (i.e. $\theta=90°$), they are, respectively, within 2 and 5 times of the tunnel radius. The asymmetry of deformation zone and plastic zone demonstrates why the predicted hydraulic conductivities approach k_0 more slowly in the horizontal direction than in the vertical direction. The changes in hydraulic conductivities resulted from strain redistribution in the disturbed rock mass indicate that a different hydraulic conductivity tensor should be associated to each geological sub-domain or even each element of the rock mass, which is important for hydro-mechanical coupling analyses.

Hydraulic Conductivity of a Cubic Block of Rock Mass with Three Orthogonal Sets of Identical Fractures

In this section, a numerical simulation is conducted to evaluate hydraulic behavior of a cubic block of rock mass containing three orthogonal sets of identical fractures under isotropic triaxial compression and shear loading. The primary goal is to investigate the change in the hydraulic conductivity of the rock mass with increasing shear load, which is obviously not achievable through any elastic models considering only the deformation of fractures under normal stresses, e.g. in Liu et al. (1999).

The underlying rock mass block model for examination, with a size of $10 \times 10 \times 10$ m (a scale that can represent both the initial mechanical and hydraulic REVs (Min et al., 2004)), is assumed to contain three orthogonal sets of identical fractures, as sketched in Fig. 22. The spacing, s, of each set of fractures and the initial aperture, b_0, of each fracture are assumed to be identical, with s=1 m and b_0=1 mm. The mechanical properties of each fracture are also regarded identical and for simplicity, both the normal stiffness and the shear stiffness of the fractures are assumed to be constant during shear loading. All parameters used in this simulation are listed in Table 6, and such parameter settings enable us to demonstrate how the hydraulic conductivity evolves from initial isotropy to anisotropy in the shearing process.

The examined rock mass block model is divided into 1000 brick elements, and the resultant mesh is shown in Fig. 22. The loading condition is as follows. First, triaxial compressive stresses are applied on the surfaces of the cubic block, with $\sigma_x = \sigma_y = \sigma_z = 20$ MPa. Then, a shearing load , τ, is applied on the upper and lower surfaces of the block model step by step, increasing at an increment of 1 MPa until a maximum shear load, 20 MPa, is reached. At each step of shear loading, numerical divergence may occur. If numerical divergence does occur, the simulation program terminates after 1000 iterations with a modified NewtonRaphson method.

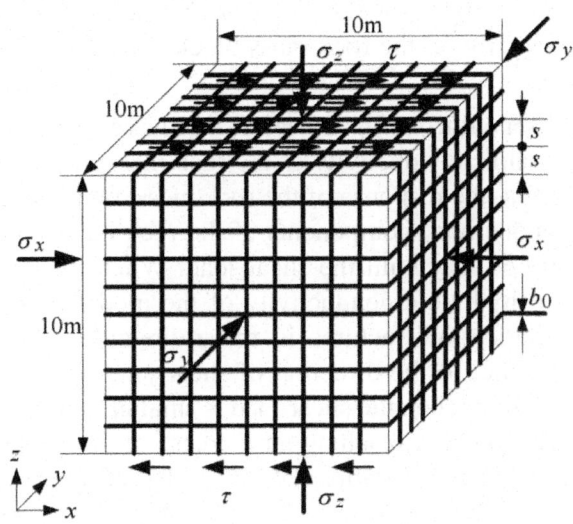

Figure 22. Sketch of a cubic block of rock mass with three orthogonal sets of identical fractures.

Table 6. Geometrical and mechanical parameters for a cubic block of fractured rock mass

Category	Parameter	Setting
Intact rock matrix	Elastic modulus, E	6 GPa
	Poisson's ratio, v	0.25
	Cohesion, c_R	1 MPa
	Friction angle, φ_R	46°
	Peak dilatancy angle, ψ_R^{peak}	35°
	Decay parameter of dilatancy, r_R	100
Fractures	Initial mechanical aperture, b_0	1 mm
	Spacing, s	1 m
	Normal stiffness, k_n	30 GPa/m
	Shear stiffness, k_s	10 GPa/m
	Dimensionless constant, ς	0.0067
	Cohesion, c_f	0.4 MPa
	Friction angle, φ_f	40°
	Peak dilatancy angle, ψ_f^{peak}	26°
	Decay parameter of dilatancy, r_f	100

Clearly, before the rock mass is loaded, its initial hydraulic properties are isotropic, with $k_{x0}=k_{y0}=k_{z0}=1.30\times10^{-2}$ cm/s by Eq. (84). Under the condition of isotropic compression, the rock mass remains elastic, the isotropic property of hydraulic conductivity is maintained, and the magnitude of the hydraulic conductivity reduces by 2 orders of magnitude due to compression of fractures, with $k_x=k_y=k_z=4.82\times10^{-4}$ cm/s by Eq. (80). When shear stress is added incrementally on the rock mass block model from 0 to 20 MPa, the proposed method

As can be observed from Fig. 23, shear load has a substantial impact on the evolution of hydraulic conductivity of the rock mass model. Before the shear load reaches 4 MPa, the response of the rock mass model remains elastic, and the hydraulic conductivity components of the rock mass model are basically identical and do not vary with the shear load. When the shear load exceeds 4 MPa, however, hydraulic conductivity of the model becomes anisotropic. Due to shear dilation of fractures in the z-direction, the major hydraulic conductivities parallel to the direction of shear load in x-y plane, k_x and k_y, increase mildly at first when the shear load is smaller than 8 MPa. Afterwards, they increase dramatically, reaching an increase of 3-4 orders of magnitude. They approach a relatively stable state after the shear load increases up to 14 MPa. Obviously, the increase of k_x and k_y is resulted from the dilatancy behavior of the fractures related to equivalent plastic strain, as shown in Fig. 24, where the mobilized dilatancy angle approaches zero as the shear load approaches 14 MPa. When the shear load exceeds 14 MPa, shear dilatancy of the related fractures becomes trivial and hence k_x and k_y become steady. From Table 7 and Fig. 23, we can further see that k_x and k_y are very close to each other in values and they generally have the same varying trend with the increasing shear load.

Table 7. Major hydraulic conductivities of a cubic block of rock mass under isotropic compression and increasing shear loading

τ (MPa)	k_x (cm/s)	k_y (cm/s)	k_z (cm/s)	τ (MPa)	k_x (cm/s)	k_y (cm/s)	k_z (cm/s)
-	0.013016	0.013016	0.013016	10	0.279373	0.279350	0.000020
0	0.000482	0.000482	0.000482	11	1.088835	1.088816	0.000056
1	0.000482	0.000482	0.000482	12	2.204162	2.204158	0.000375
2	0.000482	0.000482	0.000482	13	3.171558	3.171559	0.001374
3	0.000483	0.000483	0.000482	14	3.676801	3.697449	0.022811
4	0.000494	0.000486	0.000474	15	3.915193	4.137786	0.224877
5	0.000543	0.000509	0.000444	16	4.063688	4.696511	0.635383
6	0.000657	0.000576	0.000372	17	4.243447	5.407600	1.167070
7	0.000742	0.000643	0.000282	18	4.635512	6.233203	1.600997
8	0.000704	0.000581	0.000207	19	5.390907	7.316177	1.928768
9	0.012562	0.012459	0.000106	20	6.462514	8.618240	2.159053

With the increase of shear load from 4 to 20 MPa, the change in the major hydraulic conductivity vertical to the direction of shear load, k_z, is even more interesting. Before the shear load reaches 10 MPa, k_z decreases significantly with increasing shear load and manifests a shear contraction-like behavior. When the shear load further increases, shear dilatancy occurs and k_z increases drastically, with changes in as high as 4-5 orders of magnitude. k_z reaches a relatively stable state after the shear load increases up to 17 MPa, which is actually a critical loading point that numerical instability may occur.

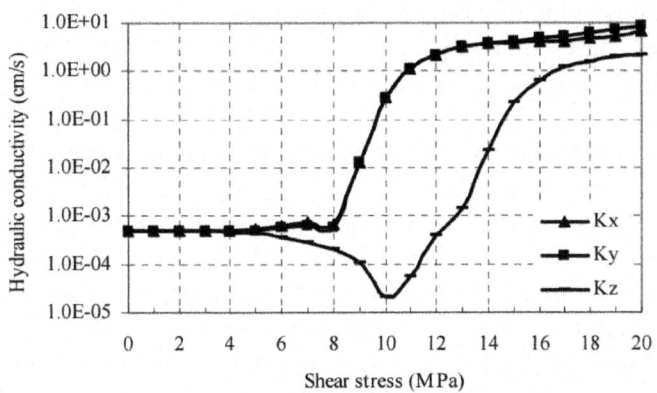

Figure 23. Major hydraulic conductivities of a cubic block of rock mass with increasing shear load.

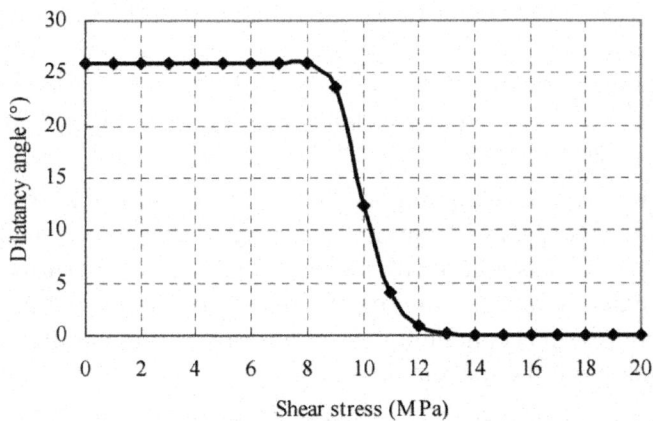

Figure 24. A typical case of mobilized dilatancy angle of a fracture with increasing shear load.

CONCLUSIONS

In this chapter, mathematical models were developed to estimate the hydraulic conductivity tensor for fractured rock masses subjected to mechanical loading or engineering disturbance. Emphases are placed on the investigation of the geological characteristics of rock masses as well as the coupling between fluid flow and stress/deformation, especially the effect of shear dilation or shear contraction on the hydraulic behavior of rock fractures.

The stress-dependent hydraulic conductivity tensor was formulated by using the superposition principle of flow dissipation energy on the basis of the concept of representative elementary volume (REV) and the assumption that rock masses can be treated as equivalent continuum media. The deformation behaviors of rock fractures subjected to normal and shear loadings are described with an elastic constitutive model, in which the pre-peak shear dilation or contraction of the fractures is empirically modelled. The validity of using the superposition principle of flow dissipation energy for development of the model is supported by the functional equivalence between the current formulation and the Snow's and Oda's models. This model is best suited for estimation of the hydraulic properties of rock masses at low stress level and with overall elastic response, and can be used to determine the applicability of the continuum approach to coupling analysis. The latter is achieved by performing numerical experiments to test the existence of the REV, and if exists, to further estimate the REV by gradually increasing the cubic volume of flow region, V_p, to see whether the hydraulic conductivity of the rock mass can eventually approach a steady point. The hydraulic properties and the REV size of the fractured rock mass at the construction site of the Laxiwa Hydropower Project were evaluated with the proposed model, and the calculation results were compared with the predictions of the Snow's model and validated by in-situ hydraulic tests, hence the feasibility of the proposed model in rock engineering practices is demonstrated.

The strain-dependent hydraulic conductivity tensor, on the other hand, was developed for disturbed rock masses under excavation or loading. In the model, a non-associative elasticperfectly plastic constitutive model was integrated to describe the deformation behaviors of the rock masses by characterizing them as equivalent continua containing one or multiple sets of parallel fractures. The clear advantages of the formulation are:

- The proposed hydraulic conductivity tensor is related to strains rather than stresses, hence enabling easier hydro-mechanical coupling analysis to include the effect of material nonlinearity of fractured rock masses.
- Beneficial from the equivalent non-associative elastic-perfectly plastic constitutive model, the hydraulic conductivity tensor considers the impact of shear dilatancy of fractures on fluid flow properties via mobilized dilatancy angles.
- When reduced to one dimensional case with a single fracture under normal and shear loadings, a closed-form solution to the hydraulic

conductivity can be obtained, enabling validation of the model by laboratory coupled shear-flow tests of rock fractures.

- The proposed model is easy to be implemented in a FEM code, particularly suitable for numerical analysis of coupled hydro-mechanical processes in rock engineering.

The closed-form solution was validated by an existing coupled shear-flow test, and the evaluation results show that the proposed solution can closely describe the hydraulic behavior of a hard rock fracture under a wide range of normal and shear loads. The results of the simulation conducted to predict the excavation-induced hydraulic conductivities around a circular tunnel in a biaxial stress field at the Stripa mine are justified by in-situ experimental observations and compared with an existing elastic strain-dependent model, which show that engineering disturbance such as underground excavations may dramatically alter the hydraulic conductivities of the rock mass surrounding the excavations and change the isotropic pattern of the initial hydraulic conductivities. The numerical simulation on a cubic block model of a rock mass with three orthogonal sets of identical fractures under isotropic triaxial compression and shear loading further demonstrates that shear loading may drastically change the hydraulic properties of fractured rocks, in the magnitude of as high as 4-5 orders, and lead to high anisotropy of the hydraulic properties.

Despite all these efforts, characterizing the hydraulic properties for fractured rock masses remains one of the most difficult research topics in rock mechanics. In the proposed models presented in this chapter, rock masses are assumed with rather regular distribution patterns of fractures, and the existence of a hydraulic conductivity tensor of the rock masses with any distribution of fractures is not discussed. The interaction between the fractures in the rock masses is also out of the scope of this chapter, and its effect on the hydraulic properties remains an open issue. Furthermore, the proposed models are established with a rather intuitive upscaling approach, and more rigorous homogenization schemes should be developed. All of there issues should be addressed in the future research.

ACKNOWLEDGEMENTS

The financial support from the National Natural Science Foundation of China (No. 51079107) and the National Natural Science Fund for Distinguished Young Scholars of China (No. 50725931), and the Program for New Century Excellent Talents in University (No. NCET-09-0610) for this study is gratefully acknowledged.

REFERENCES

1. Alejano, L. R. & Alonso, E. (2005). Consideration of the dilatancy angle in rocks and rock masses. International Journal of Rock Mechanics and Mining Sciences, Vol. 42, No. 4, 481–507

2. Barton, N. (1976). Rock mechanics review: the shear strength of rock and rock joints. International Journal of Rock Mechanics and Mining Sciences & Geomechanical abstracts, Vol. 13, No. 9, 255-279

3. Barton, N. R. & Bandis, S. C. (1982). Effects of block size on the shear behavior of jointed rocks. In: Proc 23rd US Symp Rock Mechanics, Berkeley

4. Barton, N.; Bandis, S. & Bakhtar, K. (1985). Strength, deformation and conductivity coupling of rock joints. International Journal of Rock Mechanics and Mining Sciences & Geomechanical abstracts, Vol. 22, No. 2, 121-140

5. Bear, J. (1972). Dynamics of fluids in porous media. American Elsevier, New York

6. Castillo, E. (1972). Mathematical model for two-dimensional percolation through fissured rock. In: Proc Int Symp Percolation through Fissured Rock, T1±D1–7, Stuttgart, Germany

7. Chen, Y. F.; Sheng, Y. Q. & Zhou, C. B. (2006). Strain-dependent permeability tensor for coupled M-H analysis of underground opening. Proceedings of the 4th Asian Rock Mechanics Symposium, pp. 271, Singapore, Nov 2006, World Scientific Publishing

8. Chen, Y. F.; Zhou, C. B. & Sheng, Y. Q. (2007). Formulation of strain-dependent hydraulic conductivity for fractured rock mass. International Journal of Rock Mechanics and Mining Sciences, Vol. 44, No. 7, 981-996

9. Chen, S. H. & Egger, P. (1999). Three dimensional elasto-viscoplastic finite element analysis of reinforced rock masses and its application. International Journal for Numerical and Analytical Methods in Geomechanics, Vol. 23, No. 1, 61–78

10. Chiarelli, A. S.; Shao, J. F. & Hoteit, N. (2003). Modeling of elastoplastic damage behavior of a claystone. Int J Plasticity, Vol. 19, 23–45

11. Esaki, T.; Du, S.; Mitani, Y.; Ikusada, K. & Jing, L. (1999). Development of a shear-flow test apparatus and determination of coupled properties for a single rock joint. International Journal of Rock Mechanics and Mining Sciences, Vol. 36, 641–50

12. Hoek, E.; Wood, D. & Shah, S. (1992). A modified Hoek-Brown criterion for jointed rock masses. In: Proc Rock Characterization Symp ISRM:

Eurock 92, Hudson, J. A. (Ed.), 209–214, British Geotechnical Society, London

13. Hsieh, P. A. & Neuman, S. P. (1985). Field determination of the three-dimensional hydraulic conductivity tensor of anisotropic media. Water Resource Research, Vol. 21, No. 11, 1655–1665.

14. Huang, T. H.; Chang, C. S. & Chao, C. Y. (2002). Experimental and mathematical modeling for fracture of rock joint with regular asperities. Eng Fract Mech, Vol. 69, 1977– 1996

15. Indelman, P. & Dagan, G. (1993). Upscaling of permeability of anisotropic heterogeneous formations. Water Resources Research, Vol. 29, No. 4, 917-923

16. Jing, L. (2003). A review of techniques, advances and outstanding issues in numerical modeling for rock mechanics and rock engineering. International Journal of Rock Mechanics and Mining Sciences, Vol. 40, No., 283-353

17. Jing, L.; Stephansson, O. & Nordlund, E. (1993). Study of rock joints under cyclic loading conditions. Rock Mechanics and Rock Engineering, Vol. 26, No. 3, 215–32

18. Kelsall, P. C.; Case, J. B. & Chabannes, C. R. (1984). Evaluation of excavation-induced changes in rock permeability. Int J Rock Mech Min Sci & Geomech Abstr, Vol. 21, No. 3, 123–35

19. Lai, T. Y. (2002). Multi-scale finite element modeling of strain localization in geomaterials with strong discontinuity. Ph.D. thesis, Stanford University

20. Liu, C. H.; Chen, C. X. & Fu, S. L. (2002). Testing study on seepage characteristics of single fracture with sand under shearing displacement. Chinese Journal of Rock Mechanics and Engineering, Vol. 21, No. 10, 1457-1461

21. Liu, J.; Elsworth, D. & Brady, B. H. (1999). Linking stress-dependent effective porosity and hydraulic conductivity fields to RMR. International Journal of Rock Mechanics and Mining Sciences, Vol. 36, 581-596

22. Liu, S. H. (1996). Generation of flow network and field tests on hydraulic conductivity for fractured rock mass. Northwestern Hydropower, Vol. 55, No. 1, 21-27

23. Lomize, G. M. (1951). Flow in fractured rocks. Gosenergoizdat, Moscow

24. Long, J. C. S.; Remer, J. S.; Wilson, C. R. & Witherspoon, P. A. (1982). Porous media equivalents for networks of discontinuous fractures. Water Resource Research, Vol. 18, No. 3, 645–58

25. Louis, C. (1971). A study of groundwater flow in jointed rock and its influence on the stability of rock masses. Rock Mechanics Research Report, No. 10, Imperial College of Science and Technology, London, Maini, YNT

26. Min, K. B. & Jing, L. (2003). Numerical determination of the equivalent elastic compliance tensor for fractured rock masses using the distinct element method. International Journal of Rock Mechanics and Mining Sciences, Vol. 40, No. 6, 795-816

27. Min, K. B.; Rutqvist, J.; Tsang, C. F. & Jing, L. (2004). Stress-dependent permeability of fractured rock masses: a numerical study. International Journal of Rock Mechanics and Mining Sciences, Vol. 41, No. 7, 1191-1210

28. Oda, M. (1985). Permeability tensor for discontinuous rock masses. Geotechnique, Vol. 35, No. 4, 483-195

29. Oda, M. (1986). An equivalent continuum model for coupled stress and fluid flow analysis in jointed rock masses. Water Resources Research, Vol. 22, No. 13, 1845-1856

30. Olsson, R. & Barton, N. (2001). An improved model for hydromechanical coupling during shearing of rock joints. International Journal of Rock Mechanics and Mining Sciences, Vol. 38, No. 3, 317-329

31. Pande, G. N. & Xiong, W. (1982). An improved multilaminate model of jointed rock masses. In: Numerical Models in Geomechanics, Dungar, R.; Pande, G. N. & Studer, J. A. (Ed.), 218–226, Bulkema, Rotterdam

32. Patir, N. & Cheng, H. S. (1978). An average flow model for determining effects of threedimensional roughness on hydrodynamic lubrication. ASME Journal of Lubrication Technology, Vol. 100, 12-17

33. Plesha, M. E. (1987). Constitutive models for rock discontinuities with dilatancy and surface degradation. International Journal for Numerical and Analytical Methods in Geomechanics, Vol. 11, 345–62

34. Pusch, R. (1989). Alteration of the hydraulic conductivity of rock by tunnel excavation. Int J Rock Mech Min Sci & Geomech Abstr, Vol. 26, No. 1, 79–83

35. Snow, D. T. (1969). Anisotropic permeability of fractured media. Water Resources Research, Vol. 5, No. 6, 1273-1289

36. Vajdova, V. (2003). Failure mode, strain localization and permeability evolution in porous sedimentary rocks. Ph.D. thesis, Stony Brook University

37. Wang, M. & Kulatilake, P. H. S. W. (2002). Estimation of REV size and

three dimensional hydraulic conductivity tensor for a fractured rock mass through a single well packer test and discrete fracture fluid flow modeling. International Journal of Rock Mechanics and Mining Sciences, Vol. 39, 887-904

38. Yuan, S. C. & Harrison, J. P. (2004). An empirical dilatancy index for the dilatant deformation of rock. International Journal of Rock Mechanics and Mining Sciences, Vol. 41, 679–86

39. Zimmerman, R. W.; Kumar, S. & Bodvarsson, G. S. (1991). Lubrication theory analysis of the permeability of rough-walled fractures. International Journal of Rock Mechanics and Mining Sciences, Vol. 28, No. 4, 325-331

40. Zhou, C. B.; Chen, Y. F. & Sheng, Y. Q. (2006). A generalized cubic law for rock joints considering post-peak mechanical effects. In: Proc GeoProc2006, 188–197, Nanjing, China

41. Zhou, C. B.; Sharma, R. S.; Chen Y. F. & Rong, G. (2008). Flow-Stress Coupled Permeability Tensor for Fractured Rock Masses. International Journal for Numerical and Analytical Methods in Geomechanics, Vol. 32, 1289-1309

42. Zhou, C. B. & Xiong, W. L. (1996). Permeability tensor for jointed rock masses in coupled seepage and stress fields. Chinese Journal of Rock Mechanics and Engineering, Vol. 15, No. 4, 338-344

43. Zhou, C. B. & Xiong, W. L. (1997). Influence of geostatic stresses on permeability of jointed rock masses. Acta Seismologica Sinica, Vol. 10, No. 2, 193-204

44. Zhou, C. B.; Ye, Z. T. & Han, B. (1997). A study on configuration and hydraulic conductivity of rock joints. Advances in Water Science, Vol. 8, No. 3, 233-239

45. Zhou, C. B. & Yu, S. D. (1999). Representative elementary volume (REV): a fundamental problem for selecting the mechanical parameters of jointed rock mass. Chinese Journal of Engineering Geology, Vol. 7, No. 4, 332-336

Chapter 7

LINKING XYLEM HYDRAULIC CONDUCTIVITY AND VULNERABILITY TO THE LEAF ECONOMICS SPECTRUM—A CROSS-SPECIES STUDY OF 39 EVERGREEN AND DECIDUOUS BROADLEAVED SUBTROPICAL TREE SPECIES

Wenzel Kröber[1], Shouren Zhang[2], Merten Ehmig[1], Helge Bruelheide[1,3]

[1] Institute of Biology, Geobotany and Botanical Garden, Martin-Luther-University Halle-Wittenberg, Halle (Saale), Germany

[2] State Key Laboratory of Vegetation and Environmental Change, Institute of Botany, the Chinese Academy of Sciences, Beijing, China

[3] German Centre for Integrative Biodiversity Research (iDiv) Halle-Jena-Leipzig, Leipzig, Germany

ABSTRACT

While the fundamental trade-off in leaf traits related to carbon capture as described by the leaf economics spectrum is well-established among plant species, the relationship of the leaf economics spectrum to stem hydraulics is much less known. Since carbon capture and transpiration are coupled, a close connection between leaf traits and stem hydraulics should be expected. We thus asked whether xylem traits that describe drought tolerance and vulnerability to cavitation are linked to particular leaf traits. We assessed xylem vulnerability, using the pressure sleeve technique, and anatomical xylem characteristics in 39 subtropical tree species grown under common garden conditions in the BEF-China experiment and tested for correlations with traits related to the leaf economics spectrum as well as to stomatal control, including maximum stomatal conductance, vapor pressure deficit at maximum stomatal conductance and vapor pressure deficit at which stomatal conductance is down-regulated. Our results revealed that specific xylem hydraulic conductivity and cavitation resistance were closely linked to traits represented in the leaf economic spectrum, in particular to leaf nitrogen concentration, as well as to log leaf area and leaf carbon to nitrogen ratio but not to any parameter of stomatal conductance. The study highlights the potential use of well-known leaf traits from the leaf economics spectrum to predict plant species' drought resistance.

INTRODUCTION

The worldwide leaf economics spectrum (LES) represents an important framework of trade-offs between key functional leaf traits [1]. It describes different strategies of carbon capture among vascular land plants, from that of short-lived leaves with high photosynthetic capacity per leaf mass, to long-lived leaves with low mass-based carbon assimilation rates. Mass-based photosynthetic capacity is positively related to mass-based leaf nitrogen concentration (LNC) and to specific leaf area [2] and is negatively related to leaf life span [1]. Many studies have confirmed the global validity of these trade-off patterns [3]–[5], and Osnas *et al.* [6] recently demonstrated that such relationships between traits in the LES result from relationships to leaf area and from normalizing area-proportional traits by leaf mass. Kröber & Bruelheide [7] have demonstrated that there are additional dimensions to plants' functional traits that are orthogonal to the LES. They found parameters of stomatal regulation (derived from stomatal conductance - vapor pressure deficit relationships), stomatal density and stomatal size to be independent from the LES.

So far, reported relationships between traits of the LES and those of plant organs other than leaves, such as wood or roots, are equivocal. For example, Baraloto *et al.* [8] found the main axes in leaf and wood traits to be decoupled, while Freschet *et al.* [9] provided evidence for a tight relationship between the main dimensions of the leaf, stem and root economics spectra. However, from a 'whole plant' perspective, tight relationships would be expected for those leaf and wood traits that determine a plant's tolerated minimum water potential, because the plant's water status links a multitude of physiological processes [10]. Choat *et al.* [11] and Poorter *et al.*[12] reported that leaves with high specific leaf area (SLA) were linked to stems with low wood density. Similar to wood density, hydraulic xylem properties would also be expected to be correlated across roots, stems and leaves. In particular, a high photosynthetic capacity of leaves, as expressed by high SLA, should be associated with high xylem hydraulic conductivity to facilitate sufficient water supply required for high stomatal conductance. Accordingly, in a study on ten tropical tree species in Panama, Sack and Frole [13] reported that leaf hydraulic resistance was strongly linked to leaf venation and mesophyll structure. The relationship between hydraulic conductivity and leaf venation was also confirmed across

43 species worldwide [14]. However, Sack *et al.* [15] argued that leaf hydraulic conductance might be mechanistically independent from the LES, but might be linked statistically as both hydraulic conductance and LES traits affect mass-based photosynthesis.

Besides being hydraulically efficient, another required feature of the vascular plumbing network is drought resistance. In this regard, species with stress-resistant leaves, as indicated by low SLA values, should be expected to have stress-resistant wood. Wood stress resistance is reflected in high wood density, which is thought to confer a higher tolerance from shade, wind, herbivores and drought [16]. In particular, drought resistance determined by measuring xylem vulnerability to cavitation should be correlated between leaves and wood, because cavitation is a persistent hazard under drought stress and affects leaves and wood [17]. Sustaining low water potentials requires high cavitation resistance of conduits, as derived from xylem vulnerability curves [18]. These curves allow quantifying the specific xylem hydraulic conductivity of the xylem (K_S) and the xylem pressure at which 50% loss of the maximum specific xylem hydraulic conductivity occurs (Ψ_{50}). Ψ_{50} is mainly determined by pit size and structure [19],[20]. However, low water potentials are transmitted throughout the whole plant, from the point where the water-pathway ends and the regulation of the water flow takes place to the xylem, where cavitation occurs. Thus, Ψ_{50} should be reflected in functional leaf traits and parameters of stomatal regulation.

Tree species that are able to endure severe drought periods have characteristic leaves. The leaves are tough and have a high leaf dry matter content (LDMC), allowing them to sustain low water potentials [21],[22]. Such species should likewise be characterized by low Ψ_{50} values. SLA is inversely related to LDMC, in that it decreases with drought resistance [23],[24], and it would be expected to scale negatively with Ψ_{50}. In addition, cavitation vulnerability should also be related to stomatal regulation, because cavitation-sensitive and -insensitive species would be expected to close their stomata at low and high vapor pressure deficits, respectively[25]–[27]. Such parameters of stomatal closure have recently been provided by Kröber & Bruelheide [7] for the same 39 species used also in the current study. The authors measured daily courses of stomatal conductance (g_s) with porometry in the same plots as in the present study, and modeled the species-specific $g_s \sim$ vapor pressure deficit (VPD) relationships. They found that mean g_s can be predicted from leaf traits that reflect the LES, with a positive relationship to LNC and a negative relationship to leaf carbon to nitrogen ratio. In contrast, the maximum of the $g_s \sim$ VPD curve was unrelated to traits of the LES and increased with leaf carbon concentration (LCC) and vein length. The VPD at which g_s was down-regulated, characterized by the point of inflection of the g_s ~ VPD curve at high VPD, was lower for species with higher stomatal density and lower leaf carbon concentration. In addition to leaf trait measurements, we use these parameters of stomatal control from Kröber & Bruelheide [7] to predict xylem hydraulics.

The objective of our study was to quantify hydraulic conductivity and Ψ_{50} from xylem vulnerability curves, making use of the common garden situation of the BEF-China experiment. Comparing 39 broad-leaved tree species, we hypothesized that (1) leaf traits describing the leaf economics spectrum are related to specific xylem hydraulic conductivity and cavitation resistance. Accordingly, we expected that (2) evergreen species characterized by low SLA and high LDMC are more resistant to cavitation, i.e. have lower Ψ_{50} values than deciduous species, Finally, (3) we tested the hypothesis that parameters of stomatal regulation, such as maximum stomatal conductance, the vapor pressure deficit (VPD) at maximum stomatal conductance and VPD at which stomatal conductance is down-regulated, are related to high xylem hydraulic conductivity.

MATERIALS AND METHODS

Study Site

The study was conducted in the BEF-China project, which is a biodiversity-ecosystem functioning experiment based in Jiangxi Province, southeast China (http://www.bef-china.de; 29.08–29.11 N, 117.90–117.93 E). The climate at the experimental site is subtropical with moderately cold and dry winters and warm summers. Based on data of meteorological stations established at the sites, mean annual temperature was 17.4°C and mean annual precipitation was 1635 mm (Fig. 1). Across an area of 38 ha, 219,000 trees were planted at different levels of species richness [29]. The diversity gradient spans from monoculture to two, four, eight, 16 and up to 24 species per plot. The 39 tree species included in the study (see Table 1) are representative of the local natural broadleaved subtropical forest community [30],[31], and the trees assessed had already reached an age of four or five years at the time when our study was carried out. Using young even-aged trees in a common garden situation allowed for controlling for confounding factors, such as different ontological stage, but also allowed to sample leaves and branches at a standardized height above ground. No specific permissions were required for these locations and activities. The field studies did involve neither endangered nor protected species.

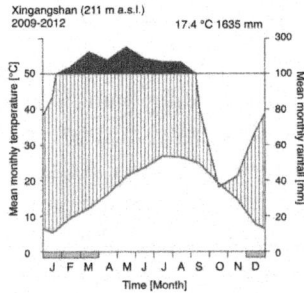

Figure 1. Climate diagram according to Walter & Lieth [28] **of** Xingangshan, the location of the experimental sites. Elevation: 211 m above sea level. Observation period was March 2009 to October 2012. Mean annual temperature was 17.4°C and total annual precipitation was 1635 mm. Monthly precipitation below 100 mm is scaled 2:1 with mean monthly temperature (vertically hatched) and above 100 mm 15:1. Turquoise bars below the x-axis show the months where frosts can occur (when absolute monthly minimums are equal or lower than 0°C). Climate data were recorded by a meteorological station established at the very center of the experimental site (Kühn, unpublished). doi:10.1371/journal.pone.0109211.g001

Table 1. Tree species planted in the BEF-China experiment and included in this study.

Species name	Family	Abbreviation	Leaf habit
Acer davidii Franch.	Aceraceae	Ad	d
Ailanthus altissima (Miller) Swingle	Simaroubaceae	Aa	d
Alniphyllum fortunei (Hemsl.) Makino	Styracaceae	Af	d
Betula luminifera Winkl.	Betulaceae	Bl	d
Castanopsis eyrei (Champion ex Bentham) Tutcher	Fagaceae	Ce	e
Castanopsis fargesii Franch.	Fagaceae	Cf	e
Castanea henryi (Skan) Rehd. et Wils.	Fagaceae	Ch	d
Castanopsis sclerophylla (Lindley & Paxton) Schottky	Fagaceae	Cs	e
Celtis biondii Pamp.	Cannabaceae	Cb	d
Choerospondias axillaris (Roxb.) Burtt et Hill	Anacardiaceae	Ca	d
Cinnamomum camphora (Linn.) Presl	Lauraceae	Cc	e
Cyclobalanopsis glauca (Thunberg) Oersted	Fagaceae	Cg	e
Cyclobalanopsis myrsinifolia (Blume) Oersted	Fagaceae	Cm	e
Daphniphyllum oldhamii (Hemsl.) Rosenthal	Daphniphyllaceae	Do	e
Diospyros japonica Siebold & Zuccarini	Ebenaceae	Dj	d
Elaeocarpus chinensis (Gardn. et Chanp.) Hook. f. ex Benth.	Elaeocarpaceae	Ec	e
Elaeocarpus glabripetalus Merr.	Elaeocarpaceae	Eg	e
Elaeocarpus japonicus Sieb. et Zucc.	Elaeocarpaceae	Ej	e
Idesia polycarpa Maxim.	Flacourtiaceae	Ip	d
Koelreuteria bipinnata Franch.	Sapindaceae	Kb	d
Liquidambar formosana Hance	Altingiaceae	Lf	d
Lithocarpus glaber (Thunb.) Nakai	Fagaceae	Lg	e
Machilus grijsii Hance	Lauraceae	Mg	e
Machilus leptophylla Hand.-Mazz.	Lauraceae	Ml	e
Machilus thunbergii Sieb. et Zucc.	Lauraceae	Mt	e
Manglietia fordiana (Oliver) Hu Y.W.Law	Magnoliaceae	Manf	e
Melia azedarach Linn.	Meliaceae	Ma	d
Meliosma flexuosa Blume	Sabiaceae	Mf	d
Nyssa sinensis Oliver	Nyssaceae	Ns	d
Phoebe bournei (Hemsl.) Yen C. Yang.	Lauraceae	Pb	e
Quercus acutissima Carruthers	Fagaceae	Qa	d
Quercus fabri Hance	Fagaceae	Qf	d
Quercus phillyreoides A. Gray	Fagaceae	Qp	e
Quercus serrata Murray	Fagaceae	Qs	d
Rhus chinensis Mill.	Anacardiaceae	Rc	d
Sapindus saponaria Linn.	Sapindaceae	Sd	d
Triadica cochinchinensis Loureiro	Euphorbiaceae	Tc	d
Triadica sebifera (L.) Small	Euphorbiaceae	Ts	d
Schima superba Gardn. et Champ.	Theaceae	Schs	e

Species names are in accordance with nomenclature in The Flora of China (http://flora.huh.harvard.edu/china). d= deciduous, e= evergreen.

doi:10.1371/journal.pone.0109211.t001

Tree Species and Vulnerability Curves

We randomly chose three individuals per species in the high-diversity plots, with one individual per species being sampled per plot. This enabled us to minimize time between sample cutting and lab procession of samples because different species grew in close proximity to each other. Xylem conductivity and vulnerability and leaf stomatal conductance (see below) were measured on the same plots, but not explicitly on the same individuals. The sampling and measurements on xylem hydraulics were conducted in August - October 2012. This period was characterized by monthly mean temperatures of about 20°C and a monthly precipitation of 40 mm (Fig. 1), which involved dry spells of several weeks, typically resulting in midday depressions of stomatal conductance. Samples were always taken in the early morning hours between 6 and 8 am, when relative humidity was still high (70–95% Rh) and temperatures were around 20°C. Measurements of leaf water potentials were made in spring 2012, using a PMS M1000 Scholander pressure chamber. These data showed that water potentials were well above −2 MPa, and for many species>−1 MPa. A twig with no leaves, buds or branches, around 15 cm in length and 5–15 mm diameter was cut and immediately immersed in water. We are aware that maximum vessel length of some of the species might be larger than 15 cm [32], which would result in overestimating specific xylem hydraulic conductivity (K_S) and the absolute value of the xylem pressure at which 50% loss of the maximum specific xylem hydraulic conductivity occurs (Ψ_{50}). However, it has been shown that extreme vessel lengths are very rare [33]. In any case, obtaining non-ramified twigs longer than 15 cm would have been impossible in most species. After transportation to the lab, the stem pieces were then placed into a double-ended pressure sleeve (PMS M1000 Scholander pressure chamber) in the laboratory following established protocols [18],[34] (Fig. 2). Xylem vulnerability was measured within at maximum four hours after cutting. Increasing the air pressure in the cavitation chamber was used to simulate increasingly negative xylem sap pressures [35]. Before the measurements were taken, each twig segment was treated for one hour with perfusion solution pressurized at 0.15 MPa in order to flush out air from older embolism events and any potential air entry into the xylem during the cutting and handling of samples. We used 10 mM citric acid perfusion solution, using filtered and demineralized water to prevent any blockages caused by microorganisms.

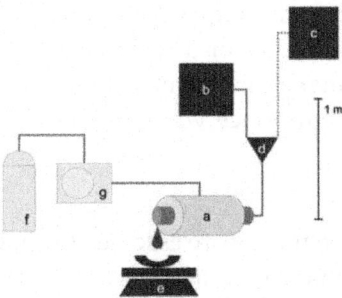

Figure 2. Simplified overview of the xylem hydraulics measurement installation. A) Pressure chamber with the stem segment, B) tank with the perfusion solution, C) flush tank, D) three-way stopcock, E) electronic scale, F) nitrogen pressure cylinder, and G) pressure gauge. doi:10.1371/journal.pone.0109211.g002

Specific xylem hydraulic conductivity (K_S) was measured as the mass of flow-through perfusion solution through the piece of wood per unit of time and per cross-sectional area of the twig. The measurements were started at ambient air pressure and repeated as pressure was increased in increments of 0.5 MPa up to 7 MPa, or to the pressure at which no more perfusion solution flow was encountered.

Measurements of Stomatal Conductance

Data on stomatal conductance were taken from [7]. For the measurements of stomatal conductance the authors had randomly selected twelve high-diversity plots (with 16 or 24 species). Eleven to 23 individuals of all species were measured per plot. In total, 3,290 measurements of stomatal conductance were made in August to October 2010, May/June 2011 and August to October 2011. Each of the 39 species was represented between at least three and at maximum 13 individuals, resulting in 218 individuals in total. This non-balanced sampling design resulted from measuring daily course of stomatal conductance, which required walking time to be minimal. The repeated measurements were taken on the same leaf, which was fully developed, undamaged and fully exposed to the sun. Daily courses of stomatal conductance were produced for all species in every plot. Stomatal conductance was recorded with an SC1 porometer (Decagon). Air temperature and relative humidity was measured simultaneously using a T2 thermo-hygrometer (Trotec). Vapor pressure deficit (VPD) was calculated following the August-Roche Magnus formula. The daily courses of all different individuals from all different daily courses were then aggregated to one g_s ~ VPD relationship which included all data for one species. Mean and maximum stomatal conductance (g_{smax}) could then be estimated per species. The species-specific g_s ~ VPD relationships were modeled by regressing the logits of g_s/g_{smax} to VPD and the quadratic term of VPD using a generalized linear model with a binomial error distribution. The parameters of the model allowed calculating the maximum stomatal conductance and the VPD at which the modeled stomatal conductance was maximal.

Trait Measurements

A total of 34 leaf and wood traits were assessed to analyze possible relationships with Ψ_{50} and K_s (see Table 2). To accomplish this, four total sets of samples were taken: 1) A set of five individuals with five leaves being sampled per individual for the traditional leaf traits, such as absolute area per leaf, leaf fresh-weight, leaf dry-weight, leaf nitrogen concentration (LNC) and leaf carbon concentration (LCC). The data were used to calculate specific leaf area (SLA), leaf dry matter content (LDMC) and carbon to nitrogen ratio (CN). We also determined leaf habit (deciduous/evergreen), leaf pinnation (pinnate or simple), leaf margin (entire or serrate) and recorded the presence or absence of extrafloral nectaries. 2) Another 30 leaves were sampled from three individuals per species to determine leaf tensile strength as a measure of leaf toughness. Leaf tensile strength was measured with a tearing apparatus modified after Hendry[36]. 3) The same leaves on which stomatal conductance was measured (see above), were taken to analyze the stomatal related traits. Stomatal traits were analyzed after Gerlach [37], with stomata being counted on a minimum area of 50,000 μm^2 on three leaves from three individuals per species. Stomata were counted on nail polish impressions made on leaf samples, which had been stored in 70% ethanol. Length and width of three stomata per replicate were measured, and stomatal density was expressed as stomatal number per area. The analysis was performed with a light-optical microscope (Zeiss Axioskop 2 plus) and using the Axio Vision (Version 3.0) software. 4) A sample was taken from each twig used in the cavitation sensitivity analysis for further xylem anatomical investigation. Twig sections were prepared for light microscopic inspection and, from an area of 4.4 mm2 per sample, every xylem vessel was analyzed. To determine xylem traits, we made use of XylemDetector that was implemented as part of the open-source package MiToBo (http://www.informatik.uni-halle.de/mitobo), an extension of the Java image processing software ImageJ. We measured the mean lumen area of conducting vessels (MEANAREA) and the mean roundness of conducting vessels (MEANROUND), which is a measure of how close the vessel shape is to a perfect circle, and ranges from 0 to 1. MEANROUND was calculated as:

$$meanround = \frac{4\pi a}{p^2}$$ where a is the area and p the perimeter of the lumen.

Table 2. List of the leaf traits measured across the 39 tree species.

Code	Trait	Analytical technique	Type	Units/categories
Ψ_{50}	Loss of 50% initial conductivity	Pressure Chamber	Continuous	MPa
K_S	Maximum conductivity	Pressure Chamber	Continuous	kg m^{-1} s^{-1} MPa^{-1}
b	Parameter b (Sigmoid Regression)	Pressure Chamber	Continuous	nondimensional
CONMEAN	Average stomatal condutance	Porometer	Continuous	mmol m^2 s^{-1}
CONMAX	Maximum stomatal condutance	Porometer	Continuous	mmol m^2 s^{-1}
VPDMAX	VPD at CONMAX	Porometer, Hygrometer, Thermometer	Continuous	hPa
CONMAXFIT	Fitted Max. stomatal conductance	Porometer, Hygrometer, Thermometer	Continuous	nondimensional
VPDMAXFIT	VPD at CONMAXFIT	Porometer, Hygrometer, Thermometer	Continuous	hPa
VPDPOI	VPD at point of inflection of fitted stomatal conductance	Porometer, Hygrometer, Thermometer	Continuous	hPa
SLA	Specific leaf area	Scanner, Balance	Continuous	m^2 kg^{-1}
LOG10LA	Decadic log (Leaf Area)	Scanner	Continuous	mm^2
LDMC	Leaf dry matter content	Balance	Continuous	mg g^{-1}
LT	Leaf toughness	Leaf toughness device	Continuous	N mm^{-1}
LEAFHABIT	Leaf habit	Literature	Binary	(0) evergreen; (1) deciduous
LNC	Leaf nitrogen concentration	CN Analyzer	Continuous	mg g^{-1}
LCC	Leaf carbon concentration	CN Analyzer	Continuous	mg g^{-1}
CN	Carbon-nitrogen ratio	CN Analyzer	Continuous	ratio
CA	Leaf calcium concentration	AAS Analyzer	Continuous	mg g^{-1}
K	Leaf potassium concentration	AAS Analyzer	Continuous	mg g^{-1}
MG	Leaf magnesium concentration	AAS Analyzer	Continuous	mg g^{-1}
LEAFPIN	Leaf pinnation	Field Observation	Binary	(0) pinnate; (1) simple
LEAFMAR	Leaf margin	Field Observation	Binary	(0) dentate; (1) entire
EXTRAFLO	Extrafloral nectaries	Field Observation	Binary	(0) no; (1) yes
STODENS	Stomata density	Microscopy	Continuous	1 mm^{-2}
STOSIZE	Stomata size	Microscopy	Continuous	μm^2
STOIND	Stomata index	Microscopy	Continuous	ratio
DIAMVEIN1	Diameter veins 1st order	Scanner	Continuous	cm
DIAMVEIN2	Diameter veins 2nd order	Scanner	Continuous	cm
VEINDENS	Length of veins per unit leaf area	Scanner	Continuous	cm cm^{-2}
WPOT	Water potential	Pressure Chamber	Continuous	MPa
WOODDENS	Wood density	Balance	Continuous	g cm^3
MEANAREA	Mean area of conducting vessels	Microscopy	Continuous	μm^2
MEANROUND	Mean roundness of conducting vessels	Microscopy	Continuous	nondimensional
DHYD	Hydraulically weighted diameter of conducting vessels	Microscopy	Continuous	μm

All traits were assessed on the individuals planted in the experiment. The table includes the trait abbreviations (Code) used throughout the text.

doi:10.1371/journal.pone.0109211.t002

Following Sperry [38], we calculated hydraulically weighted conduit diameter (DHYD) from the lumen area data according to: $dhyd = 2\dfrac{\sum r^5}{\sum r^4}$ where r are the circle radii calculated from the lumen areas.

Statistical Analyses

We plotted vulnerability curves that show the flow rates of perfusion solution through stem segments as a function of water potential [35]. A sigmoid, three-parameter regression was applied to the vulnerability data [39],[40]:

$$K_S = \frac{a}{1+e^{-\left(\frac{\Psi-\Psi_{50}}{b}\right)}}$$

where K_s is the specific hydraulic conductivity of the xylem [kg m^{-1} s^{-1} MPa^{-1}], Ψ is the xylem pressure at which water flow was measured [MPa], a is the original maximum specific xylem hydraulic conductivity, b is the slope of the regression and Ψ_{50} is the xylem pressure at which 50% loss of the original maximum specific xylem hydraulic conductivity occurs. Fig. 3 shows an example of how the sigmoid model was fitted to predict the loss of specific xylem hydraulic conductivity (K_s) from water potential (please refer to the Figure S1 and Table S1 for raw data on xylem vulnerability). We made use of the FactoMineR package in R (http://cran.r-project.org/web/packages/FactoMineR/index.html) to correlate the physiological parameters to the species› traits in a PCA. K_s, Ψ_{50} and the slope b of the $K_s \sim \Psi$ relationship were tested for differences between deciduous and evergreen leaf habit by analysis of variance (ANOVA) and for bivariate relationships to leaf traits by linear regression models. As these tests performed multiple testing, they run the risk of error inflation and cannot be used to infer statistical significances. These tests were exploratory and had the purpose to identify possible candidate predictors and to show the direction of their effects. To further investigate the emerging significant relationships, we rerun all significant linear regressions by additionally including the interaction with leaf habit. For all statistics, R software version 3.0.2 was used.

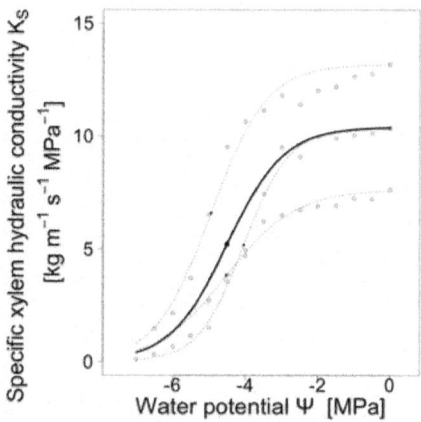

Figure 3. Conductivity rates as a function of decreasing water potential, taking *Castanea henryi* **as an example.** Outline dots show the measured values of three independent vulnerability curves. Dotted lines show the single regression functions obtained from the measured value per vulnerability curve, obtained from a 3-parametric sigmoid function. The bold line shows the regression lines constructed from the means of the parameters from the three single vulnerability curves. Filled dots represent Ψ_{50} values.

doi:10.1371/journal.pone.0109211.g003

RESULTS

Fig. 4 shows the vulnerability curves for all 39 species included in the study. Values of Ψ_{50}ranged between -1.08 MPa and -6.6 MPa for *Celtis biondii* and *Lithocarpus glaber*, respectively, with an overall mean of -3.78 MPa (SD=1.48). Specific xylem hydraulic conductivity (K_S) was highest in *Melia azedarach, Triadica sebifera* and *Castanea henryi*(17.52, 11.01 and 10.40 kg m^{-1} s^{-1} MPa^{-1}, respectively) and lowest in *Machilus grijsii* (0.036 kg m^{-1} s^{-1} MPa^{-1}), with a overall mean of 2.44 kg m^{-1} s^{-1} MPa^{-1} (SD=3.31). K_S and Ψ_{50}were not correlated across all species (p=0.512). Evergreen species had significantly lower values of maximum hydraulic conductivity and lower Ψ_{50} values than deciduous species (Fig. 5).

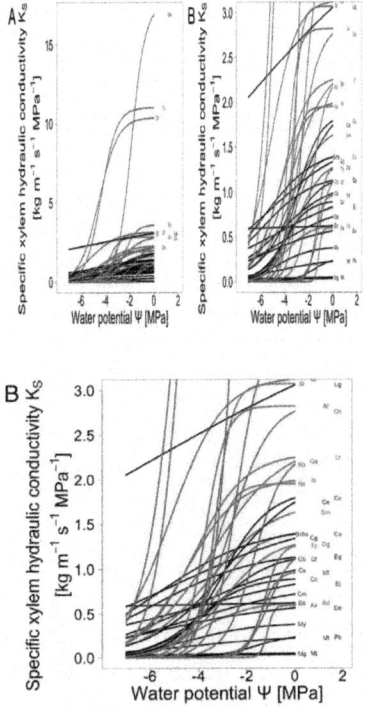

Figure 4. Plots of modeled specific xylem hydraulic conductivity K_S **versus water potentials for all 39 species included in the study.** A) and B) show the same data at different scale of K_S. Deciduous species are shown in red, evergreen species shown in black. For species abbreviations see Table 1. For details of calculation of regression lines, see Fig. 2 and Methods. doi:10.1371/journal.pone.0109211.g004

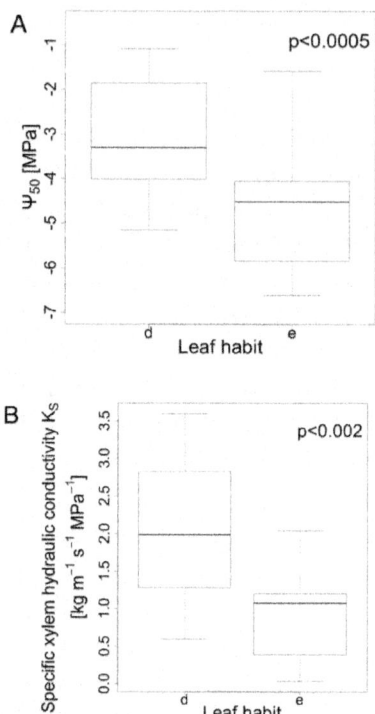

Figure 5. Boxplots characterizing the species set by leaf habit. A) xylem pressure at 50% loss of the maximum specific xylem hydraulic conductivity (Ψ_{50}) and B) specific xylem hydraulic conductivity of the xylem (K_s) as a function of leaf habit. d = deciduous, e = evergreen. The probability values were derived from an ANOVA. doi:10.1371/journal.pone.0109211.g005

Fig. 6 shows the principal components analysis (PCA) of all 34 leaf traits, including the parameters of stomatal control and xylem vulnerability for all 39 study species (Fig. 6 a, b). The species mean values of all traits are provided in Table S1. The first three PCA axes explained 43.3% of the total variance, with eigenvalues of 7.03, 4.2 and 3.48, respectively. While evergreen species tended to score higher on the first PCA axes than deciduous species, there was a large overlap between the two leaf habits (Fig.6 c, d). Positive scores on the first PCA axis reflected both decreasing xylem vulnerability and increasing values of traits of the leaf economics spectrum, such as leaf nitrogen concentration (LNC) and specific leaf area (SLA), as well as evergreen leaf habit and the logarithm of the area of a single leaf (Log10LA), while leaf toughness (LEAFT) and leaf carbon to nitrogen ratio (CN) showed negative loadings. Parameters of stomatal control were correlated with the second PCA axis, with positive loadings being recorded for stomatal index (STOIND), stomatal density (STOMDENS) and wood density (WOODDENS), and negative ones for the point of inflection of the g_s ~ VPD curve (VPDPOI) and maximum stomatal conductance (CONMAX).

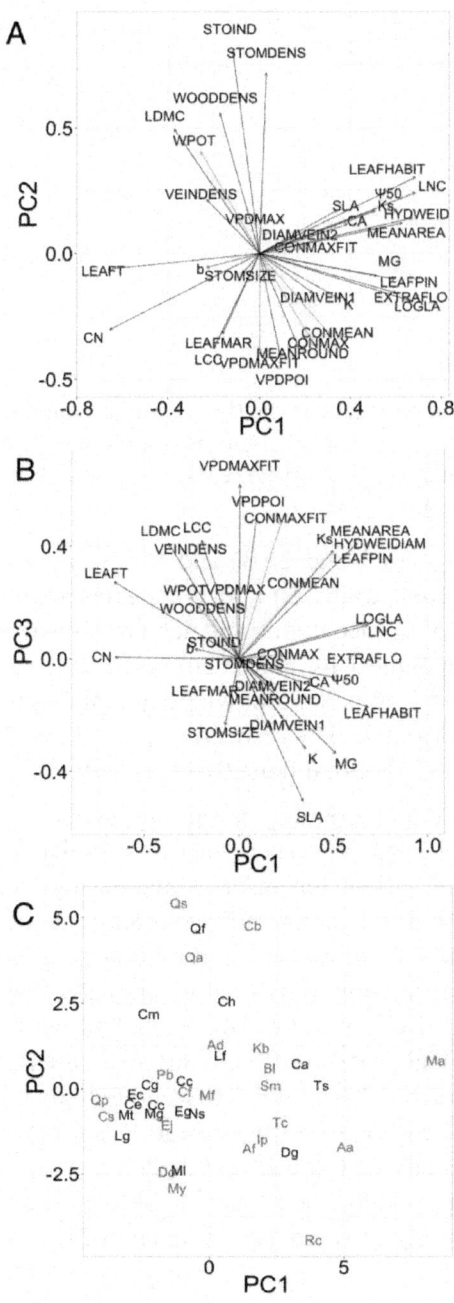

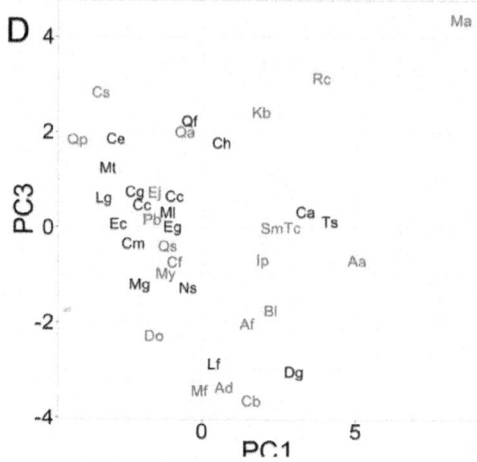

Figure 6. Principal component analysis (PCA) biplots showing the relationships between the mean values of all chemical and morphological leaf traits (black arrows), all parameters of stomatal regulation (green arrows) and all parameters of cavitation sensitivity (blue arrows).

A) and C) PCA axes 1 and 2, B) and D) PCA axes 1 and 3. A) and B) loadings of the different traits. C) and D) species scores in the PCA shown separately by leaf habit. Deciduous species are shown in red and evergreen species in black. See Table 1 for abbreviations of species names. Eigenvalues: axis 1=7.03, axis 2=4.2, axis 3=3.48, with cumulative proportion of explained inertia 20.1%, 33.0% and 43.3%, respectively. SeeTable 2 for abbreviations of trait names. doi:10.1371/journal.pone.0109211.g006

We found xylem vulnerability to be significantly related to numerous traits (Table 3, Fig. 7 & 8). Ψ_{50} was positively correlated with leaf area, LNC, hydraulically weighted diameter and leaf magnesium concentration, while it was only marginally significantly related to leaf calcium concentration, and negatively related to leaf toughness and carbon to nitrogen ratio. A similar pattern was found for maximum hydraulic conductivity (K_s), which showed a positive relationship to leaf area (Log10LA), leaf nitrogen concentration (LNC) and two morphological wood traits (i.e. the mean area of conducting vessels (MEANAREA), and the hydraulically weighted diameter) and a negative correlation with leaf carbon to nitrogen ratio (CN). The regression parameter b was not related to any of the traits studied. The regression equations from the significant linear models are shown in Table 4. There were no significant correlations between Ψ_{50} or K_s to any parameter of stomatal control.

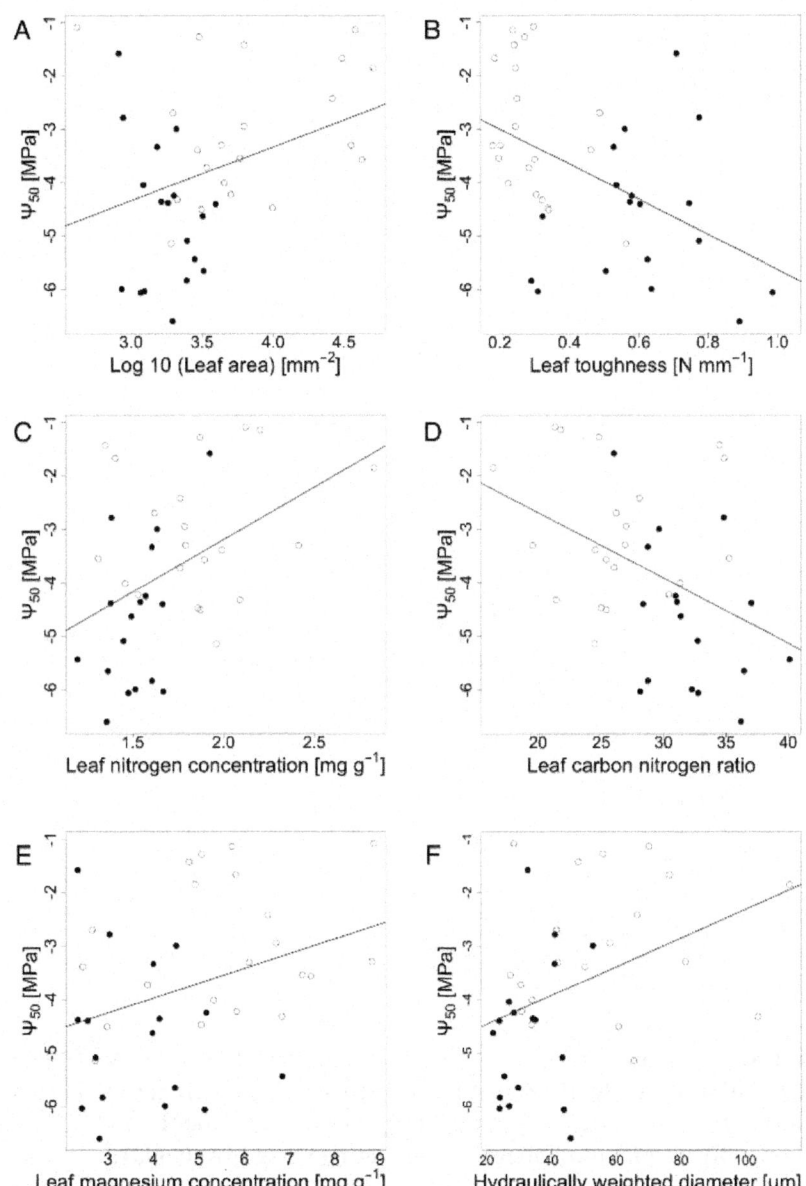

Figure 7. Ψ_{50} **as a function of A) leaf area (p=0.0315, r=0.34), B) leaf toughness (p<0.0003, r=−0.47), C) leaf nitrogen concentration (p<0.0075, r=0.43), D) leaf carbon to nitrogen ratio (p<0.0078, r=−0.43), E) leaf magnesium concentration (p<0.042, r=0.33) and F) hydraulically weighted conduit diameter (p<0.01, r=0.39).** Filled black dots represent species of evergreen leaf habit; empty dots represent species of deciduous leaf habit.

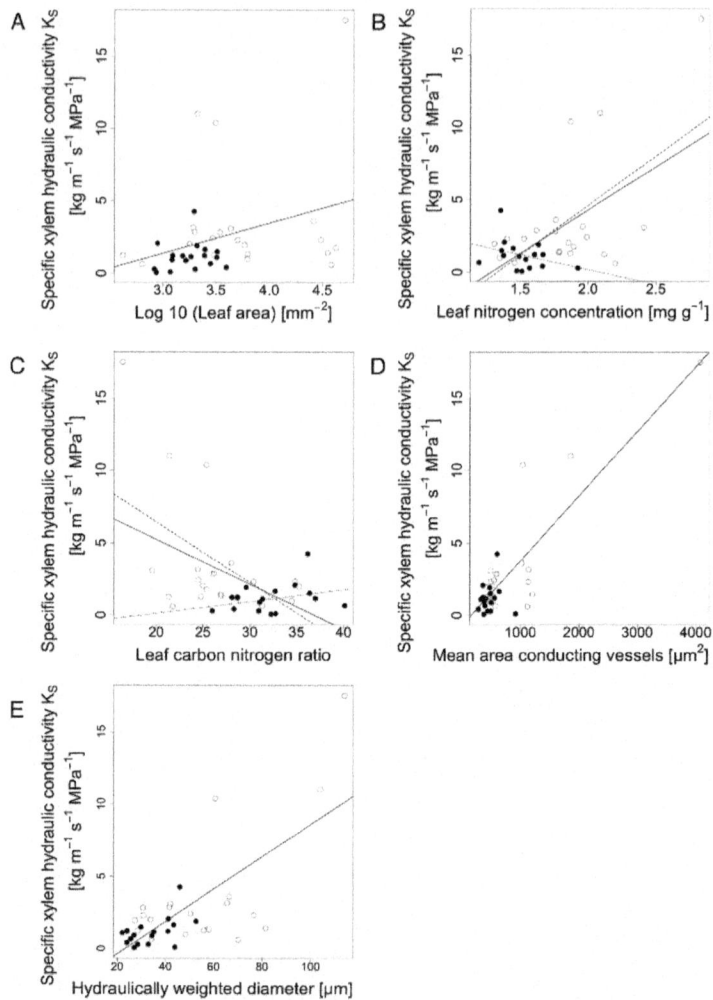

Figure 8. K_S **as a function of A) leaf area (p<0.044, r=0.1), B) leaf nitrogen concentration (p<0.00001, r=0.59), C) leaf carbon to nitrogen ratio (p<0.0019, r=0.48), D) mean area of conducting vessels (p<0.0001, r=0.85), and E) hydraulically weighted diameter of conducting vessels (p<0.0001, r=0.72).**

Filled black dots represent species of evergreen leaf habit and empty dots represent species of deciduous leaf habit. In B) and C) K_S also showed significant interactions with leaf habits. Dotted lines represent species of evergreen and broken lines represent species of deciduous leaf habit. B) leaf habit p=0.061, interaction leaf nitrogen concentration and leaf habit p=0.049, C) leaf habit p=0.026, interaction leaf carbon to nitrogen ratio and leaf habit p=0.031. doi:10.1371/journal.pone.0109211.g008

Table 3. Correlation matrix of hydraulic xylem characteristics with numerical leaf traits and parameters of stomatal control.

	Ψ_{50}		K_S		b	
	r	p	r	p	r	p
CONMEAN	0.12	0.49	0.13	0.42	−0.12	0.49
CONMAX	0.02	0.92	0.02	0.91	−0.06	0.73
VPDMAX	−0.04	0.81	0.15	0.36	0.14	0.39
CONMAXFIT	0.14	0.41	0.26	0.12	−0.11	0.50
VPDMAXFIT	−0.13	0.43	0.18	0.26	−0.01	0.96
VPDPOI	0.13	0.44	0.13	0.44	−0.10	0.53
SLA	0.27	0.09	0.05	0.78	−0.17	0.30
LOG10LA	0.34	0.03	0.10	0.04	−0.20	0.22
LDMC	−0.02	0.91	−0.02	0.91	−0.03	0.85
LEAFT	−0.47	0.00	−0.23	0.16	0.02	0.90
LNC	0.43	0.01	0.59	0.00	−0.08	0.62
LCC	−0.23	0.17	−0.12	0.49	0.05	0.76
CN	−0.43	0.01	−0.48	0.00	0.05	0.75
CA	0.32	0.05	0.09	0.58	−0.16	0.35
K	0.16	0.33	0.05	0.75	−0.13	0.44
MG	0.33	0.04	0.02	0.91	−0.24	0.15
STOMDENS	0.09	0.58	0.07	0.65	−0.08	0.64
STOMSIZE	0.06	0.74	−0.06	0.71	−0.01	0.96
STOIND	0.11	0.52	0.01	0.69	−0.09	0.59
DIAMVEIN1	0.20	0.23	−0.09	0.59	−0.23	0.17
DIAMVEIN2	0.17	0.33	0.01	0.93	−0.14	0.42
VEINLENGTH	−0.15	0.37	−0.09	0.61	0.19	0.24
WPOT	0.09	0.57	−0.04	0.79	0.01	0.95
WOODDENS	−0.06	0.73	0.06	0.71	−0.13	0.42
MEANAREA	0.24	0.15	0.85	0.00	−0.11	0.52
MEANROUND	0.04	0.82	−0.01	0.94	−0.02	0.92
DHYD	0.39	0.01	0.72	0.00	−0.20	0.23

The correlations were calculated from linear model analyses between functional leaf traits and the extracted physiological parameters. Significant relationships are shown in bold letters. Ψ_{50}= xylem pressure at which 50% loss of the original maximum specific xylem hydraulic conductivity occurred, K_S= specific hydraulic conductivity of the xylem, b= slope of the three-parametric sigmoid model of the vulnerability curve, CONMEAN = mean g_s; CONMAX = mean $g_{s\,max}$; VPDMax = VPD at $g_{s\,max}$; CONMAXFIT = modeled $g_{s\,max}$; VPDMaxfit = VPD at modeled $g_{s\,max}$, and VPDPoi = VPD at the point of inflexion of the g_s ~ VPD curve, MEANAREA = the mean area of conducting vessels, MEANROUND = the mean roundness of conducting vessels, and DHYD the hydraulically weighted diameter of conducting vessels.

doi:10.1371/journal.pone.0109211.t003

Table 4. Regression equations for all significant correlations of Ψ_{50} and K_S to the functional traits.

Xylem parameter	Functional trait	Equation
Ψ_{50}	LOG10LA	y=1.0090x−7.3717
Ψ_{50}	LEAFT	y=−3.2808x−2.3454
Ψ_{50}	LNC	y=1.9556x−7.1038
Ψ_{50}	CN	y=−0.12210x−0.25799
Ψ_{50}	MG	y=0.2763x−5.0792
Ψ_{50}	DHYD	y=0.0136x−5.005
K_S	LOG10LA	y=2.0699x−4.8375
K_S	LNC	y=6.003x−7.734
K_S	CN	y=−0.31132x+11.44974
K_S	MEANAREA	y=0.0011481x−0.7089940
K_S	DHYD	y=0.0569x−2.611

The correlations were calculated from linear model analyses. Ψ_{50}= xylem pressure at which 50% loss of the original maximum specific xylem hydraulic conductivity occurred, K_S= specific hydraulic conductivity of the xylem, b= slope of the three-parametric sigmoid model of the vulnerability curve, CONMEAN = mean g_s; CONMAX = mean $g_{s\,max}$; VPDMax = VPD at $g_{s\,max}$; CONMAXFIT = modeled $g_{s\,max}$; VPDMaxfit = VPD at modeled $g_{s\,max}$, and VPDPoi = VPD at the point of inflexion of the g_s ~ VPD curve, MEANAREA = the mean area of conducting vessels, and DHYD the hydraulically weighted diameter of conducting vessels.

doi:10.1371/journal.pone.0109211.t004

Including leaf habit in the significant models resulted in two models with significant interactions with leaf habit, i.e. for leaf nitrogen concentration (LNC) and the leaf carbon to nitrogen ratio (CN). While with increasing CN and decreasing LNC, K_S increased in evergreen species, deciduous species showed a decrease (Fig. 8 a, b).

DISCUSSION

We found a large variation in specific xylem hydraulic conductivity and xylem vulnerability among our study species, which points to different complementary strategies of the species in the same subtropical forest community. Thus, our results conform to the findings of Böhnke et al. [41], who revealed a high and temporally constant level of functional diversity in the course of succession in these forests. In particular, variation in cavitation resistance could offer an explanation for species coexistence in the same community [42]. Our values on specific xylem hydraulic conductivity with a range of K_S between 0.036 and 17.52 kg m^{-1} s^{-1} MPa^{-1} and xylem vulnerability to cavitation Ψ_{50} between −1.08 and −6.6 MPa cover a large part of the total range recorded for such variables in other studies [43]. For example, Cavender-Bares et al.[44] described a range of maximum specific xylem hydraulic conductivity of between 1.75 and 5 kg m^{-1} s^{-1} MPa^{-1} for 17 oak species, while Maherali et al. [45] reported a mean of 1.36 kg m^{-1} s^{-1} MPa^{-1} of maximum specific xylem hydraulic conductivity (n species =87) and a mean of −3.15 MPa for Ψ_{50} values (n species =167). Slightly higher Ψ_{50} values of −1.2 up to −2.76 MPa were encountered for eight tree species from a tropical dry forest [46]. However, we have to consider that extreme values of K_S of more than 10 kg m^{-1} s^{-1} MPa^{-1} might be methodological artifacts, caused by some open vessels in these samples. However, the three species with extreme K_S also ranked high in values of the predictor traits for K_S, indicating that the relative rank in K_S in these species might be realistic. As vessel length of the species was not measured, and thus, in some species, specific xylem hydraulic conductivity might have been overestimated, comparisons to other studies should be done with caution. However, as Melcher et al. [33] pointed out, such an overestimation is probably not severe, as long vessels are also very rare.

Important leaf traits such as leaf nitrogen concentration (LNC) and leaf area (Log10LA) were highly correlated to Ψ_{50} and K_S. Thus, we can fully confirm our first hypothesis that leaf traits describing the leaf economics spectrum are related to specific xylem hydraulic conductivity and cavitation resistance. Interestingly, there were no significant correlations for some of the traits of the leaf economics spectrum, such as SLA and LDMC. Thus, leaf thickness and water content of leaves seem to have less importance for hydraulic characteristics than the leaves' absolute size and protein content. This contrasts with the findings reported by Willson et al. [47], who described a significant relationship between SLA and Ψ_{50} for the genus *Juniperus*. Thus, comparative studies confined to certain taxonomic levels, such as congeneric comparisons, might arrive at different conclusions than studies covering a wider range of taxa. Alternatively, SLA may have a differing level of importance with regard to the physiology of gymnosperms and angiosperms. Interestingly, our study did also not support a link between K_S and leaf vein density, which was predicted by the flux trait network suggested by Sack et al. [15]. However, there was only equivocal evidence for a significant relationship between K_S and leaf vein density in their reviewed studies [15]. Our results of a strong relationship of Ψ_{50} and K_S to leaf area, conform to those encountered for eight southern African tree species of a seasonally dry tropical forest by Vinya *et al.* [48], except that they reported a link to leaf area only for K_S, but not for Ψ_{50}. Our findings of a relationship of Ψ_{50} to Mg concentration and a marginal one to Ca concentration might indicate that the non-vein and non-sclerenchyma mesophyll density in the leaf are more relevant for cavitation resistance than overall leaf tissue density. As a central component of chlorophyll, Mg concentration is directly related to photosynthetic capacity, and thus, might capture this proportion of actively assimilating tissue in the leaf. In addition, as a cofactor of many enzymatic processes, Mg can be considered an indicator for the plant›s nutrition status [49].

The absence of any direct relationship between SLA or LDMC with Ψ_{50} or K_S which is in accordance with Sack *et al.* [50], in combination with the large overlap in Ψ_{50} and K_S detected between deciduous and evergreen species, is not a conclusive result. It appears that evergreen and deciduous subtropical forest species form two ends of a gradient from cavitation resistance to cavitation avoidance, respectively. This view is supported by a recent study by Fu *et al.* [17], who investigated the relationship of stem hydraulics and leaf phenology in Asian tropical dry forest species. In particular, they found a negative relationship between leaf life span and K_S but no significant relationship between leaf life span and Ψ_{50}. In accordance with our results, Maherali *et al.*, Choat *et al.* and Chen *et al.* [27],[45],[51] also reported significant differences in Ψ_{50} and K_S between the different leaf habit groups. Such differences in hydraulic characteristics also translate to higher growth rates, as shown by Fan *et al.* [52] for 40 Asian tropical trees. However, some other studies failed to detect any differences, such as that of Markesteijn *et al.* [42], who attributed the substantial differences they encountered in Ψ_{50} and K_S to shade tolerance. They also argued that the distinction between pioneer vs. shade-tolerant species predicts hydraulic properties better than leaf habit, because there are considerable overlaps in strategies along the gradient of leaf longevity. As Givnish [53] pointed out, evergreen leaves can be advantageous under a wide range of ecological conditions, and the relationship of leaf habit with Ψ_{50} and K_S may therefore strongly depend on the system considered. We can also confirm our second hypothesis that evergreen species characterized by low SLA and high LDMC show lower Ψ_{50} and K_S values than deciduous species. Interestingly, we found leaf habit to significantly influence the relationships of K_S to leaf nitrogen concentration and carbon to nitrogen ratio, which might be explained by differences in basic leaf constructing principles. Deciduous species tend to invest high amounts of nitrogen to maximize photosynthetic assimilation per leaf mass, whereas in evergreen species, the focus is on increased leaf lifespan, which is reflected in higher leaf carbon concentration [54]. Furthermore, deciduous species show a more conservative stomatal control to avoid embolism, whereas evergreen species tend to have more cavitation-resistant vessels [7,25],[7],[25]–[27],[55]. The underlying reason is that evergreen species are mostly diffuse-porous, which also explains the strong impact of the hydraulically weighted conduit diameter and mean area of conducting vessels on K_S, which is also well-known from the literature [19].

Contrary to expectations, this study did not find a significant link between xylem hydraulic conductivity and parameters of stomatal regulation. Neither the maximum stomatal conductance, the vapor pressure deficit at maximum stomatal conductance nor the vapor pressure deficit at which stomatal conductance is down-regulated was related to any parameter of the xylem vulnerability curves. As such, our third hypothesis has to be rejected, which implies that the ability of a very precise and fast stomatal regulation versus a retarded and inert stomatal regulation does not translate into cavitation resistance. Additional insights into the relationship between K_S and leaf stomatal regulation might be gained by calculating leaf-specific xylem hydraulic conductivity K_L, which would directly refer to the capacity of the vascular system of a stem to supply the water to that stem [56]. However, our results confirm those of Brodribb *et al.* [46], who found no correlation between Ψ_{50} and the leaf water potential at stomatal closure in eight tropical, dry forest trees. They concluded that xylem cavitation and stomatal closure are linked through complex indirect regulatory mechanisms and argue that this potential linkage is considerably flexible, especially with regard to different leaf phenology strategies, and that there may be carry-over effects of preceding embolism events on stomatal control. A further explanation of a lacking relationship between xylem vulnerability and stomatal regulation may be the different scale at which stomatal regulation is considered. At the level of whole trees, Litvak *et al.* [55] found a strong linear relationship between the sensitivity of tree-level sap flow to VPD and Ψ_{50} both within diffuse- and ring-porous species, which was not encountered for leaf-level transpiration rates. The authors argue that the tree-level transpiration sensitivity, in addition to stomatal regulation, also directly responds to drought-induced embolisms. Several studies showed a trade-off between high hydraulic conductivity and cavitation resistance [57]–[60]. In contrast to these studies, we found Ψ_{50} and K_S to be unrelated. According to our current understanding of the causes of xylem embolism under drought conditions, there may indeed be no mechanistic link between these two hydraulic characteristics. As such, diameter and length of vessels may differ autonomously from pit structure and size [19]. While K_S is mainly driven by vessel diameter, Ψ_{50} depends on pit size and structure [19],[20]. The pit area hypothesis states that cavitation resistance is linked to the total area of inter-vessel pits per vessel [61],[62]. Thus, the risk of an embolism expanding between vessels rises with the maximum size of the pit membrane pore, which in turn is dependent on the associated pit membrane area per vessel. This was demonstrated by Hacke et al. [57], who reported a strong negative link between xylem vulnerability and pit membrane area per vessel, resulting in small pits potentially increasing hydraulic resistance and decreasing K_S. However, pit size may be of minor importance to K_S compared to that of vessel diameter, and the relationship of hydraulic conductivity and cavitation resistance might depend on the specific ecosystem considered. Tyree et al. [59] distinguished between frost- and drought-induced cavitation. In their meta-analysis, the trade-off between hydraulic conductivity and cavitation resistance was mainly related to frost-induced cavitation events. Although frosts occur in the Chinese subtropics, they are neither very strong, nor long-lasting [63]. Thus, cavitation in the forests of our study area will mainly be brought about by drought events, which may result in far fewer, or insignificant, trade-offs. Since all our individuals have the same age, our species set provides a high comparability usually not found in comparative studies. We expect that some of our response variables will change with tree age, such as specific xylem hydraulic conductivity [64],[65]. In addition, future comparisons should take the sustained leaf area into account, as whole-tree leaf-specific hydraulic conductance (K_L) is known to decrease with tree age [66],[67].

CONCLUSION

For the studied subtropical forest community, we demonstrated a clear link of K_S and Ψ_{50} with functional traits, and particularly with leaf nitrogen concentration, log leaf area and leaf carbon to nitrogen ratio. Thus, easily measured leaf traits from the LES have the potential to predict plant species' drought resistance. However, current knowledge on xylem vulnerability and traits from other ecosystems do not allow generalizing from these results. In addition, our finding of an absence of any relationship between parameters of stomatal control raises the question whether stomatal control as characteristics that are an independent axis of the LES might be related to an axis of xylem characteristics that are independent of specific xylem hydraulic conductivity and xylem vulnerability.

Supporting Information

Figure S1: Raw data for the vulnerability curves of the 39 study species analyzed. Filled dots represent measured data, empty dots show estimated Ψ_{50} values and the broken lines represent the fitted models of xylem vulnerability.

doi:10.1371/journal.pone.0109211.s001

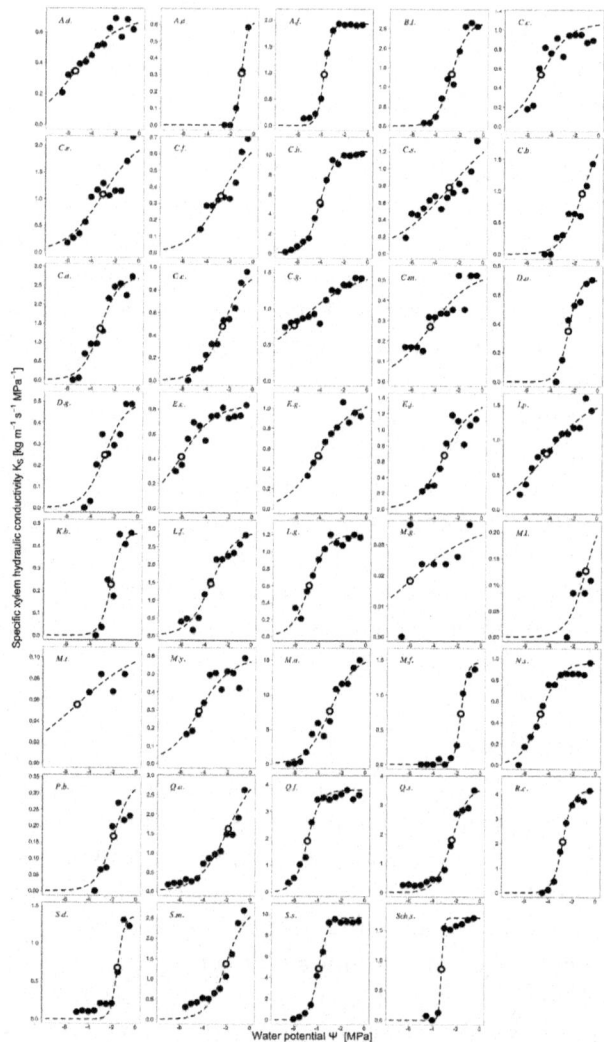

Acknowledgments

We are very grateful for the help of Xuefei Yang, Sabine Both, Lin Chen and Xiaojuan Liu in coordinating the fieldwork for the BEF-China experiment. We also thank the whole BEF-China research group for their general support. In addition, we would like to thank Ricardo Schöps, Ricarda Pohl and Michael Staab for their support in the lab. Special thanks are due to Tim Langhammer and Stefan Posch for providing data on the xylem structure and to Peter Kühn for providing climate data.

Author Contributions

Conceived and designed the experiments: WK SZ ME HB. Performed the experiments: WK ME. Analyzed the data: WK HB. Contributed reagents/materials/analysis tools: WK SZ ME HB. Wrote the paper: WK SZ ME HB.

REFERENCES

1. Wright IJ, Reich PB, Westoby M, Ackerly DD, Baruch Z, et al. (2004) The worldwide leaf economics spectrum. Nature 428: 821–827.

2. Reich PB, Ellsworth DS, Walters MB, Vose JM, Gresham C, et al. (1999) Generality of leaf trait relationships: A test across six biomes. Ecology 80: 1955–1969.

3. Shipley B, Lechowicz MJ, Wright I, Reich PB (2006) Fundamental trade-offs generating the worldwide leaf economics spectrum. Ecology 87: 535–541.

4. Cornelissen JHC, Wright IJ, Reich PB, Falster DS, Garnier E, et al. (2005) Assessing the generality of global leaf trait relationships. New Phytol 166: 485–496.

5. Santiago LS (2007) Extending the leaf economics spectrum to decomposition: evidence from a tropical forest. Ecology 88: 1126–1131.

6. Osnas JLD, Lichstein JW, Reich PB, Pacala SW (2013) Global leaf trait relationships: mass, area, and the leaf economics spectrum. Science 340: 741–744.

7. Kro¨ber W, Bruelheide H (2014) Transpiration and stomatal control: A cross-species study of leaf traits in 39 evergreen and deciduous broadleaved subtropical tree species. Trees 28: 901–914.

8. Baraloto C, Paine C, Poorter L, Beauchene J, Bonal D, et al. (2010) Decoupled leaf and stem economics in rain forest trees. Ecol Letters 13: 1338–1347.

9. Freschet GT, Cornelissen JHC, van Logtestijn RSP, Aerts R (2010)

Evidence of the "plant economics spectrum" in a subarctic flora. J Ecol 98: 362–373.

10. Lambers H, Chapin F III, Pons T (2008) Plant Physiological Ecology. 2nd ed. Berlin: Springer. 623 pp.

11. Sperry J, Choat B, Jansen S, Brodribb TJ, Cochard H, et al. (2012) Global convergence in the vulnerability of forests to drought. Nature 491: 752–755.

12. Poorter L, Mcdonald I, Alarco´ n A, Fichtler E, Licona J-C, et al. (2010) The importance of wood traits and hydraulic conductance for the performance and life history strategies of 42 rainforest tree species. New Phytol 185: 481–492.

13. Sack L, Frole K (2006) Leaf structural diversity is related to hydraulic capacity in tropical rain forest trees. Ecology 87: 483–491.

14. Brodribb TJ, Feild TS, Jordan GJ (2007) Leaf maximum photosynthetic rate and venation are linked by hydraulics. Plant Physiol 144: 1890–1898.

15. Sack L, Scoffoni C, John GP, Poorter H, Mason CM, et al. (2013) How do leaf veins influence the worldwide leaf economic spectrum? Review and synthesis. J Exp Bot 64: 4053–4080.

16. Chave J, Coomes D, Jansen S, Lewis SL, Swenson NG, et al. (2009) Towards a worldwide wood economics spectrum. Ecol Letters 12: 351–366.

17. Fu P-L, Jiang Y-J, Wang A-Y, Brodribb TJ, Zhang J-L, et al. (2012) Stem hydraulic traits and leaf water-stress tolerance are co-ordinated with the leaf phenology of angiosperm trees in an Asian tropical dry karst forest. Ann Bot-London 110: 189–199.

18. Sperry J, Donelly J, Tyree MT (1988) A method for measuring hydraulic conductivity and embolism in xylem. Plant Cell Environ 11: 35–40.

19. Hacke U, Sperry J (2000) Functional and ecological xylem anatomy. PPEES 4: 97–115.

20. Tyree MT, Zimmermann MH (2002) Xylem Structure and the Ascent of Sap. 2nd ed. Berlin: Springer. 302 pp.

21. Kursar TA, Engelbrecht BM, Burke A, Tyree MT, El Omari B, et al. (2009) Tolerance to low leaf water status of tropical tree seedlings is related to drought performance and distribution. Funct Ecol 23: 93– 102.

22. Engelbrecht BM, Kursar TA (2003) Comparative drought-resistance of seedlings of 28 species of co-occurring tropical woody plants. Oecologia 136: 383–393.

23. Fonseca CR, Overton JM, Collins B, Westoby M (2000) Shifts in trait-

combinations along rainfall and phosphorus gradients. J Ecol 88: 964–977.

24. Skarpe C (1996) Plant functional types and climate in a southern African savanna. J Veg Sci 7: 397–404.

25. Brodribb TJ, Jordan GJ (2008) Internal coordination between hydraulics and stomatal control in leaves. Plant Cell Environ 31: 1557–1564.

26. Cochard H, Bre´da N, Granier A (1996) Whole tree hydraulic conductance and water loss regulation in Quercus during drought: evidence for stomatal control of embolism? Ann For Sci 53: 197–206.

27. Choat B, Ball MC, Luly JG, Holtum JAM (2005) Hydraulic architecture of deciduous and evergreen dry rainforest tree species from north-eastern Australia. Trees 19: 305–311.

28. Walter H, Lieth H (1967) Klima-Diagramm Weltatlas. Jena: Gustav Fischer Verlag. 250 pp.

29. Bruelheide H, Nadrowski K, Assmann T, Bauhus J, Both S, et al. (2014) Designing forest biodiversity experiments: general considerations illustrated by a new large experiment in subtropical China. Methods Ecol Evol 5: 74–89.

30. Yang X, Bauhus J, Both S, Fang T, Ha¨rdtle W, et al. (2013) Establishment success in a forest biodiversity and ecosystem functioning experiment in subtropical China (BEF-China). Eur J Forest Res 132: 593–606.

31. Bruelheide H, Bo¨hnke M, Both S, Fang T, Assmann T, et al. (2011) Community assembly during secondary forest succession in a Chinese subtropical forest. Ecol Monogr 81: 25–41.

32. Ennajeh M, Simo˜es F, Khemira H, Cochard H (2011) How reliable is the double-ended pressure sleeve technique for assessing xylem vulnerability to cavitation in woody angiosperms? Physiologia Plant 142: 205–210.

33. Melcher PJ, Michele Holbrook N, Burns MJ, Zwieniecki MA, Cobb AR, et al. (2012) Measurements of stem xylem hydraulic conductivity in the laboratory and field. Methods Ecol Evol 3: 685–694.

34. Perez-Harguindeguy N, Dı´az S, Garnier E, Lavorel S, Poorter H, et al. (2013) New handbook for standardised measurement of plant functional traits worldwide. Aust J Bot 61: 167–234.

35. Cochard H, Badel E, Herbette S, Delzon S, Choat B, et al. (2013) Methods for measuring plant vulnerability to cavitation: a critical review. J Exp Bot 64: 4779–4791.

36. Hendry G, Grime J (1993) Methods in comparative Plant Ecology: a laboratory Manual. London: Chapman & Hall. 252 pp.

37. Gerlach D (1984) Botanische Mikrotechnik. Stuttgart, New York: Thieme. 289 pp.

38. Sperry J, Nichols KL, Sullivan JEM, Eastlack SE (1994) Xylem embolism in ring-porous, diffuseorous, and coniferous trees of Northern Utah and interior Alaska. Ecology 75: 1736.

39. Vander Willigen C, Sherwin HW, Pammenter NW (2000) Xylem hydraulic characteristics of subtropical trees from contrasting habitats grown under identical environmental conditions. New Phytol 145: 51–59.

40. Domec J, Gartner BL (2001) Cavitation and water storage capacity in bole xylem segments of mature and young Douglas-fir trees. Trees 15: 204–214.

41. Bo" hnke M, Kro" ber W, Welk E, Wirth C, Bruelheide H (2013) Maintenance of constant functional diversity during secondary succession of a subtropical forest in China. J Veg Sci 25: 897–911.

42. Markesteijn L, Poorter L, Paz H, Sack L, Bongers F (2011) Ecological differentiation in xylem cavitation resistance is associated with stem and leaf structural traits. Plant Cell Environ 34: 137–148.

43. Sperry J, Tyree MT (1989) Vulnerability of xylem to cavitation and embolism. Annu Rev Plant Biol 40: 19–36.

44. Cavender-Bares J, Kitajima K, Bazzaz FA (2004) Multiple trait associations in relation to habitat differentiation among 17 Floridian oak species. Ecol Monogr 74: 635–662.

45. Maherali H, Pockman WT, Jackson RB (2004) Adaptive variation in the vulnerability of woody plants to xylem cavitation. Ecology 85: 2184–2199.

46. Brodribb TJ, Holbrook NM, Edwards E, Gutie´ rrez MRVA (2003) Relations between stomatal closure, leaf turgor and xylem vulnerability in eight tropical dry forest trees. Plant Cell Environ 26: 443–450.

47. Willson CJ, Manos PS, Jackson RB (2008) Hydraulic traits are influenced by phylogenetic history in the drought-resistant, invasive genus Juniperus (Cupressaceae). Am J Bot 95: 299–314.

48. Vinya R, Malhi Y, Brown N, Fisher JB (2012) Functional coordination between branch hydraulic properties and leaf functional traits in miombo woodlands: implications for water stress management and species habitat preference. Acta Physiol Plant 34: 1701–1710.

49. Bell P (2000) Green Plants: their Origin and Diversity. Cambridge: Cambridge Univ Pr. 349 pp.

50. Sack L, Cowan PD, Jaikumar N, Holbrook NM (2003) The "hydrology"

of leaves: co-ordination of structure and function in temperate woody species. Plant Cell Environ 26: 1343–1356.

51. Chen J-W, Zhang Q, Cao K-F (2009) Inter-species variation of photosynthetic and xylem hydraulic traits in the deciduous and evergreen Euphorbiaceae tree species from a seasonally tropical forest in southwestern China. Ecol Res 24: 65–73.

52. Fan ZX, Zhang S-B, Hao GY, Ferry Slik JW, Cao K-F (2012) Hydraulic conductivity traits predict growth rates and adult stature of 40 Asian tropical tree species better than wood density. J Ecol 100: 732–741.

53. Givnish T (2002) Adaptive significance of evergreen vs. deciduous leaves: solving the triple paradox. Silva Fenn 36: 703–743.

54. Aerts R (1995) The advantages of being evergreen. Trends Ecol Evol 10: 402–407.

55. Litvak E, McCarthy HR, Pataki DE (2012) Transpiration sensitivity of urban trees in a semi-arid climate is constrained by xylem vulnerability to cavitation. Tree Physiol 32: 373–388.

56. Choat B, Sack L, Holbrook NM (2007) Diversity of hydraulic traits in nine Cordia species growing in tropical forests with contrasting precipitation. New Phytol 175: 686–698.

57. Hacke UG, Sperry J, Wheeler JK, Castro L (2006) Scaling of angiosperm xylem structure with safety and efficiency. Tree Physiol 26: 689–701.

58. Markesteijn L, Poorter L, Bongers F, Paz H, Sack L (2011) Hydraulics and life history of tropical dry forest tree species: coordination of species' drought and shade tolerance. New Phytol 191: 480–495.

59. Tyree MT, Davis SD, Cochard H (1994) Biophysical perspectives of xylem evolution: is there a tradeoff of hydraulic efficiency for vulnerability to dysfunction? IAWA J 15: 335–360.

60. Martı́nez-Vilalta J, Prat E, Oliveras I, Pin˜ol J (2002) Xylem hydraulic properties of roots and stems of nine Mediterranean woody species. Oecologia 133: 19–29.

61. Jarbeau JA, Ewers FW, Davis SD (1995) The mechanism of water-stress-induced embolism in two species of chaparral shrubs. Plant Cell Environ 18: 189–196.

62. Hargrave KR, Kolb KJ, Ewers FW, Davis SD (1994) Conduit diameter and drought-induced embolism in Salvia mellifera Greene (Labiatae). New Phytol 126: 695–705.

63. Box E, Peet R, Masuzawa T, Yamada I, Fujiwara K, et al, editors (1995) Vegetation Science in Forestry. Dordrecht: Kluwer Academic Publishers.

664 pp.

64. Domec JC, Gartner BL (2003) Relationship between growth rates and xylem hydraulic characteristics in young, mature and old-growth ponderosa pine trees. Plant Cell Environ 26: 471–483.

65. Rosner S, Klein A, Muller U, Karlsson B (2008) Tradeoffs between hydraulic and mechanical stress responses of mature Norway spruce trunk wood. Tree Physiol 28: 1179–1188.

66. McDowell N, Phillips N, Lunch C, Bond B, Ryan M (2002) An investigation of hydraulic limitation and compensation in large, old Douglas-fir trees. Tree Physiol 22: 763–774.

67. Delzon S, Sartore M, Burlett R, Dewar R, Lousteau D (2004) Hydraulic responses to height growth in maritime pine trees. Plant Cell Environ 27: 1077–1087.

Chapter 8

ARSENIC IN GROUNDWATER: A SUMMARY OF SOURCES AND THE BIOGEOCHEMICAL AND HYDROGEOLOGIC FACTORS AFFECTING ARSENIC OCCURRENCE AND MOBILITY

Julia L. Barringer and Pamela A. Reilly

U.S. Geological Survey, USA

INTRODUCTION

World-Wide Occurrences of Arsenic–Contaminated Groundwater – Forms and Toxicity

Arsenic (As) is a metalloid element (atomic number 33) with one naturally occurring isotope of atomic mass 75, and four oxidation states (-3, 0, +3, and +5) (Smedley and Kinniburgh, 2002). In the aqueous environment, the +3 and +5 oxidation states are most prevalent, as the oxyanions arsenite (H_3AsO_3 or $H_2AsO_3^-$ at pH ~9-11) and arsenate ($H_2AsO_4^-$ and $HAsO_4^{2-}$ at pH ~4-10) (Smedley and Kinniburgh, 2002). In soils, arsine gases (containing As^{3-}) may be generated by fungi and other organisms (Woolson, 1977).

The different forms of As have different toxicities, with arsine gas being the most toxic form. Of the inorganic oxyanions, arsenite is considered more toxic than arsenate, and the organic (methylated) arsenic forms are considered least toxic (for a detailed discussion of toxicity issues, the reader is referred to Mandal and Suzuki (2002)). Arsenic is a global health concern due to its toxicity and the fact that it occurs at unhealthful levels in water supplies, particularly groundwater, in more than 70 countries (Ravenscroft et al., 2009) on six continents.

Health Effects and Standards

Despite its use in medicines for nearly 2,500 years (Mandal and Suzuki, 2002; Cullen, 2008) As has long been recognized as a toxic and often lethal substance. Chronic exposure to As can cause harm to the human cardiovascular, dermal, gastrointestinal, hepatic, neurological, pulmonary, renal and respiratory systems (ATSDR, 2000) and reproductive system (Mandal and Suzuki, 2002). Research on health effects is summarized and discussed by Mandal and Suzuki (2002) and Ng et al. (2003). A compilation of their reviews is found in Table 1.

Table 1. Summary of effects of chronic arsenic exposure on human health. (Data from Mandal and Suzuki, 2002, and Ng et al., 2003, and references therein.)

System	Health effects
Cardiovascular	Heart attack, cardiac arrhythmias, thickening of blood vessels, loss of circulation leading to gangrene of extremities, hypertension
Dermal	Hyperpigmentation, abnormal skin thickening, narrowing of small arteries leading to numbness (Raynaud's Disease), squamous and basal-cell cancer
Gastrointestinal	Heartburn, nausea, abdominal pain
Hematological	Anemia, low white-blood-cell count (leucopenia)
Hepatic	Cirrhosis, fatty degeneration, abnormal cell growth (neoplasia)
Neurological	Brain malfunction, hallucinations, memory loss, seizures, coma, peripheral neuropathy
Pulmonary	Chronic cough, restrictive lung disease, cancer
Respiratory	Laryngitis, tracheal bronchitis, rhinitis, pharyngitis, shortness of breath, perforation of nasal septum
Renal	Hematuria, proteinuria, shock, dehydration, cortical necrosis, cancer of kidneys and bladder
Reproductive	Spontaneous abortions, still-births, congenital malformations of fetus, low birth weight

The carcinogenic properties of As were suspected as early as the late 19th Century (Smith et al., 2002). Arsenic is now widely recognized and regulated as a carcinogen (ATSDR, 2000; National Research Council, 1999; USEPA, 2001). Consequently, the occurrence of As in waters at concentrations that exceed existing standards for drinking-water supplies has become of increasing concern, leading to recommended or legislated decreases in concentrations of As in drinking water in many countries. In 1993, the World Health Organization provisionally recommended a decrease from 50 μg/L to 10 μg/L (WHO, 1993). The United States (USA) federal standard, the

European Union (EU) Drinking Water Directive (98/83/EC), the New Zealand Drinking Water Standard, the Japanese standard, and recent laws in many Latin American countries (Argentina, Bolivia, Brazil, Chile, Colombia, Costa Rica, El Salvador, Guatemala, Honduras, Nicaragua, and Panama) now place 10 µg/L as the drinking water maximum contaminant level (MCL) (Bundschuh et al., 2012; Robinson et al., 2004; Rowland et al., 2011; Smedley and Kinniburgh, 2002; USEPA, 2001). Mexico has adopted 25 µg/L as a standard (Bundschuh et al., 2012), whereas Australia has instituted a standard of 7 µg/L (NHMRC, 1996) and the State of New Jersey in the USA adopted an As MCL of 5 µg/L in 2006 (NJDEP, 2009). Some developing countries (Bangladesh, for example) have maintained the earlier 50 µg/L MCL standard (Ng et al., 2003). Many instances of As concentrations in groundwater that far exceed standards have been reported throughout much of the world (Smedley, 2008) and the number of countries in which groundwater is found to be contaminated by As has increased substantially over the past 80 years. This chapter presents a brief overview of the history of groundwater As contamination and summarizes information about the sources, occurrence and mobility of As in groundwater. A compilation of worldwide hazardous waste sites is beyond the scope of the chapter, and only a few examples will be presented. Information on As occurrence reported in previous important summaries and discussions (e.g., Bhattacharya et al., (eds) 2007; Smedley and Kinniburgh, 2002; Welch and Stollenwerk, (eds.) 2003) is presented in addition to recent findings from the past decade.

Chronology of Discoveries of Geogenic Arsenic Contamination

Currently (2012), As contamination of groundwater resources has been identified in many parts of the world, although recognition of the widespread nature of the problem has been advanced only relatively recently. Despite localized inputs of As from human activities, much of the contamination of groundwater with As is shown to arise from geogenic sources and affected groundwater has been found in countries on nearly every continent or major land mass. To date, none has been reported for Greenland, and Antarctica (Figure 1).

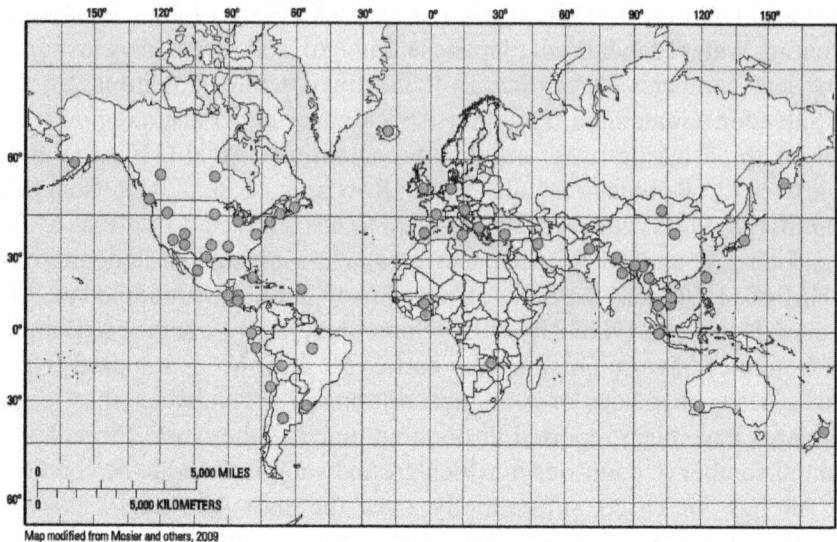

Map modified from Mosier and others, 2009

Figure 1. Countries, states, provinces, or areas mentioned in this chapter in which arsenic concentrations in groundwater, including geothermal waters and water contaminated by mining, exceed 10 micrograms per liter (the World Health Organization (WHO) recommended standard (WHO, 1993)). Arsenic concentrations at most locations shown exceed 50 micrograms per liter. Locations of dots are spatially generalized and do not indicate precise locations of arsenic-contaminated waters.

The discoveries of As contamination of groundwater have occurred over a span of nearly 100 years, the most recent within the last decade. Observations of health problems first led to the realization that As was being inadvertently ingested. Arsenic poisoning in humans in Argentina was recognized as early as 1913, and attributed to the drinking of groundwater (Bundschuh et al., 2012, and reference therein). A possible connection between skin cancers and drinking water was recognized in Taiwan during the 1930s (Chen et al., 1994). In the 1940s, As contamination of well water in the Pannonian Basin in Romania and adjacent Hungary was discovered (Gurzau and Pop, 2012; Mukherjee et al., 2006). Recognition of similar occurrences in other European countries such as southwestern England, Germany, Greece, and Spain followed, and groundwaters with As concentrations that exceed standards are now observed in more than 70 countries worldwide (Nordstrom, 2002; Ravenscroft et al., 2009).

During the mid-20th century, instances of As contamination of groundwater were reported for the western USA and Alaska (Welch et al., 1988; Mueller et al., 2001), but were not fully recognized in the states of Oklahoma, Texas, and Arkansas and in the Midwest and Northeastern parts of the country until the 1980s (Ayotte et al., 2003; Ayotte et al., 1999; Haque et al., 2008; Peters, 2008; Peters and Burkert, 2007; Scanlon et al., 2009; Sharif et al., 2008; Welch et al., 2000 and references therein). Most recently, groundwater containing As in excess of USA Federal and State MCLs was found in parts of the Atlantic Coastal Plain (Barringer et al., 2010; Drummond and Bolton, 2010; Haque et al., 2008; Mumford et al., 2012; Pearcy et al., 2011). In Canada, As-contaminated groundwaters in New Brunswick and Nova Scotia were noted in the 1970s (Bottomley, 1984), and instances in western Canada were noted in the 1960s and 1980s (Wang and Mulligan, 2006a).

In West Bengal, India, cases of arsenic poisoning were first noted in 1983-84, according to Rahman et al. (2005), although Mandal and Suzuki place 1978 as the time when arsenicosis (skin lesions) and groundwater contamination were first noticed in West Bengal. Since the 1980s, extensive sampling of well water in West Bengal has revealed levels of As that exceed 50 µg/L —concentrations in some samples exceeding 1000 µg/L (Rahman et al., 2005). In 1983-84, several patients treated for arsenicosis in West Bengal came from neighboring Bangladesh. Sampling in Bangladesh during the early 1990s of waters from tube wells (installed two decades earlier to provide what was thought to be safe, pathogen-free drinking water) revealed elevated concentrations of As (Smith et al., 2000).

Continued sampling throughout West Bengal showed concentrations ranging from <1 to 2,500 µg/L (Nordstrom, 2002). Although the population exposed to As-contaminated water in West Bengal was ultimately estimated to have been about 6,000,000, the exposed population in Bangladesh has been estimated to be about five times that number (Nordstrom, 2002), making the contamination in Bangladesh the arsenic-related public health problem of the greatest magnitude yet observed. Following the early discoveries in West Bengal and Bangladesh, well-water sampling in Vietnam, Cambodia, and Pakistan also identified contamination of groundwater with As (Agusa et al., 2006; Berg et al., 2001;Chanpiwat et al., 2011; Hoang et al., 2010; Luu et al., 2009; Nickson et al., 2005; Smedley, 2008;Sthiannopkao et al., 2008;), as well as in 1999, Nepal (Gurung et al., 2005), and in 2000, Myanmar (Tun, 2003).

Health problems attributed to As exposure were first noted within a district in a province of Thailand in 1987. Sampling indicated As at concentrations in groundwater that exceeded 5,000 µg/L with about 15,000 people thought to have been exposed (Nordstrom, 2002; Smedley, 2008). Sampling of well waters during the 1990s and later in northern parts of China, including Inner Mongolia, have shown groundwater As concentrations that range widely, from <1 to about 2,400 µg/L (Nordstrom, 2002;Smedley, 2008). In Japan, As levels were found to be high in geothermal waters and springs as early as the 1950s (Noguchi and Nakagawa, 1969), and As contamination of groundwater was noted in 1994. Exposure to As from industrial sources was noted as early as the 1950s (Mukherjee et al., 2006). In 1981 in Iran, chronic arsenic poisoning was noticed and subsequent well-water sampling revealed concentrations exceeding 1,000 µg/L (Mukherjee et al., 2006).

Arsenic contamination of both surface water and groundwater also is found in many Latin American countries, but the full extent of the problem is not yet clear (Bundschuh et al., 2012). Groundwater concentrations in Argentina are reported to range as high as about 15,000 µg/L (Bundschuh et al., and references therein). High levels of As were found in waters in Chile and Mexico in the 1950s and 60s (Bundschuh et al., 2012; Rosas et al., 1999). In northern Chile, As concentrations in groundwater ranged from 20 to 5,000 µg/L, and there is strong evidence from hair, skin, bones and funerary preparations (clay, paint) of Chinchorro culture mummies that the population there was exposed to high levels of As more than 7,000 years ago (Bundschuh et al., 2012), and, presumably, has been ever since. Concentrations of As in groundwater in Mexico generally are not reported to reach the higher levels found in Argentina except for a geothermal area in Michoacán, where 24,000 µg/L are reported (Bundschuh et al., 2012, and references therein). In mined areas of Mexico, leaching from tailings piles have contributed As to groundwater (Carillo-Chávez et al., 2000; Méndez and Armienta, 2003). Arsenic-contaminated surface water and groundwater subsequently was recognized in parts of Peru in the 1970s (Bundschuh et al., 2010). Although As contamination of waters in Argentina, Chile, Mexico and Peru was known from the early and mid-20th century, surface water and (or) groundwaters (and geothermal waters) containing geogenic As at contaminant levels were discovered in alluvial, metasedimentary, volcanic, and metavolcanic aquifers only since the late 1990s in Bolivia, Brazil, Columbia, Cuba, Ecuador, El Salvador, Guatemala, Honduras, Nicaragua, and Uruguay (Bundschuh et al., 2012).

In Africa, the occurrence of high concentrations of As in groundwater from wells in a village in Burkina Faso was first noticed in the 1970s, but more extensive sampling did not occur until several decades later (Smedley, 2008). Arsenic-related skin diseases were noted in the region, and, although many wells yielded water with As concentrations < 10 µg/L, concentrations as high as 1,600 µg/L were found (Smedley et al., 2007). Effects of mining on soils and waters in other African countries (Ghana and Zambia) have been studied only recently (e.g., Bowell et al., 1994; Nakayama et al., 2011; Smedley and Kinniburgh (2002); As contamination of groundwater has been reported in

Ghana. Arsenic contamination of well water was discovered during the drought in Perth, Australia in 2002 (Appleyard et al., 2006).

SOURCES OF ARSENIC

Anthropogenic Sources

Sources of As that arise from human activities include mining and processing of ores and manufacturing using As-bearing sulfides. Smelters in numerous countries, including Canada, Chile, Italy, South Africa, the USA, and the former USSR have processed metal ores (mainly copper, but also zinc, gold, and tin) that contain As. The smelting process, both recent and ancient, has released As to the air and soils both locally and globally (Matschullat, 2000). Elevated As levels in precipitation and in soils surrounding smelters have frequently been documented (e.g., Ball et al., 1983; Beaulieu and Savage, 2005; Carpenter et al., 1978). Disposal of mining wastes has caused As contamination of groundwater in numerous places, including in Thailand, Ghana, and Turkey (Gunduz et al., 2010;Smedley and Kinniburgh, 2002; Smedley, 2008). In southwestern England, mineral deposits and mineral processing (tin, copper, with accessory As minerals) are recognized sources of As to soils and groundwater (Brunt et al., 2004; Camm et al., 2004; Palumbo-Roe et al., 2007). In southeastern Europe (Serbia, Bosnia, Poland), and in Spain, mining activities left a legacy of arsenic that has contaminated soils and waters (Dangic, 2007; Gomez et al., 2006; Karczewska et al., 2007). Groundwater contamination in mined areas is also found in parts of the western USA and Canada (Moncur et al., 2005; Welch et al., 2000).

Arsenic compounds have been used in the manufacture of numerous products. Arsenic has been used in glass production and by the wood-preservation industry. The latter industry has been, through the end of the 20th century, the most active user of such compounds in the USA (Welch et al., 2000), but the industry's voluntary reduction of the use of chromated copper arsenate (CCA) since 2003 has resulted in a more limited availability of CCA treated wood and products (Brooks, 2008). Contamination of soils and surface-water bodies has resulted from use of CCA-treated wood (Khan et al., 2006; Rice et al., 2002).

The use of inorganic arsenical pesticides has waned in recent years owing to bans in the 1980s and 90s (Welch et al., 2000), but, in the past, manufacture and use of arsenical pesticides were important contributors of As to the environment (e.g., Barringer et al., 1998; Barringer et al., 2001; USEPA, 2011). Inorganic arsenicals have been used on a variety of crops (citrus, cotton, tobacco, and potatoes) and on fruit trees; (Walsh and Keeney, 1975; Welch et al., 2000). The use of lead-arsenate pesticides in orchards has prompted concern that additions of phosphate fertilizers could displace arsenate sorbed to soil particles, mobilizing As to groundwater (e.g. Davenport and Peryea, 1991). Manufacture of pesticides has been responsible for As contamination of soil, surface water, and groundwater.

Examples of contamination caused by former pesticide manufacture are found in India and in the USA (Barringer et al., 1998; Mukherjee et al., 2006; USEPA, 2011). Soil contamination by As may also occur in areas where soils are amended with chicken and swine manure. Such fertilizers can contain As due to use of chicken and swine feeds containing the growth additive Roxarsone, an organic arsenical (4-hydroxy-3-nitrophenylarsonic acid) (Hileman, 2007; O'Day, 2006).

There have been military uses of As, and attempts to recycle As-bearing materials. Arsenic was used in chemical warfare agents, most recently in the first half of the 20th century (Krüger et al., 2007). In the 1930s, concern over disposal of stockpiles of As compounds led to experiments with incorporating As in cement which was then used to coat pilings and other wooden structures (van Siclen and Gerry, 1936). One ounce of white arsenic was added to 12 pounds of sand and 3 pounds of cement, water was added and the slurry was applied to wood pilings by air gun. To the authors' knowledge, no studies of the effects of this practice on the environment have been published. Arsenic currently is used in various electronic devices, and improper disposal and lack of care in recycling these materials also can add As to the environment (Brooks, 2008).

Anthropogenic sources of As can affect the quality of surface water through groundwater discharge and runoff (Hemond, 1995; Martin and Pederson 2002). In the case of pesticides, the effect can be through direct applications to water bodies for control of nuisance vegetation (Kobayashi and Lee, 1978; Tanner and Clayton, 1990; Durant et al., 2004). Although groundwater contamination does exist at various sites affected by agricultural, industrial or military releases, for example (e.g., Hemond, 1995; Krüger et al. 2007; USEPA, 2011), contamination introduced at the land surface does not always move to groundwater. Owing to the affinity of As for soil constituents such as metal oxides and hydroxides (mainly iron (Fe), aluminum (Al), and manganese (Mn)) and clays (Goldberg and Glaubig, 1988; Manning and Goldberg, 1996), the As can be attenuated in the intervening soils by sorption to Fe hydroxides or clays, or by precipitation reactions, such as formation of As- or Fe-sulfides in anoxic soils (e.g., Brunt et al., 2004; Cancès et al, 2008).

Geologic Sources

Arsenic Minerals

For most known areally extensive instances of As contamination of groundwater, the sources of the As have been shown to be geogenic (Smedley and Kinniburgh, 2002). A summary table of the worldwide occurrence of As in groundwater (Nordstrom, 2002) indicates that mining of arsenic and metal ores and natural geologic sources of As dominate the environmental conditions listed for inputs of As to groundwater.

There are about 24 As-bearing minerals that are commonly found in hydrothermal veins, ore deposits, and rocks. Most primary As minerals are sulfides, of which arsenopyrite is the most common (Ehrlich and Newman, 2009). Secondary minerals tend to be less common arsenates and oxides. WHO (2001) provides a list of these minerals, which also can be found tabulated in Mandal and Suzuki (2002, p. 203). Arsenic in crustal rocks also has an affinity for, and is associated with, pyrite or Fe hydroxides and oxides (Nordstrom, 2002) for which chemical formulas are FeS_2, $FeOOH$, Fe_2O_3, and Fe_3O_4, respectively. The As content of crustal rocks varies widely; Smedley and Kinniburgh, 2002, p. 531) compile and tabulate those results. Concentrations of As in water associated with crustal rocks are described below.

Geothermal Activity, Volcanic and Plutonic Rocks, and Mineralized Zones

Geothermally active zones occur along plate boundaries, in tectonic rift areas such those in East Africa and at seafloor spreading centers, such as in Iceland, and at "hot spots" where mantle-derived plumes ascend, such as in Hawaii and Yellowstone National Park, USA. Arsenic is one of a suite of incompatible elements (these do not fit easily into the lattices of common rock-forming minerals), which include antimony (Sb), boron (B), fluoride (F), lithium (Li), mercury (Hg), selenium (Se) and thallium (Tl). Together with hydrogen sulfide, these elements are ubiquitous in high-temperature geothermal settings (Webster and Nordstrom, 2003). Concentrations of arsenic are high mainly in geothermal waters that leach continental rocks; geothermal waters in basaltic rocks, such as in Iceland, contain lesser amounts of As. Arsenic in hot geothermal fluids was shown to be derived mainly from leaching of host rocks at Yellowstone National Park, in Wyoming, USA (Stauffer and Thompson, 1984), rather than derived from magmas. The As in the hot fluid is present as As(III) in arsenious acid (H_3AsO_3); in low-sulfide fluids, the arsenite in the arsenious acid is oxidized to arsenate as the rising fluid mixes with cold oxygenated groundwater or encounters the atmosphere. In high sulfide solutions, As may be present as thioarsenate complexes (Webster and Nordstrom, 2003; Planer-Friedrich et al., 2007).

In the western USA, there are As inputs to groundwater and surface water from geothermal fluids in and near Yellowstone National Park (e.g., Ball et al., 1998, 2002; Nimick et al., 1998) and in other western mineralized areas (Welch et al., 1988). Groundwater associated with volcanics (tuffs and rhyolites) in California contain As at concentrations ranging up to 48,000 µg/L, with As-bearing sulfide minerals as the main source of As (Welch et al., 1988, and references therein). Geothermal waters on Dominica in the Lesser Antilles also contain concentrations of As >50 µg/L (McCarthy et al., 2005).

In general, because arsenic is an incompatible element, it accumulates in differentiated magmas, and is commonly found at higher concentrations in volcanic rocks of intermediate (andesites) to felsic (rhyolites) composition than in mafic (basaltic) rocks—as shown for the western USA (Welch et al., 1988; Welch et al., 2000). In Maine and New Hampshire, USA, where As-contaminated groundwater is present, pegmatites, granites and metamorphic rocks (granofels) were found to have substantial As contents—up to 60, 46 and 39 mg/kg, respectively (Peters, 2008; Peters et al., 1999; Peters and Blum, 2003). Weathering of pegmatite veins in Connecticut, USA, was thought to contribute As to groundwater (Brown and Chute, 2002). Although the As content of mafic rocks can be relatively low, fractured ultramafic rocks in Vermont, USA, contribute up to 327 µg/L of As to groundwater (Ryan et al., 2011).

Leached from surrounding rocks, As in hot springs from geothermal fields in New Zealand is found at concentrations that range to 4,800 µg/L (Brown and Simmons, 2003). Acidic (pH 1.2) geothermal springs in Japan contained As at 2,600 µg/L (Noguchi and Nakagawa, 1969). Arsenic in these springs precipitated out as As sulfides and lead (Pb) As sulfides in surface-water sediments. The As contents of the sediments ranged from about 5 to 56 wt. % (Noguchi and Nakagawa, 1969).

In Latin America, volcanic and geothermal activity along the Pacific tectonic plate boundaries produces As-rich waters and gases in springs and fumaroles. In sodium-chloride (Na-Cl)-rich waters, As concentrations can reach about 50,000 µg/L at the El Tatio geothermal field in Chile (Lopez et al, 2012). Arsenic concentrations in waters of geothermal fields in Mexico vary; 250 to 73,600 µg/L are reported, where the host rocks through which these waters have risen include sandstones and shales, lava flows and pyroclastics, and metamorphosed carbonate rocks, basalts and hornblende andesites (Lopez et al., 2012).

In the coastal volcanic areas of Central and South American countries, rocks are mainly andesitic or rhyolitic in composition. Arsenic concentrations in the geothermal springs throughout these areas vary widely—concentrations of several thousand micrograms per liter are reported, but none are reported as high as the highest concentration in a Mexican geothermal field (see above). Where springs and fumaroles discharge water and gases to a lake in the Bolivian Altiplano, however, As concentrations in the lake water are reported as high as 4,600,000 µg/L, the As apparently being contributed by oxidation of sulfide deposits (Lopez et al., 2012).

Dissolution of volcanic glasses in ash layers and leaching of loess-type deposits in the Chaco-Pampean plain of Argentina have resulted in groundwater As concentrations that range from <10 to 5,300 µg/L (Nicolli, et al., 2012) where the potentially affected rural population numbers several million people. Mining of various metals (gold, copper, silver) in Latin America has played an important role in mobilizing As from the geologic materials and exacerbating contamination of groundwater resources (McClintock et al., 2012).

In Europe, groundwaters containing As at concentrations that exceed 50 µg/L are found in geothermal fields of the Massif Central in France, and in Greece, (Brunt et al., 2004; Karydakis et al., 2005). Iceland, where As concentrations in groundwater can exceed 10 µg/L (Arnórsson, 2003) sits astride the Mid-Atlantic Ridge, and is subject to outpourings of basaltic lava that typically contain less As than do more silicic lavas (Onishi and Sandell (1955; Baur and Onishi, 1969; Ure and Berrow, 1982). In geothermal systems of northern and northeastern Spain, the As concentrations are high and deposit As-rich minerals (Navarro et al., 2011); the Caldes de Malavella field in northern Spain contributes substantial As to groundwater—50-80 µg/L in springs, and from <1 to 200 µg/L in groundwater (Piqué et al., 2010). Groundwaters (including brines) contained As concentrations ranging from 1.6 to 6,900 µg/L in the Phlegraean Fields in southern Italy (Aiuppa et al., 2006). Quaternary volcanic rocks with hydrothermal activity on the island of Ischia (offshore from Naples, Italy) impart As to groundwater at concentrations that range up to 3,800 µg/L (Aiuppa, et al., 2006; Daniele, 2004).

In the Mid-East, in northwestern Iran, As concentrations in thermal waters and hot springs are as high as 3,500 and 890 µg/L, respectively, in the area of Mt. Sabalan, a stratavolcano (Haeri et al, 2011). In western Anatolia, in Turkey, natural leaching, aided by pumping and discharge of waste geothermal fluids from an active geothermal system, has mobilized As from metamorphic, igneous and sedimentary rocks to groundwater in a shallow alluvial aquifer, where highest As concentrations in groundwater and geothermal waters were 561 and 594 µg/L, respectively (Gunduz and Simsek (2008).

Groundwater in early Proterozoic silicic volcanics and granites of the Chhattisgarh Basin of India contains As at concentrations that exceed 10 μg/L. The As is perhaps emplaced there by hydrothermal fluids (Acharyya, 2002).

Sedimentary and Meta-Sedimentary Bedrock

Arsenic is found in coals, with the content of some coals from southwestern China being highest—826 to 2,578 mg/kg is reported (Nriagu et al., 2007) and up to 32,000 mg/kg is listed by Wang et al., (2006). In Germany, the As content of bituminous shales ranges from 100 to 900 mg/kg (Smedley and Kinniburgh, 2002). The As contents of American coals are reported to range as high as 2,200 mg/kg (Wang et al., 2006), but the mean concentration for more than 7000 samples is about 24 mg/kg (Kolker et al., 2006). Pyrite is the main source of As in coals with high As content, whereas in lower As coals, the As tends to be associated with the organic material (Yudovich and Ketris, 2005). In Pennsylvania, USA, As concentrations in water discharging from abandoned anthracite mines ranged from <0.03 to 15 μg/L and from abandoned bituminous mines, from 0.10 to 64 μg/L, with 10% of samples exceeding the USEPA MCL of 10 μg/L (Cravotta, 2008).

In Wisconsin, USA, As concentrations of water in sandstone and dolomite aquifers were as high as 100 μg/L. Oxidation of pyrite hosted by these formations was the likely source of the As, the transport of which was, in some instances, retarded by its association with Fe oxyhydroxides (Burkel and Stoll, 1999; Thornburg and Sahai, 2004). In the adjacent State of Michigan, USA, As concentrations in groundwater reached 220 μg/L in another sandstone aquifer (Haack and Rachol, 2000). In Australia, a combination of increased water withdrawals during development and declining recharge due to drought caused oxidation of pyrite in sedimentary aquifers, resulting in As contamination of well water (Appleyard et al., 2006). In England, groundwater from a sandstone aquifer contained As at concentrations that spanned 10 to 50 μg/L; the As content of the sandstone ranged from 5 to 15 mg/kg. Desorption at pH of about 8 appeared to be the mechanism for As release to groundwater (Kinniburgh et al., 2006). Water from wells completed in a Mesozoic Era sandstone in northern Bavaria also contained As at concentrations from 10 to 150 μg/L (Heinrichs and Udluft, 1999), although the mineralogy contributing the As was not identified.

In the Piedmont of Pennsylvania and New Jersey, USA, groundwater in Mesozoic age aquifers of red and black shale, mudstone, and siltstone contains elevated levels of As—domestic well waters from Pennsylvania contained up to 65 μg/L (Peters and Burkert, 2007), whereas in New Jersey the highest concentration measured recently was 215 µg/L (Serfes, 2005). Pyrites in the reduced black shales and mudstones are a major source of As, with measured As contents of 3,000 and 40,000 mg/kg in some samples (Serfes, 2005). Arsenic (as arsenate) is also released from the red shales by desorption as pH rises above 6.5 (Serfes, 2005). These Piedmont rocks also contain hornfels along contacts with diabase intrusions, some of which contain mineralization by copper (Cu), As, and uranium (U). Typical As-bearing minerals in the Piedmont rocks include arsenopyrite (FeAsS), cobaltite ((Ni, Co, Fe)AsS), alloclasite ((Co, Fe)AsS), gersdorffite (NiAsS), erythrite ($Co_3(AsO_4)_2.8H_2O$), and safflorite ((Co, Fe, Ni)As_2) (Senior and Sloto, 2006).

In Taiwan, groundwater from artesian wells completed in black shales, muds and fine sands is contaminated with As. In northern Bavaria, mineralized sandstone yields As-contaminated groundwater (Smedley and Kinniburgh, 2002).

Carbonate rocks typically contain low concentrations of As (Baur and Onishi, 1969; Smedley and Kinniburgh, 2002), although some limestones may contain As-bearing pyrite (e.g. Price and Pichler, 2006). Because arsenate can substitute for phosphate in minerals, phosphorite deposits can contain substantial amounts of As—up to about 400 mg/kg is reported (Smedley and Kinniburgh, 2002).Barringer et al. (2011) report 19.5 to 56.6 mg/kg of As in phosphorite deposits in the Coastal Plain of New Jersey, USA.

The As content of meta-sedimentary rocks varies widely, with the contents of gneisses and quartzites generally < 10 mg/kg, and higher As contents for slates and phyllites —up to 143 mg/kg (Boyle and Jonasson, 1973). In the New England states of northeastern USA, groundwater contamination with As was found to be prevalent in water from wells completed in formations containing metapelite rocks (schists, phyllites, slates), particularly those rocks adjacent to intrusive bodies. In New England, meta-shales contain As-bearing minerals pyrrhotite, cobaltite, and arsenopyrite, and supergene minerals include orpiment (AsS) and loellingite ($FeAs_2$)(Foley et al., 2002). Mineralization in a Proterozoic marble in New Jersey, USA, has resulted in zinc ores and, in addition to arsenopyrite and loellingite, a variety of uncommon As-bearing minerals. Past mining activities have contributed to the release of As from the bedrock to shallow groundwater; concentrations ranged from 2.02 to 22.0 µg/L in water discharging to the area's major river, the Wallkill River (Barringer et al., 2007). In mineralized meta-sedimentary rocks containing sulfide minerals in Fairbanks, Alaska, USA, groundwater contains As at concentrations that range from <3 µg/L to 1,670 µg/L (Mueller et al., 2001).

Alluvial and Coastal Plain Unconsolidated Sediments

The sediments shed from the Himalayas have formed the extensive alluvial plain and delta through which the Ganges, Brahmaputra and Meghna Rivers flow and which form aquifers in India (West Bengal Delta Region) and adjacent Bangladesh. To the east, the deltas of the Mekong and Red Rivers form aquifers in Vietnam and adjacent Cambodia and Laos. Arsenic released from these sediments has caused the most widespread contamination in the world, with populations of many millions affected by drinking As-laden well water. The largest number of people (about 35 million) exposed to As contamination is in Bangladesh, and about 6 million in neighboring West Bengal, India (Nordstrom, 2002). The As contents of these young sediments are not extremely high— about 1 to 15 mg/kg, but vary with depth (Smedley, 2008). The concentrations of As in groundwater range, however, from undetectable to several thousand μg/L, with a survey of about 3,200 wells in Bangladesh by researchers from the United Kingdom (UK) and Bangladesh finding that As concentrations in about 27% of samples exceeded 50 μg/L, the Bangladeshi Drinking Water Standard (Smedley, 2008). In local areas, the percentage of affected wells was higher; in central Bangladesh, As in water from about 75% of 6000 wells exceeded 50 μg/L (van Geen et al., 2003). The sediments of Holocene age are micaceous sands, silts, and clays. Reduced, gray sediments of the upper aquifer, where As concentrations in groundwater are high, are underlain by brown, oxidized sediments where As concentrations in groundwater are low. In some places the two layers are separated by a thick clay layer (Harvey et al., 2002). The mineralogy of both layers, where the clay was absent, was found to be similar for part of the Bengal Basin near Dhaka, Bangladesh. Minerals included quartz, plagioclase and potassium feldspar, micas (biotite, muscovite, and phlogopite), chlorite, and amphibole; carbonaceous material was sparse. Trace amounts of siderite were present in the reduced sediments, but not in the deeper, oxidized sediments (Stollenwerk et al., 2007). In the shallow aquifer in West Bengal, organic carbon is present as petroleum-related compounds (Rowland et al., 2006).

In Cambodia, Vietnam, and Laos, As in young deltaic sediments of the Mekong and Red River basins has also contaminated groundwater, again exposing several million people to unhealthy levels of As (> 1,000 μg/L in some cases) in drinking water (Agusa et al., 2006; Berg et al., 2001; Chanpiwat et al., 2011; Hoang et al., 2010; Luu et al., 2009; Sthiannopkao et al., 2008). The aquifers are composed of quartz sands and clays that host Fe oxide and hydroxide phases, also possibly manganese (Mn) oxides, and organic matter is present. Siderite, pyrite, and orpiment are also found in sediments of the Mekong delta (Quicksall et al., 2008) and siderite, ilmenite, vivianite, gibbsite and boehmite are reported for the Red River delta sediments (Eiche et al., 2008).

In the Pannonian Basin of Hungary and Romania, Quaternary sediments of fluvial and eolian origin have contributed As to groundwater; the sediments are composed of sands and loess. Quartz, feldspar, carbonates (calcite and dolomite, muscovite, chlorite, clays, and humic substances are reported and fine particles of Fe hydroxides are indicated (Varsányi and Kovács, 2006). Other Fe-bearing minerals, from which As may be released, are reported to include goethite, limonite, pyrite, and siderite (Rowland et al., 2011).

Alluvial and lacustrine sediments in the Huhhot Basin of Inner Mongolia form two aquifers, separated by a clay confining layer; some boreholes completed in the deeper aquifer are artesian. The sediments are more fine-grained in the low-lying parts of the basin and it is in these sediments that reducing conditions are present and As concentration in groundwater are highest (1,500 µg/L). Organic matter is found in the aquifers and dissolved organic carbon concentrations in groundwater are high (Smedley et al., 2003).

The Atlantic Coastal Plain is located along the east coast of the USA. In addition to quartz-rich deposits of near-shore origin, the Coastal Plain is composed partly of sediments of marine origin that contain the mineral glauconite, the As contents of which are high (up to 130 mg/kg) in some formations (Dooley, 2001). In the state of Maryland, USA, in an aquifer composed of these marine sediments, As concentrations in groundwater exceed 10 µg/L and have been found as high as 80 µg/L (Pearcy et al, 2011). Farther north, in the state of New Jersey, As concentrations in water from an observation well in a glauconite-bearing aquifer were 110 µg/L (dePaul and Szabo, 2007), and water from several domestic wells in similar aquifers in the same region has exceeded the state MCL of 5 µg/L. In shallow groundwater discharging to New Jersey Coastal Plain streams underlain by the glauconitic sediments, As concentrations have exceeded the MCL, ranging as high as 89.2 µg/L. The sediments below the streambeds also contain other phyllosilicates (illite, smectites, muscovite, biotite, chlorite) and quartz. In addition to the glauconite sands, associated phosphorite deposits were found to contain As up to 56.6 mg/kg and siderite that precipitated in sediments beneath a streambed contained 184 mg/kg of As (Barringer et al., 2011; Mumford et al., 2012).

Glauconite sands and clays of Pliocene and Miocene age, overlain by younger non-glauconitic sands, clays, thin coal beds and peats, are also present in the lowlands of xouth Sumatra, Indonesia. Arsenic concentrations exceeded the WHO guidelines in water from several wells completed in both the glauconitic formations and the younger sediments, with the higher concentrations found in water from the youngest (Holocene) sediments

(Winkel et al., 2008). Glauconitic sediments are found on several continents (Barringer et al., 2010) and the recent findings in the Atlantic Coastal Plain and Sumatra indicate that these marine sediments, in addition to all the aquifers in other geologic settings, can now also be considered a potential source of As-contaminated groundwater.

BIOGEOCHEMICAL FACTORS

Oxidation

Oxidation of As-bearing sulfides has been proposed as a mechanism for releasing As from geologic materials (Smedley, 2008). Although originally proposed as a mechanism for As release from the alluvial sediments of West Bengal and Bangladesh, the presence of sulfide minerals in those aquifers is rare and appears limited to biogenic framboidal pyrite, and pyrite in woody peat, and on magnetite (Acharyya, 2002). Oxidation of sulfides in mined areas throughout the world is a well-known phenomenon that has led to high concentrations of As in soils, surface water and groundwater; examples of such occurrences include western Canada, the western USA, and the Bolivian Altiplano (Lopez et al., 2012; Moncur et al., 2005; Welch et al., 2000). Sulfide oxidation has also resulted in high As concentrations (up to 215 μg/L) in groundwater in the eastern USA, in parts of the Piedmont rocks in Pennsylvania and New Jersey (Peters and Burkert, 2007; Serfes, 2005), where an As-oxidizing bacterium was involved in the mobilization of As (Rhine et al., 2008). A broad diversity of microorganisms oxidizes dissolved arsenic for different reasons including dissimilatory respiration, detoxification, and energy needs (Santini and Ward, 2012).

Reduction

Because of the magnitude of the contamination in West Bengal and Bangladesh, these two regions have received substantial attention from the research community. In general, reductive dissolution of Fe hydroxides and release of sorbed As explains much of the observed mobilization of As from sediments to groundwater (e.g., Nickson et al., 2000; Zheng et al., 2004). Organic matter in the alluvial aquifers is likely an important component of the reduction process. Field and experimental studies have shown that metal-reducing microbes can enhance mobilization of As, and that oxidation of the organic matter drives the redox reactions whereby Fe hydroxides are reductively dissolved and sorbed As released (Islam et al., 2004; McArthur et al., 2004). Arsenic also is released from Mn oxides as they reductively dissolve, but may not remain in the groundwater, instead resorbing to Fe hydroxides (McArthur et al.,

2004). Other studies indicate that As could also be released from biotite into Bangladeshi groundwater (Hopf et al., 2001; Seddique et al., 2008).

Reductive dissolution of Fe hydroxides has been proposed as a viable mechanism for As release in many other affected aquifers as well—for example, in Croatia, Inner Mongolia, northern China, and the eastern and southeastern USA (Barringer et al., 2010, Barringer et al., 2011; Guo et al., 2010; Haque et al., 2008; Mumford et al, 2012; Pearcy et al., 2011; Sharif et al., 2008; Ujevic et al., 2010; Xie et al, 2008). Although less intensely studied than Bengal Basin sediments, release of As from Fe oxides appears to be an important mechanism in the Inner Mongolian sediments (Smedley, 2008).

Microbially Mediated Reactions

Given the similarities to the shallow aquifers in Cambodia, Vietnam, Laos, and Myanmar, much of what has been found in West Bengal and Bangladesh may apply to the aquifers in those countries as well. A study of indigenous bacteria in Cambodian sediments indicated arsenic-respiring bacteria that reduce As(V) (arsenate) to As(III) (arsenite) were fueled by inputs of organic carbon (Lear et al., 2007), similar to findings for bacteria in West Bengal sediments. Additionally, results of experiments showed microbially mediated reduction of As(V), and Fe, and release from minerals (clay, glauconite sands, oxides and hydroxides) (Campbell et al., 2006; Dong et al., 2003; Hopf et al., 2009; Kostka et al., 1999; McLean et al., 2006; Pearcy et al., 2011), demonstrating the involvement of microbes in the reduction-oxidation (redox) reactions.

The bacteria involved in the reactions involve several groups. Iron-reducing bacteria of the genus *Geobacter* can reduce Fe in minerals such as hydroxides (Lloyd and Oremland, 2006), thus leading to dissolution of the hydroxides and sorbed As release. Geobacter bacteria have been investigated as As reducers; although *G. uraniumreducens* contains genes for As respiration, it was not conclusively shown to respire As, and *G. sulfurreducens* did not reduce As enzymatically (Islam et al., 2005; Lear et al., 2007). Bacteria known as dissimilatory arsenate respiring prokaryotes (DARPs) are identified as arsenic reducers, which means that As(V) serves as the terminal electron acceptor in dissimilatory reduction of arsenate. In the glauconitic sediments of the New Jersey Coastal Plain, USA, amplification of the arsenic respiratory reductase gene (*arrA*) followed by alignment and gene sequencing revealed clones with close (99%) similarity to *Alkaliphilus oremlandii* (CP000453) (formerly *Clostridium* species strain OhILAs)—a known arsenate-respiring bacterium (Mumford et al., 2012). Also using molecular techniques, an arsenate-respiring proteobacterium *Sulfurospirillum* sp. strain NP4 was identified by Lear et al. (2007) in Cambodian sediments. The bacterium *Desulfotomaculum auripigmentum* reduces As(V) to As(III) as well as sulfate to sulfide, and precipitates orpiment (Newman et al. 1997; Ehrlich and Newman, 2009)

It should be noted, however, that not all microbial reduction of arsenate is the result of bacterial respiration. Many bacteria detoxify arsenate by reducing it to arsenite and expelling it (Oremland and Stolz, 2005) and, in some cases, by methylating and expelling it (Bentley and Chasteen, 2002). Presence of dimethylarsinate and monomethylarsonate in groundwater, as was found in shallow groundwater in glauconitic sediments of the New Jersey Coastal Plain, probably was indicative of such microbial activity (Mumford, et al., 2012). In experiments, As was mobilized from apatite by the bacterium*Burkholderia fungorum*, which utilizes phosphorus from the apatite (Mailloux et al., 2009). There is apatite present in the sediments released from the Himalayas to the alluvial aquifers of South East Asia, and there are apatite-rich phosphorite beds in the glauconitic New Jersey Coastal Plain. In addition to reductive dissolution of iron hydroxides releasing sorbed As, arsenate reduction by As-respirers and other As-reducing bacteria, As may also be released by the mechanism suggested by Mailloux et al. (2009) in some aquifers.

Conditions that support microbial sulfate reduction are reported for the aquifers of the West Bengal, Mekong, and Red River deltas. Where such conditions exist, there is the potential for precipitation of sulfide minerals that could remove As as well as Fe from solution. Buschmann and Berg (2009) found As concentrations to be lower in groundwater from zones where sulfate (SO_4^{2-}) and Fe reduction were occurring in the aquifers of the Bengal, Mekong, and Red River deltas. They suggest such conditions, which could result in precipitation of insoluble sulfides, are a control on As levels in groundwater. Some of the chemical reactions that likely affect As mobility are shown in Table 2.

Table 2. Reactions involved in, or affecting, reduction, oxidation, and (or) precipitation of arsenic in water and sediments.

Reaction description	Equation	Reference
Oxidation of pyrite	$FeS_2 + 15/4\, O_2 + 7/2\, H_2O \rightarrow Fe(OH)_3 + 2H_2SO_4$ $10FeS_2 + 30NO_3^- + 20\, H_2O \rightarrow 10\, Fe(OH)_3 + 15\, N_2 + 15\, SO_4^{2-} + 5\, H_2S_0\, 4$	Welch et al., (2000)
Oxidation of arsenopyrite	$FeAsS(S) + 11/4O_2(aq)\, 3/2H_2O\, (aq) \rightarrow Fe^{2+}(Aq) + SO_4^{2-}(aq) + H_3AsO_3(aq)$	Morin & Calas, (2006)
Oxidation of arsenite	$H_3AsO_3 + \tfrac{1}{2}\, O_2\, (aq) \rightarrow H_2AsO_4^- + {}^{H}+$	Morin & Calas, (2006)
Reductive dissolution of Fe hydroxides (release of sorbed arsenate not shown)	$4Fe^{III}OOH + CH_2O + 7H_2CO_3 \rightarrow 4Fe^{II} + 8HCO_3^- + 6H_2O$	Nriagu et al. (2007)
Reduction of sulfate, formation of sulfide	$2CH_2O + SO_4^{\,2-} \rightarrow 2HCO_3^{\,-} + H_2S$	Nriagu et al., (2007)
Oxidation of organic carbon (lactate) and reduction of As(V)	$CH_3\text{-}CHOH\text{-}COO^- + 2HAsO_4^{2-} + 3H^+ \rightarrow CH_3COO^- + 2HAsO_2 + 2H_2O + HCO_3^-$	Saltikov et al., (2003)
Microbially mediated precipitation of orpiment	$2HAsO_2 + 3HS^- + 3H^+ \rightarrow As_2S_3 + 4H_2O$	Ehrlich and Newman(2009)
Incongruent dissolution of glauconite (with release of arsenic not shown)	$K_2(Fe_{1-x},Mg)_2Al_x(Si_4\, {}_6O_{10})_3(OH)_{12(s)} + 3/2\, O_2 + 6H^+ \rightarrow K^+ +2xMg^{+2}\, _{(aq)} + 6SiO_{2(aq)} + 2(1\text{-}X).$ $Fe(OH)_{3(s)} + 3/2\, Al_4(Si_4\, _{10})(OH)_{8(s)}$	Chapelle and Knobel (1983)

Sources, Sinks, Electron Donors, and Competitive Ions

Aqueous sulfide was measured in As-rich groundwater discharging to a New Jersey Coastal Plain, USA, stream (Barringer et al., 2010), but sulfide minerals (pyrite) were rare or not present in cores of streambed sediments. Siderite ($FeCO_3$) was found, however, more than a meter deep in the sediments and, judging from the As content (184 mg/kg), the siderite is an effective sink for As released to the shallow groundwater (Mumford et al, 2012). Siderite apparently forms when bacteria respire organic matter, creating bicarbonate (HCO_3^-), in an Fe-rich, reducing environment with circumneutral to alkaline pH (Fredrickson et al., 1998). Siderite is reported for parts of the Southeast Asian alluvial aquifers, and presumably acts as an As sink there as well, because Islam et al., (2005) found both arsenate (As(V)) and arsenite (As(III)) sorbed effectively to the siderite in their experiments. Jönsson and Sherman (2008), however, found that the binding of As(III) with siderite is weak. Clearly, however, siderite plays a role in removing one or both of the prevalent As species from solution. Whether siderite remains a permanent sink for As is not known; it would seem possible that biogeochemical conditions in an aquifer could change such that sorbed As would be released back into solution.

In the microbial release of As from geologic materials to groundwater, the presence of organic matter is seen as a critical factor, as it is an electron donor that fuels microbial activity. Organic acids also may compete with As species, along with oxyanions such as phosphate, molybdate, sulfate, and silicate, for binding sites on solids (Wang and Mulligan, 2006b). Nevertheless, the main role of organic matter in As release appears to be that it provides the necessary substrate to bacterial communities for growth and activity—as part of the process in which it is oxidized by bacteria, the organic matter also may produce quinone-like moities that act as electron shuttles in the resulting redox reactions (Mladenov, et al., 2010).

In the studies of the biogeochemistry of the Southeast Asian aquifers, the source of the organic matter was a matter of controversy. Buried peat was suggested as the source (McArthur et al., 2004), whereasHarvey et al. (2002) indicated that young carbon from the land surface moved to depth by irrigation pumping accounted for the organic matter in the redox reactions. (It may be that both sources are operative in different places.) Rowland et al., (2006, 2007) found naturally occurring hydrocarbons in West Bengal and Cambodian aquifers that could promote the microbial activity involved in arsenic release. Héry et al., (2010) point out that very low organic carbon contents (i.e. ≤1%) in sediments is sufficient to stimulate the microbially mediated reactions that result in metal and arsenic reduction in the aquifers.

Organic matter can come from anthropogenic sources as well as natural sources. Barringer et al. (2010) indicated that the likely source of organic matter in the glauconitic system they studied came from wastewater discharged for many years from farming and other subsequent activities. The issue of whether anthropogenic inputs from agricultural practices have contributed organic matter to shallow groundwater in West Bengal and Bangladesh has received much debate (Farooq et al., 2010; Neumann et al., 2009; Sengupta et al., 2008) and it is not entirely resolved. It has been noted, however, at various contamination sites, that petroleum leaks, organic-rich leachates from landfills, as well as inputs of organic carbon for remediation purposes has led to mobilization of As (Hering et al., 2009). Thus it is clear that inputs of organic carbon from both natural and anthropogenic activities can supply electrons and stimulate the microbially mediated processes that lead to As release from geologic materials into groundwater.

HYDROGEOLOGIC FACTORS

Residence Time

Several biogeochemical processes that release As from geologic materials have been identified, as presented in the previous section. Smedley and Kinniburgh (2002) point out that whether released As remains at problematic levels in groundwater depends not only on whether there are biogeochemical reactions that retard the transport of As, but also upon the hydrologic and hydrogeologic properties of the aquifer, such as flow velocity and dispersion. If the kinetics of As release are slow, and groundwater residence time is short, then As concentrations may not increase to the point where groundwater would be considered contaminated. Conversely, if reactions that mobilize As are rapid and residence time is long, then As can accumulate in groundwater such that concentrations become hazardous—as seen in Bangladesh, for example. Eventually, if the biogeochemical conditions that lead to release and mobilization of As continue to be present (within a geologic timeframe), then the source could become exhausted.

Seasonal Changes in Recharge

Natural fluctuations can affect the fate and transport of As within groundwater systems. Seasonal fluctuations in recharge could, during periods of high precipitation, bring dilution to shallow groundwater, but also transport surficially derived materials to the aquifer. As mentioned above, the transport of dissolved organic matter from agricultural land in the Bengal Delta has been suggested to fuel bacterial activity that releases As from the aquifer materials, and, in that region, seasonal (monsoon) rainfall has an important effect on recharge rates and the transmission of land-derived substances to depth in the aquifers. On a much smaller scale, As concentrations in shallow groundwater that discharges to Coastal Plain streams in New Jersey, USA, varies with season and hydrologic conditions. Increased recharge during springtime results in more diluted shallow groundwater and low As concentrations, whereas hot, dry weather results in decreased recharge and higher As concentrations. Where clay lenses underlie the stream channel shallow groundwater levels above the clay decline during warm, dry periods, and some stream segments may ultimately lose water to groundwater. Thus, in some stream segments seasonal hydrologic conditions control As-rich groundwater discharges to the stream. In other segments, As-rich groundwater may discharge on a relatively constant basis with higher concentrations of As being present during warm dry weather (Barringer et al., 2010).

Effects of Pumping

Pumping-induced changes to hydraulic gradients can alter flow paths at regional and local scales and can lead to introduction of contaminants to otherwise potable water. For As-contaminated well water in Bangladesh, the contamination is found in the shallower of two aquifers. Although water of the deeper aquifer is generally free of As contamination, the replacement of shallow tube wells with wells in the deep aquifer was thought to have the potential to transport contamination downward to low-As ground water (Ravenscroft et al, 2001; Harvey et al., 2002; van Geen et al., 2003).

Pumping also can result in changes in redox conditions along a flow path. As suggested by Peters and Blum (2003), anoxic water can be drawn upward to oxic zones near the wellhead, resulting in disequilibrium between As species. The converse is also possible—anoxic waters could be introduced by pumping to oxic zones such that Fe(III) in ferric hydroxides is reduced, hydroxides are dissolved, and sorbed As is released. Or, in cases where As is released under anoxic conditions, introduction of oxic water through pumping could slow or terminate the reaction. An example of pumping-induced changes in redox conditions is found at the individual borehole scale in eastern Wisconsin, USA. In a field experiment, longer periods between pumping episodes allowed longer periods of anoxia, which resulted in higher concentrations of As in groundwater (Ayotte et al., 2011). Pumping also could move constituents such as organic carbon from the surface to depth, where the carbon could stimulate microbially mediated redox reactions such as Fe reduction that leads to As release from aquifer materials. Pumping could also move higher pH water into a zone of lower pH water, creating an environment in which As sorbed to aquifer materials can desorb.

Flow rates at the individual borehole scale can also be sufficiently rapid that contact time between groundwater and aquifer material is minimized, and reactions releasing As may be relatively slow. Limited contact time is thought to be the explanation for lower dissolved constituents, including As, in water from wells with high yield compared with those with low yield and higher dissolved constituents in the Hungarian Pannonian Basin (Varsányi and Kovács, 2006). Similar results are reported for a well in a sandstone aquifer in Wisconsin, USA, where reducing conditions developed with no pumping and As concentrations increased (indicating dissolution of Fe hydroxides and As release), whereas rapid well purging introduced oxic conditions to water near the well bore (Gotkowitz et al., 2004).

CONCLUSIONS

Arsenic contamination of groundwater resources has been identified in many parts of the world. In some cases, as in geothermal fields, the impact of As on drinking-water supplies may not be great, but there are parts of the world where groundwater is a major drinking-water source for millions of people, many of which are in poverty and have limited ability to solve the problem of a contaminated water source. Since the discoveries of widespread groundwater contamination with As, considerable effort has been expended to find suitable, inexpensive methods for removing As. A discussion of those efforts, and the results, is beyond the scope of this chapter; the reader is directed to Feenstra et al. (2007) for an overview of As removal methods.

Geogenic sources of As are numerous—within the various geologic materials, the most common occurrences of As appear to be in sulfides (mainly pyrite) and as sorbed species on Fe hydroxides, although As also appears within some silicate and carbonate rocks. The main processes involved in As release to groundwater are reduction of Fe hydroxides, reduction of As within minerals and as a sorbed species, competitive sorption with other oxyanions, and sulfide oxidation. Increasingly, studies show that these processes can be mediated by microbes.

Although major geologic sources of arsenic include alluvial materials, mineralized sedimentary and metasedimentary rocks and volcanic rocks and related deposits, recent findings indicating that As also is released from glauconitic sediments suggests that not all geologic sources and conditions for the release of As to groundwater have yet been identified. Further, it is apparent from some studies that human activities can increase the rates and amounts of As mobilized and dissolved in groundwater through inputs of organic carbon, and from water withdrawals and other changes to natural hydrologic systems. Thus, while there are human efforts to mitigate the As contamination of drinking-water supplies, there are also human activities that exacerbate the problem. The more fully we understand how, when, and where As is mobilized from geologic materials, or from anthropogenic releases to the environment, the more effectively we can find solutions to this major contamination problem.

REFERENCES

1.   Acharyya, S.K. (2002). Arsenic contamination in groundwater affecting major parts of southern West Bengal and parts of western Chhattisgarh: Source and mobilization process. *Current Science*, 82, 740-744.

2.   Agusa, T., Kunito, T., Fujihara, J., Kubota, R., Minh, T.B., Trang, P.T.K., Iwata, H., Subramanian, A., Viet, P.H. & Tanabe, S. (2006). Contamination by arsenic and other trace elements in tube-well water and its risk assessment to humans in Hanoi, Vietnam. *Environmental Pollution,* 139, 95-106.

3.   Aiuppa, A., Avino, R., Caliro, S., Chiodini, G., D'Alessandro, W., Favara, R., Federico. C., Ginevra, C., Inguaggiato, S., Longo, M., Pegoraino, G. & Valenza, M. (2006). Mineral control of arsenic content in thermal waters from volcanic-hosted hydrothermal systems: Insights from the island of Ischia and Phlegrean Fields (Campanian Volcanic Province, Italy). *Chemical Geology*, 229, 313-330.

4.   Appleyard, S.J., Angeloni, J., & Watkins, R. (2006). Arsenic-rich groundwater in an urban area experiencing drought and increasing population density, Perth, Australia. *Applied Geochemistry*, 21, 83-97.

5.   Arnórsson, S. (2003). Arsenic in surface-and in 90°C ground waters in a basaltic area, N-Iceland: processes controlling its mobility. *Applied Geochemistry,* 18, 1297-1312.

6.   ATSDR (2000). Toxicological profile for arsenic. U.S. Department of Health & Human Services, Public Health Service Agency for Toxic Substances and Disease Registry, 428.

7.   Ayotte, J.D., Montgomery, D.L., Flangan, S.M., & Robinson, K.W. (2003). Arsenic in groundwater in Eastern New England: occurrence, controls, and human health implications. *Environmental Science and Technology*, 37, 2075-2083.

8.   Ayotte, J.D., Neilson, M.G., Robinson, G.R., & Moore, R.B. (1999). Relation of arsenic, iron, and manganese in ground water to aquifer type, bedrock lithogeochemistry, and land use in the New England Coastal Basins. *U.S. Geological Survey Water-Resources Investigations Report 99-4162.*

9.   Ayotte, J.D., Szabo, Z., Focazio, M.J. & Eberts, S.M. (2011). Effects

of human-induced alteration of groundwater flow on concentrations of naturally-occurring trace elements at water-supply wells. *Applied Geochemistry*, 26, 747-762.

10. Ball, A.L., Rom, W.N. & Glenne, B. (1983). Arsenic distribution in soils surrounding the Utah copper smelter. *American Industrial Hygiene Association Journal*, 44, 341-348.

11. Ball, J.W., McCleskey, R.B., Nordstrom, D.K., Holloway, J.M., Verplank, P.L. & Sturtevant, S.A. (2002). Water chemistry data for selected springs, geysers, and streams in Yellowstone National Park, Wyoming, 1999-2000. *U.S. Geological Survey Open-File Report 2002-382.*

12. Ball, J.W. Nordstrom, D.K., Jenne, E.A. & Vivit, D.V. (1998). Chemical analyses of hot springs, pools, geysers, and surface waters of Yellowstone National Park, Wyoming and vicinity, 1972-1975. *U.S. Geological Survey Open-File Report 98-182.*

13. Barringer, J.L., Barringer, T.H., Lacombe, P.J. & Holmes, C.W. (2001). Arsenic in soils and sediment adjacent to Birch Swamp Brook in the vicinity of Texas road (downstream from the Imperial Oil Company Superfund Site)), Monmouth County, New Jersey. *U.S. Geological Survey Water-Resources Investigations Report 00-4185.*

14. Barringer, J.L., Bonin, J.L., Deluca, M.J., Romaga, T., Cenno, K., Alebus, M., Kratzer, T. & Hirst B. (2007). Sources and temporal dynamics of arsenic in a New Jersey watershed, USA. *Science of the Environment*, 379, 56-74.

15. Barringer, J.L., Mumford, A, Young, L.Y., Reilly, P.A., Bonin, J.L. & Rosman, R. (2010). Pathways for arsenic from sediments to groundwater to streams: Biogeochemical processes in the Inner Coastal Plain, New Jersey, USA. *Water Research*, 44, 5532-5544.

16. Barringer, J.L., Reilly, P.A., Eberl, D.D., Blum, A.E., Bonin, J.L., Rosman, R., Hirst, B., Alebus, M., Cenno, K., & Gorska, M. (2011). Arsenic in sediments, groundwater, and streamwater of a glauconitic Coastal Plain terrain, New Jersey, USA—Chemical "fingerprints" for geogenic and anthropogenic sources. *Applied Geochemistry*, 26, 763-776.

17. Barringer, J.L. Szabo, Z. & Barringer, T.H. (1998). Arsenic and metals in soils in the vicinity of the Imperial Oil Company Superfund site, Marlboro Township, Monmouth County, New Jersey. *U.S. Geological Survey Water-Resources Investigations Report 98-4016.*

18. Baur, W.H. & Onishi, B-M.H. (1969). Arsenic. In *Handbook of geochemistry*, ed. K.H. Wedepohl, 33-A-1-33-0-5. Berlin, Springer-Verlag.

19. Beaulieu, B.T. & Savage, K.S. (2005). Arsenate adsorption structures on aluminum oxide and phyllosilicate mineral surfaces in smelter-impacted soils. *Environmental Science and Technology*, 39, 3571-3579.

20. Bentley, R., & Chasteen,.G. (2002). Microbial methylation of metalloids, arsenic, antimony, and bismuth. *Microbiology and Molecular Biology Reviews*, 66, 250-271,

21. Berg, M., Tran, H.C., Nguyen, T.C., Pham, H.V., Schertenleib, R. & Giger, W. (2001). Arsenic contamination of groundwater and drinking water in Vietnam: A human health threat. *Environmental Science and Technology*, 35, 2621-2626.

22. Bhattacharya, P., Mukherjee, A.B., Bundschuh, J., Zevenhoven, R. & Loeppert, R.H. (eds.) (2007). *Trace Metals and other Contaminants in the Environment*, Volume 9. Elsevier BV.

23. Bottomley, D.J. (1984). Origins of some arseniferous groundwaters in Nova Scotia and New Brunswick, Canada. *Journal of Hydrology*, 69, 223-257.

24. Bowell, R.J., Morley, N.H. & Din, V.K. (1994). Arsenic speciation in soil porewaters from the Ashanti Mine, Ghana. *Applied Geochemistry*, 9, 15-22.

25. Boyle, R.W. & Jonasson, I.R. (1973). The geochemistry of arsenic and its use as an indicator element in geochemical prospecting. *Journal of Geochemical Exploration*, 2, 251-296.

26. Brooks, W.E. (2008). Arsenic. *2007 Minerals Yearbook, U.S. Geological Survey*.

27. Brown, C.J. & Chute, S.K. (2002). Arsenic in bedrock wells in Connecticut. (Abstract). *Arsenic in New England: A Multidisciplinary Scientific Conference*, National Institute of Environmental Health Sciences, Superfund Basic Research Program, Manchester, New Hampshire, May 29-31, 2002.

28. Brown, K.L., & Simmons, S.F. (2003). Precious metals in high-temperature geothermal systems in New Zealand. *Geothermics*, 32, 619-625.

29. Brunt, R., Vasak, L.,& Griffioen, J. (2004). Arsenic in groundwater: Probability of occurrence of excessive concentration on a global scale. *International Groundwater Resources Assessment Centre: Report nr. SP 2004-1*.

30. Bundschuh, J., Litter, M.I., Parvez, F., Román-Ross, G., Nicolli, H.B., Jean, J-S., Liu, C-W., López, D., Armienta, M.A., Guilherme, L.R.G.,

Cuevas, A.G., Cornejo, L., Cumbal, L., Toujaguez, R. (2012). One century of arsenic exposure in Latin America: a review of history and occurrence from 14 countries. *Science of the Total Environment*, 429, 2-35.

31. Burkel, R.S., & Stoll, R.C. (1999). Naturally occurring arsenic in sandstone aquifer water supply wells of Northeastern Wisconsin. *Ground Water Monitoring and Remediation*, 19, 114-121.

32. Buschmann, J., & Berg, M. (2009). Impact of sulfate reduction on the scale of arsenic contaminations in groundwater of the Mekong, Bengal, and Red River deltas. *Applied Geochemistry*, 24, 1278-1286.

33. Camm, G.. Glass, H.J., Bryce, D.W. & Butcher, A.R. (2004). Characterization of a mining-related arsenic-contaminated site, Cornwall, UK. *Journal of Geochemical Exploration*, 82, 1-15.

34. Campbell, K.M., Malasarn, D. Saltikov, C., Newman, D.K. & Hering, J.G. (2006). Simultaneous microbial reduction of iron (II) and arsenic (V) in suspensions of hydrous ferric oxide. *Environmental Science and Technology*, 40, 5950-5955.

35. Cancès. B., Juillot, F., Morin, G., Laperche, V., Polya, D., Vaughn, D.J., Hazeman, J.-L, Proux, O., Brown, G.E., Jr. & Calas, G. (2008). Changes in arsenic speciation through a contaminated soil profile: an XAS based study. *Science of the Total Environment*, 397, 178-189.

36. Carrillo-Chávez, A., Drever, J.I. & Martinez, M. (2000). Arsenic content and groundwater geochemistry of the San Antonio-El Triunfo, Carrizal and Los Planes aquifers in southernmost Baja California, Mexico. *Environmental Geology*, 39, 1295-1303.

37. Carpenter, R., Peterson, M.L. & Jahnke, R.A. (1978). Sources, sinks, and cycling of arsenic in the Puget Sound Region. In *Estuarine Interactions* ed. M.L. Wiley. New York, Academic Press.

38. Chanpiwat, P., Sthiannopkao, S., Cho, K.H., Kim, K-W, San, V., Suvathong, B. & Vongthavady, C. (2011). Contamination by arsenic and other trace elements of tube-well waters along the Mekong River in Lao PDR. *Environmental Pollution*, 159, 567-576.

39. Chapelle, F.H. & Knobel, L.L. (1983). Aqueous geochemistry and the exchangeable cation composition of glauconite in the Aquia aquifer, Maryland.*Ground Water*, 21, 343-352.

40. Chen, S-L., Dzeng, S.R., Yang, M-H., Chiu, K-H., Shieh, G-M., Wai, C.M. (1994). Arsenic species in groundwaters of the Blackfoot Disease Area, Taiwan. *Environmental Science and Technol.ogy*, 28, 877-881.

41. Cravotta, C.A., III. (2008). Dissolved metals and associated constituents in abandoned coal-mine drainages, Pennsylvania, USA: Part I. Constituent quantities and correlations. *Applied Geochemistry*, 23, 166-202.

42. Cullen, W.R. (2008). Is Arsenic an Aphrodisiac? The Sociochemistry of an Element, Cambridge, UK, RSC Publishing.

43. Daniele, L. (2004). Distribution of arsenic and other minor trace elements in the groundwater of Ischia Island (southern Italy). *Environmental Geology*, 46, 96-103.

44. Dangic, A. (2007). Arsenic in surface- and groundwater in central parts of the Balkan Peninsula (SE Europe), Chapter 5,127-156. In *Trace Metals and other Contaminants in the Environment, Volume 9*. Ed. P. Bhattacharya,, A.B. Mukherjee, J. Bundschuh, R. Zevenhoven, & R.H. Loeppert. Elsevier BV.

45. Davenport, J.R. & Peryea, F.J. (1991). Phosphate fertilizers influence leaching of lead and arsenic in a soil contaminated with lead arsenate. *Water, Air, and Soil Pollution*, 57-58, 101-110.

46. DePaul, V.T. & Szabo, Z. (2007). Occurrence of radium-224, radium-226, and radium-228 in water from the Vincentown and Wenonah-Mt. Laurel aquifers, the Englishtown aquifer system, and the Hornerstown and Red Bank Sands. *U.S. Geological Survey Scientific Investigations Report 2007-5064*.

47. Dong, H. Kukkadapu, R.K., Fredrickson, J.K., Zachara, J.M., Kennedy, D.W. & Kostanadrithes, H. (2003). Microbial reduction of structural Fe(III) in illite and goethite. *Environmental Science and Technology*, 37, 1268-1276.

48. Dooley, J.H. (2001). Baseline concentrations of arsenic, beryllium, and associated elements in glauconite and glauconitic soils in the New Jersey Coastal Plain. N.*J. Geological Survey Investigation Report*. Trenton, NJ, N.J. Department of Environmental Protection..

49. Drummond, D.D., & Bolton, D.W. (2010). Arsenic in ground water in the Coastal Plain aquifers of Maryland. *Report of Investigations No. 78. Department of Natural Resources, DNR 12-4282010-450*. Resource Assessment Service, Maryland Geological Survey, Baltimore MD.

50. Durant, J.L., Ivushkina, T., MacLaughlin, K., Lukacs, H., Gawel, J, Senn, D. & Hemond, H.F. (2004). Elevated levels of arsenic in the sediments of an urban pond: sources, distribution and water quality impacts. *Water Research*, 38, 2989-3000.

51. Ehrlich, H.L. & Newman, D.K. (2009). Geomicrobiology, Fifth ed., Boca Raton, FL, CRC Press

52. Eiche, E., Neumann, T., Berg, M., Weinman, B., van Geen, A., Norra, S., Berner, Z., Trang, P.T.K., Viet, P.H. & Stüben, D. (2008). Geochemical processes underlying a sharp contrast in groundwater arsenic concentrations in a village on the Red River delta, Vietnam. *Applied Geochemistry*, 23, 3143-3154.

53. Farooq, S.H., Chandrasekharam, D., Berner, Z., Norra, S. & Stüben, D. (2010). Influence of traditional agricultural practices on mobilization of arsenic from sediments to groundwater in Bengal delta. *Water Research*, 44, 5575-5588.

54. Feenstra, L., van Erkel, J. & Vasak, L. (2007). Arsenic in groundwater: Overview and evaluation of removal methods. *IGRAC Report nr.SP 2007-2*. Utrecht, International Groundwater Resources Assessment Centre.

55. Foley, N.K., Ayuso, R.A., Ayotte, J.D., Marvinney, R.G., Reeve, A.S. & Robinson, G.R., Jr. (2002). Mineralogical pathways for arsenic in weathering meta-shales: An analysis of regional and site studies in the northern Appalachians. (Abstract). *Arsenic in New England: A Multidisciplinary Scientific Conference*, National Institute of Environmental Health Sciences, Superfund Basic Research Program, Manchester, New Hampshire, May 29-31, 2002.

56. Fredrickson, J.K., Zachara, J.M., Kennedy, D.W., Dong. H., Onstott, T.C., Hinman, N.W. & Li, S.M. (1998). Biogenic iron mineralization accompanying the dissimilatory reduction of hydrous ferric oxide by a groundwater bacterium, *Geochimica et Cosmochimica Acta*, 62, 3239-3257.

57. Goldberg, S., & Glaubig, R.A. (1988). Anion sorption on a calcareous, montmorillonitic soil—Arsenic. *Soil Science Society of America Journal*, 52, 1297-1300.

58. Gómez, J.J., Lillo, J. & Sahún, B. (2006). Naturally occurring arsenic in groundwater and identification of the geochemical sources in the Duero Cenozoic Basin, Spain. *Environmental Geology*, 50, 1151-1170.

59. Gotkowitz, M.B., Schreiber, M.E., & Simo, J.A. (2004). Effects of water use on arsenic release to well water in a confined aquifer. *Ground Water,* 42, 568-575.

60. Gunduz, O. & Simsek, C. (2008). Mechanisms of arsenic contamination of a surficial aquifer in Turkey. *GQ07: Securing Groundwater Quality in Urban and Industrial Environments (Proceedings of the 6th International Groundwater Quality Conference, Freemantle, Australia, 2-7 December 2007, IAHS publ. no. XXX, 2008.*

61. Gunduz, O., Simsek, C. & Hasozbek, A. (2010). Arsenic pollution in the

groundwater of Simav Plain, Turkey: Its impact on water quality and human health. *Water, Air, and Soil Pollution*, 205, 43-62.

62. Guo, H., Zhang, B., Wang, G. & Shen, Z. (2010). Geochemical controls on arsenic and rare earth elements approximately along a groundwater flow path in the shallow aquifer of the Hetao Basin, Inner Mongolia. *Chemical Geology*, 270, 117-125.

63. Gurung, J., Ishiga, H. & Khadka, M.S. (2005). Geological and geochemical examination of arsenic contamination of groundwater in the Holocene Terai basin, Nepal. *Environmental Geology*, 49, 98-113.

64. Gurzau, A.E., and Pop, C. (2012). A new public health issue: Contamination with arsenic of private water sources. *Proceedings of the AERAPA Conference*. Available from aerapa.conference.ubbcluj.ro/2012/Gurzau.htm, accessed 7/6/12.

65. Haack, S.K., & Rachol, C.M. (2000). Arsenic in groundwater in Washtenaw County, Michigan. *U.S. Geological Survey Fact Sheet FS 134-00*.

66. Haeri, A., Strelbitskaya, S., Porkhial, S., & Ashayeri, A. (2011). Distribution of arsenic in geothermal waters from Sabalan geothermal field, N-W Iran. *Proceedings 36th Workshop on Geothermal Reservoir Engineering, Standford University, Stanford, California, January 31-February 2, 2011*.

67. Haque, S., Ji, J. & Johannesson, K.H. (2008). Evaluating mobilization and transport of arsenic in sediments and groundwaters of the Aquia aquifer, Maryland, USA. *Journal of Contaminant Hydrology*, 99, 68-84.

68. Harvey, C.F., Swartz, C.H., Badruzzaman, A.B.M., Keon-Blute, N., Yu, W., Ali, M.A.l, Jay, J., Beckie, R., Niedan, V., Brabander, D., Oates, R.M., Ashfaque, K.N., Islam, S., Hemond, H.F. & Ahmed, M.F. (2002). Arsenic mobility and groundwater extraction in Bangladesh. *Science*, 298, 1602-1606.

69. Heinrichs, G. & Udluft, P. (1999). Natural arsenic in Triassic rocks: a source of drinking-water contamination in Bavaria, Germany. *Hydrogeology Journal*, 7, 468-476.

70. Hemond, H.F. (1995). Movement and distribution of arsenic in the Aberjona watershed. *Environmental Health Perspectives, Supplement 1*, 103, 35-40.

71. Hering, J.G., O'Day, P., Ford, R.G., He, Y.T., Bilgin, A., Reisinger, H.J. & Burns, D.R. (2009). MNA as a remedy for arsenic mobilized by anthropogenic inputs of organic carbon. *Ground Water Monitoring and Remediation*, 29, 84-92.

72. Héry, M., van Dongen, B.E., Gill, F., Mondal, D., Baughan, D.J., Pancoast, R.D., Polya, D.A. & Lloyd, J.R. (2010). Arsenic release and attenuation in low organic carbon aquifer sediments from West Bengal. *Geobiology*, 8, 155-168..

73. Hileman, B. (2007). Arsenic in chicken production. *Chemical Engineering News*, 85, 34-35.

74. Hoang, T.H., Bang, S., Kim, K-W, Nguyen, M.H. & Dang, D.M. (2010). Arsenic in groundwater and sediment in the Mekong River delta, Vietnam. *Environmental Pollution*, 158, 2648-2658.

75. Hopf, J., Langenhorst, F., Pollok, K., Merten, D. & Kothe E. (2009). Influence of organisms on biotite dissolution: an experimental approach. *Chemie der Erde*, 69, 45-56.

76. Islam, F.S., Gault, A.G., Boothman, C., Polya, D.A., Charnock, J.M., Chatterjee, D. & Lloyd, J.R. (2004). Role of metal-reducing bacteria in arsenic release from Bengal delta sediments. *Nature*, 430, 68-71.

77. Islam, F.S., Pederick, R.L., Gault, A.G., Adams, L.K., Polya, D.A., Charnock, J.M. & Lloyd, J.R. (2005). Interactions between the Fe(III)-reducing bacterium *Geobacter sulfurreducens* and arsenate, and capture of the metalloid by biogenic Fe(II). *Applied and Environmental Microbiology*, 71, 8642-8648.

78. Jönsson, J. & Sherman, D.M. (2008). Sorption of As(III) and As(V) to siderite, green rust (fougerite) and magnetite: implications for arsenic release in anoxic groundwaters. *Chemical Geology*, 255, 173-181.

79. Kahn, B.I., Solo-Gabriele, H.M., Townsend, T.G. & Cai, Y. (2006). Release of arsenic to the environment from CCA-treated wood: 1. Leaching and speciation during service. *Environmental Science and Technology*, 40, 988-993.

80. Karczewska, A., Bogda, A. & Kryasiak, A. (2007). Arsenic in soils in areas of former mining and mineral processing in Lower Silesia, southwestern Poland, Chapter 16, 411-440. In *Trace Metals and other Contaminants in the Environment, Volume 9*. Ed. P. Bhattacharya, A.B. Mukherjee, J. Bundschuh, R. Zevenhoven & R.H. Loeppert. Elsevier BV.

81. Karydakis, G., Arvanitis, A., Andritsos, N. & Fytikas, M. (2005). Low enthalpy geothermal fields in the Strymon Basin (Northern Greece). *Proceedings World Geothermal Congress 2005*, Antalya, Turkey, 24-29 April 2005.

82. Kinniburgh, D.G., Newell, A.J., Davies, J., Smedley, P.L., Milodowski, A.E., Ingram, J.A. & Merrin, P.D. (2006). The arsenic concentrations on groundwater from the Abbey Arms Wood observation borehole,

Delamere, Cheshire, UK. 265-284. In *Fluid flow and solute movement in sandstone: the onshore UK Permo-Triassic redbed sequence. Ed. R.D.* Barker & J.H. Tellam. London, Geological Society of London, Special Publication 263.

83. Kobayashi, D.S. & Lee, G.F. (1978). Accumulation of arsenic in sediments of lakes treated with sodium arsenite. *Environmental Science and Technology*, 12, 1195-2000.

84. Kolker, A., Palmer, C.A., Bragg, L.J. & Bunnell, J.E. (2006). Arsenic in Coal. *U.S. Geological Survey Fact Sheet FS 2005-3152.*

85. Kostka, J.E., Haefele, E., Viehweger, R. & Stucki, J.W. (1999). Respiration and dissolution of iron (III)-containing clay minerals by bacteria.*Environmental Science and Technology*, 33, 3127-3133.

86. Krüger, T., Holländer, H.M., Boochs, P-W., Billib, M., Stummeyer, J., Harazim, B. (2007). *In situ* remediation of arsenic at a highly contaminated site in Northern Germany. GQ07 Securing groundwater quality in urban and industrial environments, *Proc. 6th International Groundwater Quality Conference,* Freemantle, Western Australia, 2-7 December 2007.

87. Lear, G., Song B., Gault, A.G., Polya, D.A. & Lloyd, J.R. (2007). Molecular analysis of arsenate reducing bacteria within Cambodian sediments following amendment with acetate. *Applied and Environmental Microbiology,* 73, 1041-1048.

88. Lloyd, J.R. & Oremland, R.S. (2006). Microbial transformations of arsenic in the environment: From soda lakes to aquifers. *Elements,* 2, 85-90.

89. Lopez, D.L., Bundschuh, J., Birkle, P., Armienta, M.A., Cumbal, L., Sracek, O., Cornejo, L. & Ormachea, M. (2012). Arsenic in volcanic geothermal fluids of Latin America. *Science of the Total Environment,* 429, 57-75.

90. Luu, T.T.G., Sthiannopkao, S. & Kim, K-W. (2009). Arsenic and other trace elements contamination in groundwater and a risk assessment study for the residents of the Kandal Province of Cambodia. *Environment International,* 35, 455-460.

91. Mailloux, B.J., Aleandrova, E., Keimowitz, A.R., Wovkulich, K.,Freyer, G.A., Herron, M., Stolz, J.F., Kenna, T.C., Pichler, T., Polizzotto, M.L., Dong, H., Bishop, M. & Knappett, P.S.K (2009). Microbial mineral weathering for nutrient acquisition releases arsenic. *Applied and Environmental Microbiology,* 75, 2558-2565.

92. Mandal, B.K., & Suzuki, K.T., 2002. Arsenic round the world: a Review. *Talanta,* 58, 201-235.

93. Manning, B.A. & Goldberg, S. (1996). Modeling competitive adsorption of arsenate with phosphate and molybdate on oxide minerals. *Soil Science Society of America Journal*, 60, 121-131.

94. Martin, A.J. & Pedersen, T.F. (2002). The seasonal and interannual mobility of arsenic in a mine-impacted lake. *Environmental Science and Technology*, 36, 1516-1523.

95. Matschullat, J. (2000). Arsenic in the geosphere—a review. *Science of the Total Environment*, 249, 297-312.

96. McArthur, J.M., Banergee, D.M., Hudson-Edwards, K.A., Mishra, R., Purohit, R., Ravenscroft, P., Cronin, A., Howarth, R.J., Chatterjee, A., Talukder, T., Lowry, D., Houghton, S.M & Chadha, D.K. (2004). Natural organic matter in sedimentary basins and its relation to arsenic in anoxic ground water: the example of West Bengal and its worldwide implications. *Applied Geochemistry*, 19, 1255-1293.

97. McCarthy, K.T., Pichler, T. & Price, R.E. (2005). Geochemistry of Champagne Hot Springs shallow hydrothermal vent field and associated sediments, Dominica, Less Antilles. *Chemical Geology*, 224, 55-68.

98. McClintock, T.R., Chen, Y., Bundschuh, J., Oliver, J.T., Navoni, J., Olmos, V., Lepori, E.V., Ahsan, H., Parvez, F. (2012). Arsenic exposure in Latin America: Biomarkers, risk assessments and related health effects. *Science of the Total Environment*, 429, 76-91.

99. McLean, J.E., DuPont, R.R. & Sorensen, D.L. (2006). Iron and arsenic release from aquifer solids in response to biostimulation. *Journal of Environmental Quality*. 35, 1193-1203.

100. Méndez, M., & Armienta, M.A. (2003). Arsenic phase distribution in Zimapan mine tailings, Mexico. *Geofisica Internacional*, 42, 131-140.

101. Mladenov, N., Zheng, Y., Miller, M.P., Nemergut, D.R., Legg, T., Simone, B., Hageman, C., Rahman, M.M., Ahmed, K.M. & McKnight, D.M. (2010). Dissolved organic matter sources and consequences for iron and arsenic mobilization in Bangladesh aquifers. *Environmental Science and Technology*, 44, 123-128.

102. Moncur, M.C., Ptacek, C.J., Blowes, D.W. & Jambor, J.L. (2005). Release, transport and attenuation of metals from an old tailings impoundment. *Applied Geochemistry*, 20, 639-659.

103. Morin, G. & Calas, G. (2006). Arsenic in soils, mine tailings, and former industrial sites. *Elements*, 2, 97-101.

104. Mosier, D.L., Beger, V.I. & Singer, D.A. (2009). Volcanogenic massive sulfide deposits of the world: database and grade and tonnage models. *U.S.*

Geological Survey Open-File Report 2009-1034. Available from http://pubs.usgs.gov/of/2009/1034.

105. Mueller, S., Verplanck, P. & Goldfarb, R. (2001). Ground-water studies in Fairbanks, Alaska—A better understanding of some of the United States' highest natural arsenic concentrations. *U.S. Geological Survey Fact Sheet FS-111-01.*

106. Mukherjee, A., Sengupta, M.K., Hossain, M.A., Ahamed, S., Das, B., Nayak, B., Lodh, D., Rahman, M.M. & Chakraborti, D. (2006). Arsenic contamination in groundwater: A global perspective with emphasis on the Asian scenario. *Journal of Health and Population Nutrition*, 24, 142-163.

107. Mumford, A.C., Barringer, J.L., Benzel, W.M., Reilly, P.A. & Young, L.Y. (2012). Microbial transformations of arsenic: Mobilization from glauconitic sediments to water. *Water Research*, 46, 2859-2868.

108. Nakayama, S.M.M., Ikenaka, Y., Hamada, K., Muzandu, K., Choongo, K., Teraoka, H., Mizuno, N. & Ishizuka, M. (2011). Metal and metalloid contamination in roadside soils and wild rats around a Pb-Zn mine in Kabwe, Zambia. *Environmental Pollution*, 159, 175-181.

109. National Research Council (1999). Arsenic in Drinking Water. Washington, D.C. National Academy Press.

110. Navarro, A., Font, X., Viladevall, M. (2011). Geochemistry and groundwater contamination in the La Selva geothermal system (Girona, Northeast Spain). *Geothermics,* 40, 275-285.

111. Neumann, R.B., Ashfaque, K.N., Badruzzaman, A.B.M, Ali, M.A., Shoemaker, J.K & Harvey, C.F. (2009). Anthropogenic influences on groundwater arsenic concentrations on Bangladesh. *Nature Geoscience* 1-7. DOI:10.1038/NGEO685.

112. Newman, D.K., Beveridge, T.J. & Morel, F.M. (1997). Precipitation of arsenic trisulfide by *Desulfotomaculum auripigmentum. Applied and Environmental Microbiology*, 63, 2022-2028.

113. Ng, J.C., Wang J., Shraim, A. (2003). A global health problem caused by arsenic from natural sources. *Chemosphere,* 52, 1353-1359.

114. NHMRC (1996). National Health and Medical Research Council. Australian drinking water guidelines.

115. Nickson, R.T, McArthur, J.M., Ravenscroft, P., Burgess, W.G. & Ahmed, K.M. (2000). Mechanism of arsenic release to groundwater, Bangladesh and West Bengal. *Applied Geochemistry,* 15, 403-413.

116. Nickson, R.T, McArthur, J.M., Shrestha, B., Kyaw-Myint, T.O., Lowry,

D. (2005). Arsenic and other drinking water quality issues, Muzaffargarh District, Pakistan. *Applied Geochemistry*, 20, 55-68.

117. Nicolli, H.B., Bundschuh, J., Blanco, M, del, C., Tujchneider, O.C., Panarello, H.O., Dapeña, C., Rusnasky, J.E. (2012). Arsenic and associated trace-elements in groundwater from the Chaco-Pampean plain, Argentina: Results from 100 years of research. *Science of the Total Environment*, 429, 36-56.

118. Nimick, D.A., Moore, J.N., Dalby, C.E. & Savka, M.W. (1998). The fate of arsenic in the Madison and Missouri Rivers, Montana and Wyoming. *Water Resources Research*, 34, 3051-3067.

119. NJDEP (2009). NJ Department of Environmental Protection Standards for Drinking Water. Ground Water, Soil, and Surface Water—Arsenic (Total). Available from www.state.nj.us/dep/standards/pdf/7440-38-2. pdf.

120. Noguchi, K. & Nakagawa, R. (1969). Arsenic and arsenic-lead sulfides in sediments from Tamagawa hot springs, Akita Prefecture. *Proceedings of Japan Academy*, 45, 45-50.

121. Nordstrom, D.K. (2002). Worldwide occurrences of arsenic in ground water. *Science,* 296, 2143-2145.

122. Nriagu, J.O., Bhattacharya, P., Mukherjee, A.b., Bundschuh, J., Zevenhoven, R., & Loeppert, R.H. (2007). Chapter 1. Arsenic in soil and groundwater: an overview. 3-60. In *Trace Metals and other Contaminants in the Environment, Volume 9.* Ed. P. Bhattacharya, A.B.Mukherjee, J. Bundschuh, R. Zevenhoven & R.H.Loeppert 3-60. Elsevier BV

123. O'Day, P.A. (2006). Chemistry and mineralogy of arsenic. *Elements,* 2, 77-83.

124. Onishi, H & Sandell, E.B. (1955). Geochemistry of arsenic. *Geochimica et Cosmochimica Acta,* 7, 1-33.

125. Oremland R.,S., & Stolz, J.F. (2005). Arsenic, microbes, and contaminated aquifers. *Trends in Microbiology,* 13, 45-49.

126. Palumbo-Roe, B., Klinck, B. & Cave, M. (2007). Arsenic speciation and mobility in mine wastes from a copper-arsenic mine in Devon, UK: a SEM, XAS, sequential chemical extraction study, Chapter 17. 441-471. In *Trace Metals and other Contaminants in the Environment, Volume 9.* Ed. P. Bhattacharya, A.B.Mukherjee, J.Bundschuh,, R. Zevenhoven & R.H. Loeppert. Elsevier BV.

127. Pearcy, C.A., Chevis, D.A., Haug, T.J., Jeffries, H.A., Yang, N., Tang, J., Grimm. D.A., & Johannesson, K.H. (2011). Evidence of microbially

mediated mobilization from sediments of the Aquia aquifer, Maryland, USA. *Applied Geochemistry*, 26, 575-586.

128. Peters, S.C. (2008). Arsenic in groundwaters in the Northern Appalachian Mountain belt: A review of patterns and processes. *Journal of Contaminant Hydrology*, 99, 8-21.

129. Peters, S.C. & Burkert, L. (2007). The occurrence and geochemistry of arsenic in groundwaters of the Newark Basin of Pennsylvania. *Applied Geochemistry*, 23, 85-98.

130. Peters, S.C. & Blum, J.D. (2003). The source and transport of arsenic in a bedrock aquifer, New Hampshire, USA. *Applied Geochemistry*, 18, 1773-1787.

131. Peters, S.C., Blum, J.D., Klaue, B. & Karagas, M.R. (1999). Arsenic occurrence in New Hampshire drinking water. *Environmental Science and Technology*, 33. 1328-1333.

132. Piqué, A., Grandia, F., & Canals, A. (2010). Processes releasing arsenic to groundwater in the Caldes de Malavella geothermal area, NE Spain. *Water Research,* 44, 5618-5630.

133. Planer-Friedrich, B., London, J., McCleskey, R.B., Nordstrom, D.K., & Wallschläger. (2007). Thioarsenates in geothermal waters of Yellowstone National Park: Determination, preservation, and geochemical importance. *Environmental Science and Technology*, 41, 5245-5251.

134. Price, R.E. & Pichler, T. (2006). Abundance and mineralogical association of arsenic in the Suwannee Limestone (Florida): Implications for arsenic release during water-rock interaction. *Chemical Geology*, 228, 44-56.

135. Quicksall, A.N., Bostick, B.C. & Sampson, M.L. (2008). Linking organic matter deposition and iron mineral transformations to groundwater arsenic levels in the Mekong delta, Cambodia. *Applied Geochemistry*, 23, 3088-3098.

136. Rahman, M.M., Sengupta, M.K., Ahamed, S., Lodh, D., Das, B., Hossain, M.A., Nayak, B., Mukherjee, A., Mukherjee, S.C., Pati, S., Saha, K.C., Palit, S.K., Kaies, I., Barua, K., Asad, K.A. (2005). Murshidabad—One of the nine groundwater arsenic-affected districts of West Bengal, India, Part I: Magnitude of contamination and population at risk. *Clinical Toxicology,* 43, 823-834.

137. Ravenscroft, P., Brammer, H., & Richards, K. (2009) Arsenic Pollution: A Global Synthesis. Wiley-Blackwell, 588 pp.

138. Ravenscroft, P., McArthur, J.M. & Hoque, B.A. (2001). Geochemical and palaeohydrological controls on pollution of groundwater by arsenic.

53-78. In. *Arsenic exposure and health effects IV*. Ed. W.R. Chappell, C.O. Abernathy & R. Calderon.. Oxford, Elsevier Science, Ltd.

139. Rhine, E.D., Onesios, K.M., Serfes, M.E., Reinfelder, J.R. & Young, L.Y. (2008). Arsenic transformation and mobilization from minerals by the arsenite oxidizing strain WAO. *Environmental Science and Technology,* 42, 1423-1429.

140. Rice, K.C., Conko, K.M. & Hornberger, G.M. (2002). Anthropogenic sources of arsenic and copper to sediments in a suburban lake, northern Virginia. *Environmental Science and Technology,* 36, 4962-4967.

141. Robinson, B., Clothier, B., Bolan, N.S., Mahimairaja, S., Grevn, M., Moni, C., Marchetti, M., van den Dijssel, C. & Milne, G. (2004). Arsenic in the New Zealand environment. *SuperSoil 2004. 3rd Australian New Zealand Soils Conference,* 509, December 2004, University of Sydney, Australia. Available from www.regional.org.au/au/asssi.

142. Rosas, I., Belmont, R., Armienta, A., Boaz, A. (1999). Arsenic concentrations in water, soil, milk, and forage in Comarca Lagunera, Mexico. *Water, Air, and Soil Pollution*, 112, 133-149.

143. Rowland, H., Omoregie, E., Millot, R., Jimenez C., Mertens, J., Baciu, C., Hug, S.J. & Berg, M. (2011). Geochemistry and arsenic mobilization to groundwaters of the Pannonian Basin (Hungary and Romania). *Applied Geochemistry*, 26, 1-17.

144. Rowland, H.A.L., Pederick, R.L., Polya, D.A., Pancost, R.D., van Dongen, B.E., Gault, A.G., Vaugh, D.J., Bryant, C., Anderson, B., Lloyd, J.R. (2007). The control of organic matter on microbially mediated iron reduction and arsenic release in shallow alluvial aquifers, Cambodia. *Geobiology,* 5, 281-292.

145. Rowland, H.A.L., Polya, D.A,, Lloyd, J.R. & Pancost, R.D. (2006). Characterization of organic matter in a shallow, reducing, arsenic-rich aquifer, West Bengal. *Organic Chemistry*, 37, 1101-1114.

146. Ryan, P.C., Kim, J., Wall, A.J., Moen, J.C., Corenthal, L.G., Chow, D.R., Sullivan, C.M., & Bright, K.S. (2011). Ultramafic-derived arsenic in a fractured rock aquifer. *Applied Geochemistry*, 26, 444-457.

147. Saltikov, C.W., Cifuentes, A., Venkateswaran, K., & Newman, D.K. (2003). The *ars* detoxification system is advantageous but not required for As(V) respiration by the genetically tractable *Shewanella* strain ANA-3. *Applied and Environmental Microbiology*, 69, 2800-2809.

148. Santini, J.M. and Ward, S.A. (eds.) (2012) Metabolism of Arsenite. CRC Press.

149. Scanlon, B.R., Nicot, J.P., Reedy, R.C., Kurtzman, D., Mukherjee, A., & Nordstrom, D.K. (2009). Elevated naturally occurring arsenic in a semiarid oxidizing system, Southern High Plains aquifer, Texas, USA. *Applied Geochemistry*, 24, 2061-2071.

150. Seddique, A.A., Masuda, H., Miamura, M., Shinoda, K., Yamanaka, T., Itai, T., Maruoka, T., Uesugi, K., Ahmed, K.M. & Biswas, D.K. (2008). Arsenic release from biotite into a Holocene groundwater aquifer in Bangladesh. *Applied Geochemistry*, 23, 85-98.

151. Sengupta, S., McArthur, J.M., Sarkar, A., Leng, M.j., Ravenscroft, P., Howarth, R.J., & Banerjee, D.M. (2008). Do ponds cause arsenic pollution of groundwater in the Bengal Basin? An answer from West Bengal. *Environmental Science and Technology*, 42, 5156-5164.

152. Senior, L.A. & Sloto, R.A. (2006). Arsenic, boron, and fluoride concentrations in ground water in and near diabase intrusions, Newark Basin, Southeastern Pennsylvania. *U.S. Geological Survey Scientific Investigations Report 2006-5261.*

153. Serfes, M. (2005). Arsenic occurrence, sources, mobilization, transport and prediction in the major bedrock aquifers of the Newark Basin, Ph.D. Dissertation, Rutgers University, 122 p.

154. Sharif, M.U., Davis, R.K., Steele, K.F., Kim, B., Hays, P.D., Kresse, T.M. & Fazio, J.A. (2008). Distribution and variability of redox zones controlling spatial variability of arsenic in the Mississippi River Valley alluvial aquifer, southeastern Arkansas. *Journal of Contaminant Hydrology*, 99, 49-67.

155. Smedley, P.L. (2008). Sources and distribution of arsenic in groundwater and aquifers. Chapter 4, In *Arsenic in Groundwater: a World Problem*, Ed. C.A.J. Appelo. *Proceedings of the IAH Seminar*, Utrecht, Netherlands, Nov. 2006. NERC Open Research Archive. Available from http://nova.nerc.ac.uk, accessed 7/6/12.

156. Smedley P.L., & Kinniburgh, D.G. (2002). A review of the source, behavior and distribution of arsenic in natural waters. *Applied Geochemistry*, 17, 517-568.

157. Smedley, P.L., Knudsen, J. & Maiga, D.(2007). Arsenic in groundwater from mineralized Proterozoic basement rocks of Burkina Faso. *Applied Geochemistry*, 22, 1074-1092.

158. Smedley, P.L., Zhang, M-Y, Zhangm G-Y & Luo, Z-D. (2003). Mobilization of arsenic and other trace elements in fluviolacustrine aquifers of the Huhhot Basin, Inner Mongolia. *Applied Geochemistry*, 18, 1453-1477.

159. Smith, A.H., Lingas, E.O., & Rahman, M. (2000). Contamination of drinking-water by arsenic in Bangladesh: a public health emergency. *Bulletin of the World Health Organization*, 78, 1093-1103.

160. Smith, A.H., Lopipero, P.A., Bates, M.N., & Steinmaus, C.M. (2002). Arsenic epidemiology and drinking water standards. *Science*, 296, 2145-2146.

161. Stauffer, R.E. & Thompson, J.M. (1984). Arsenic and antimony in geothermal waters of Yellowstone National Park, Wyoming, USA. *Geochimica et Cosmochimica Acta*, 48, 2547-2561.

162. Sthiannopkao, S., Kim, K.W., Sotham, S. & Choup, S. (2008). Arsenic and manganese in tube well waters of Prey Veng and Kandal Provinces, Cambodia. *Applied Geochemistry*, 23, 1086-1093.

163. Stollenwerk, K.G., Breit, G.N., Welch, A.H., Yount, J.C., Whitney, J.W., Foster, A.L., Uddin, M.N., Majumder, R.K. & Ahmed, N. (2007). Arsenic attenuation by oxidized aquifer sediments in Bangladesh. *Science of the Total Environment*, 379, 133-150.

164. Tanner, C.C. & Clayton, J.S. (1990). Persistence of arsenic 24 years after sodium arsenite herbicide application to Lake Rotoroa, Hamilton, New Zealand. New Zealand Journal of Marine & Freshwater Research, 24, 173-179.

165. Thornburg, K. & Sahai, N. (2004). Arsenic occurrence, mobilization, and retardation in sandstone and dolomite formations of the Fox River Valley in Eastern Wisconsin. *Environmental Science and Technology*, 38, 5087-5094.

166. Tun, T.N. (2003). Arsenic contamination of water sources in rural Myanmar. Proceedings, 29th WEDC International Conference, Abuja Nigeria, 219-221.

167. Ujevic, M., Duic, Z, Casiot, C., Sipos, L., Santo, V., Dadic, Z. & Halamic, J. (2010). Occurrence and geochemistry of arsenic in the groundwater of eastern Croatia. *Applied Geochemistry*, 25, 1917-1029.

168. Ure, A. & Berrow, M., (1982). Chapter 3. The elemental constituents of soils. In *Environmental Chemistry* Ed. H.J.M Bowen. 94-203. London, Royal Society of Chemistry.

169. USEPA (2001). Drinking Water Standard for Arsenic, USEPA Fact Sheet 815-F-00-105.

170. USEPA (2011). Five-Year Review Report, Vineland Chemical Company Superfund Site, Vineland Township, Cumberland County, New Jersey. U.S. Environmental Protection Agency, Region 2, New York.

171. van Geen, A., Zheng, Y., Versteeg, R., Stute, M., Horneman, A., Dhar, R., Steckler, M., Gelman, A., Small, C., Ahsan, H., Graziano, J.H., Hussain, I. & Ahmed, K.M. (2003). Spatial variability of arsenic in 6000 tube wells in a 25 km2 area of Bangladesh. *Water Resources Research*, 39, (HWC 3) 1-16.

172. van Siclen, A.P. & Gerry, C.N. (1936). Arsenic. *U.S. Minerals Yearbook*. 495-500.

173. Varsányi, I. & Kovács, L.Ó. (2006). Arsenic, iron, and organic matter in sediments and groundwater in the Pannonian Basin, Hungary. *Applied Geochemistry*, 21, 949-963.

174. Walsh, L.M., & Keeney, D.R. (1975). Behavior and phytotoxicity of inorganic arsenicals in soils. Chapter 3, In *Arsenical Pesticides,* Ed. E.A. Woolson. 35-52. Washington, D.C., ACS Symposium series 7, American Chemical Society.

175. Wang, S., & Mulligan, C.N. (2006a). Occurrence of arsenic contamination in Canada: Sources, behavior and distribution. *Science of the Total Environment*, 366, 701-721.

176. Wang, S., & Mulligan, C.N. (2006b). Effect of natural organic matter on arsenic release from soils and sediments to groundwater. *Environmental Geochemistry and Health*, 28, 197-214.

177. Wang, M., Zheng, B., Wang, B, Li, S., Wu, D. & Hu, J. (2006). Arsenic concentrations in Chinese coals. *Science of the Total Environment*, 357, 96-102.

178. Webster, J.G., & Nordstrom, D.K. (2003). Geothermal Arsenic; Chapter 4, 101-125. In *Arsenic in Groundwater*, Ed. A.H. Welch & K.G. Stollenwerk. Boston, Kluwer Academic Publishers.

179. Welch, A.H., Lico, M.S., Hughes, J.L. (1988). Arsenic in groundwater of the western United States. *Ground Water*, 26, 333-347.

180. Welch, A.H. & Stollenwerk, K. G. (eds.) (2003). Arsenic in Groundwater. Boston, Kluwer Academic Publishers.

181. Welch, A.H., Westjohn, D.B., Helsel, D.R., Wanty, R.B. (2000). Arsenic in groundwater of the United States: Occurrence and geochemistry. *Ground Water*, 38, 589-604.

182. WHO (1993). Guidelines for Drinking Water Quality. Recommendations, 2nd Ed., Geneva, World Health Organization.

183. WHO (2001). Arsenic compounds: Environmental health criteria, 224, 2nd ed. Geneva, World Health Organization.

184. Winkel, L., Berg, M., Stengel, C. & Rosenberg, T. (2008). Hydrogeological

survey assessing arsenic and other groundwater contaminants in the lowlands of Sumatra, Indonesia. *Applied Geochemistry*, 23, 3019-3028.

185. Woolson, E.A. (1977). Fate of arsenicals in different environmental substrates. *Environmental Health Perspectives*. 19, 73-81

186. Xie, X., Wang, Y., Duan, M. & Liu, H. (2008). Sediment geochemistry and arsenic mobilization in shallow aquifers of the Datong basin, northern China. *Environmental Geochemistry and Health,* DOI 10.1007/s10653-008-9204-7.

187. Yudovich, Ya. E. & Ketris, M.P. (2005). Arsenic in coal: a review. *International Journal of Coal Geology*, 61, 141-196.

188. Zheng, Y., Stute, M., van Geen, A., Gavrieli, I., Dhar, R., Simpson, H.J., Schlosser, P. & Ahmed, K.M. (2004). Redox control of arsenic mobilization in Bangladesh groundwater. *Applied Geochemistry*, 19, 201-214.

CITATION

CHAPTER 1

Mansoor Zoveidavianpoor, Ariffin Samsuri and Seyed Reza Shadizadeh (2012). Overview of Environmental Management by Drill Cutting Re-Injection Through Hydraulic Fracturing in Upstream Oil and Gas Industry, Sustainable Development - Authoritative and Leading Edge Content for Environmental Management, Dr. Sime Curkovic (Ed.), ISBN: 978-953-51-0682-1, InTech, DOI: 10.5772/45828.

CHAPTER 2

Chen L, Zhang Z, Ewers BE (2012) Urban Tree Species Show the Same Hydraulic Response to Vapor Pressure Deficit across Varying Tree Size and Environmental Conditions. PLoS ONE 7(10): e47882. doi:10.1371/journal.pone.0047882.

CHAPTER 3

Wirtz S, Seeger M, Zell A, Wagner C, Wagner J-F, Ries JB (2013) Applicability of Different Hydraulic Parameters to Describe Soil Detachment in Eroding Rills. PLoS ONE 8(5): e64861. doi:10.1371/journal.pone.0064861.

CHAPTER 4

Schymanski SJ, Or D, Zwieniecki M (2013) Stomatal Control and Leaf Thermal and Hydraulic Capacitances under Rapid Environmental Fluctuations. PLoS ONE 8(1): e54231. doi:10.1371/journal.pone.0054231.

CHAPTER 5

Barnes RT, Gallagher ME, Masiello CA, Liu Z, Dugan B (2014) Biochar-Induced Changes in Soil Hydraulic Conductivity and Dissolved Nutrient Fluxes Constrained by Laboratory Experiments. PLoS ONE 9(9): e108340. doi:10.1371/journal.pone.0108340.

CHAPTER 6

Yifeng Chen and Chuangbing Zhou (2011). Stress/Strain-Dependent Properties of Hydraulic Conductivity for Fractured Rocks, Developments in Hydraulic Conductivity Research, Dr. Oagile Dikinya (Ed.), ISBN: 978-953-307-470-2, InTech, DOI: 10.5772/16007.

CHAPTER 7

Kröber W, Zhang S, Ehmig M, Bruelheide H (2014) Linking Xylem Hydraulic Conductivity and Vulnerability to the Leaf Economics Spectrum—A Cross-Species Study of 39 Evergreen and Deciduous Broadleaved Subtropical Tree Species. PLoS ONE 9(11): e109211. doi:10.1371/journal.pone.0109211.

CHAPTER 8

Julia L. Barringer and Pamela A. Reilly (2013). Arsenic in Groundwater: A Summary of Sources and the Biogeochemical and Hydrogeologic Factors Affecting Arsenic Occurrence and Mobility, Current Perspectives in Contaminant Hydrology and Water Resources Sustainability, Dr. Paul Bradley (Ed.), ISBN: 978-953-51-1046-0, InTech, DOI: 10.5772/55354.

INDEX